OCEANOGRAPHY

Editor

Geghard Torosian

scitus
academics

Oceanography

Edited by **Geghard Torosian**

This Book is Printed on Acid-Free Paper. ♾

ISBN: 9781681179025

Library of Congress Control Number: 2017948930

SCITUS Academics LLC,

616, Corporate Way, Suite 2, 4766,

Valley Cottage, NY 10989

www.scitusacademics.com

Notice

Printed in United States of America.

Preface

Oceanography, the study of the ocean, is a field that requires a broad understanding of many disciplines, from biology and ecology, to physics and chemistry, to history and geology. The major disciplines of oceanography are geological oceanography, physical oceanography and chemical oceanography. Oceanographers and others involved in these disciplines often work together to unravel the mysteries and unknowns of ocean science. In reading about each of these sub-fields, keep in mind that some of the most important oceanographic discoveries have been made as a result of an integrated, multidisciplinary approach, often involving geologists, chemists, biologists, physical oceanographers and engineers. As a growing global population stresses the ability of our society to produce food, water and shelter, we will continue to look to the oceans to help sustain our basic needs. Advances in technology, combined with demand, will improve our ability to derive food, drinking water, energy sources, waste disposal and transportation from the ocean. It will be up to this and future generations to build upon our existing knowledge of the ocean and its potential to help meet the needs of the world and its inhabitants.

This book covers a wide range of topics, including marine life and ecosystems, ocean circulation, plate tectonics and the geology of the seafloor, and the chemical and physical properties of the ocean. It provides chapters on very different topics under very different settings, some with a focused angle, others with a wider approach, yet all sharing the inspiration that we need to understand the small pieces to put collectively the big picture for a much larger mechanism, the functioning of the ocean as a whole. The modern oceanographic research represents one of the last frontiers of the knowledge of our planet, it depends on the oceans exploration and so it is strictly connected to the development of new technologies. Furthermore, other scientific and social disciplines can provide many fundamental inputs to complete the description of the entire ocean ecosystem.

Table of Contents

CHAPTER 1

Estimation of extremes in the southern Arabian Gulf

Thomaskutty Varghese, Shailesh Nayak, Gangadhara Bhat

ABSTRACT

Extreme all-direction wave conditions and joint probabilities of high waves and high water levels are derived at selected locations in the southern Arabian Gulf. The wave conditions are obtained using the spectral wave model SWAN with wind and offshore wave information procured from ocean weather, supplemented with local water level records. Wave transformations from offshore to nearshore and waves generated by local wind conditions are derived using wind speed and direction, offshore wave height, peak period and direction and water level inputs to the SWAN model. The model predicted extreme 3-hourly wind speeds for return periods ranging from 0.06 – 200 years and the wave conditions for each storm. The study indicates that future sea level rise to 2058 gives 1:50, 1:100 and 1:250 return period water levels of 2.64, 2.71 and 2.79 mCD respectively at Jebel Ali.

Keywords: Joint probability, extreme water levels, extreme waves, return period, wave modelling.

1. INTRODUCTION

The knowledge of extreme water level is essential in many design and operational activities in the marine environment, for the safety of public and for design of coastal and marine engineering structures. Careful assessment of the probabilities of extreme sea levels employing good data and reliable analysis techniques is necessary for the design of all modern coastal infrastructure systems (Mendez et al., 2007). The coastal sea in the southern Gulf is presently undergoing significant modifications resulting from various developmental activities. The establishment of a power and desalination complex at the southern extend of Dubai Municipality and the waterfront developments immediately adjacent to its north necissitated studies on extreme sea level and wave climate on the southern Gulf. Though the power and desalination plants are on a relatively open coast at the southern Gulf, influence of the marine forcing functions is of importance in the context of their economic lifetime. Because these forcing functions are essentially random, appropriate statistical procedures need be developed for an accurate assessment.

Extreme water levels result from superposition of the extremes of tidal level, surge level and wave height. Heights of the highest waves encountered at sea vary widely. Information on freak ocean waves and extreme wave conditions is needed in the design of offshore structures and shipping activities (Draper, 1964, 1973). Efficacy of marine structures and their cost analysis require good estimates of the extreme water levels (Draper, 1970; Thom, 1971). Extreme water levels have always been a challenge to modellers and sailors. Li and Song (2005) presented a procedure to correlate extreme wave heights and extreme water levels in coastal waters using numerical models together with joint probability analysis. They calibrated and validated their model with wind, wave and water level data from the coastal waters of Hong Kong. Using hourly values of sea level record for the years 2000 and 2005, Abdullah (2010) studied tide and sea level characteristics at Juaymah on the west coast of the Arabian Gulf. Predictive models on extreme sea level have also been validated at certain regions in the Indian Ocean (Kurup et al., 2007; Muraleedharan et al., 2012).

Extremely high or low water levels at coastal locations are an important public concern and a factor in coastal hazard assessment, navigational safety, coastal management and planning, and ecosystem management. This paper presents the results of an investigation on extreme water levels in the coastal environs of the southern Gulf. Computational wave modelling studies, together with a desk assessment of extreme water levels, are carried out to derive extreme wave and water level conditions. Scope of this paper is limited to the estimation of extreme water level and design wave conditions for the future developments at Jebel Ali.

1.1. Probabilistic approach on estimation of extremes recent studies

Accurate estimates of water level and its frequency are essential to coastal flood risk studies. Many research articles for example, Haigh et al. (2013), Irish et al. (2011) and Jain et al. (2010) has focused on water level variation and joint probability techniques to combine multiple parameters. There are big uncertainties with estimates pertaining to extreme values. When obtaining level estimates, the variances of the estimates can be obtained from the expected information matrix or from the observed information matrix. An alternative, and usually more accurate, method is the profile likelihood method (Coles, 2001), which is based on the deviance function and yields asymmetric confidence intervals. Other alternatives are based on bootstrap procedures with adjustments as suggested by Coles and Simiu (2003). According to a recent study on the coverage rate of confidence intervals of extreme value estimates based on various methods Weibull method turns out to produce the best confidence intervals from of the point of view of coverage rates. The peaks-over-threshold method approach on Weibull distribution is generally thought to be somewhat superior we shall not consider the r-largest method in this study (Coles, 2001).

2. DATA AND METHODOLOGY

The area of study consists of the coastal waters in the southern Gulf. Figure 1 shows locations of the output points PT1- PT12 selected for joint probability analysis. Extreme all-direction wave conditions and joint probabilities of waves and water levels are investigated at these locations. Rashed Chowdhury et al. (2010) presented a perspective on extremes of sea-level variability and predictability for the U.S.-Affiliated Pacific Islands on seasonal time-scales. Based on the Generalized Extreme Value Model, they used the L-moments method to estimate the model parameters. In the present study, desk assessment of water levels is carried out using ocean weather model. This includes an assessment of future rise in sea level based on local records and archive data. The wind and wave hindcasts are made with basin-wide fine grids (7 km in lat. and long.), nested within coarser grid systems covering the Gulf of Oman and Arabian Sea. 2-D surge and currents are derived for storms with a grid spacing of 0.035° x 0.035°. The results are merged with the storm wave hindcasts. The wave model incorporates third-generation physics (OWI3G version 52, half friction). All the hindcast time series are available in either OWI's OSMOSIS format or ASCII.

2.1 Extreme sea levels

Ocean weather provides continuous time series of hourly wave and wind conditions for 20 years from 1983 to 2002. It also contains hourly data on 108 individual storms from 1960 to 2002 covering each storm for a period of 2 - 8 days. In the present study, relevant high sea level records are selected using 'Peaks over Threshold criteria', taking an average of about five per year from the 20 years' time series data and an average of about two per year from the 42 years' storm data. Weibull and Gumbel distributions are fitted to each of the two data sets and extrapolated to extreme values. As the storm data are selected by wind speed, some of the high sea levels in the time series data do not appear in the storm data, and so extreme values derived from the time series data shall be the higher of the two. However, four highest sea levels in the storm data occurred during an earlier period, and so one might expect extreme values for higher return periods to be higher than those derived from the time series data. The results (average of Weibull and Gumbel in each case) are summarised in Table 3. Return periods of 50 years and above based on storm data give slightly higher values.

2.2 Joint probability of high waves and high sea levels

Damage to marine structures is often associated strong waves with high tidal levels. The combination of extreme waves and high water levels results in increased coastal damages (Hawkes et al., 2010). The simultaneous occurrence of high waves and high water level is important in their combined effect on structures. Wave period can also be important in assessing run-up and overtopping, and so it is useful to have information on the joint distribution of wave height and period. These multivariate extremes are difficult to predict directly from observational data, as there may be only a few events among the observations. Fitting a joint probability distribution for variables that have a degree of correlation, such as waves and sea levels, is problematic. This makes the accurate estimation of design parameters difficult. Inaccurate estimations of the joint probability relationships are detrimental in the design of coastal and marine structures.

In the present study, the joint probability assessment is carried out following the procedures set out in Defra Environment Agency Guidelines (2005a, b). Visual assessment of wave heights against simultaneous sea levels from 42 years' storm data shows only slight dependence between large waves and high sea levels (Figure 2). Figure 2 shows monthly mean sea levels, relative to the local land level, downloaded from the website of Permanent Service for Mean Sea Level, tide gauges in the Gulf with at least 15 years of data available are only considered.

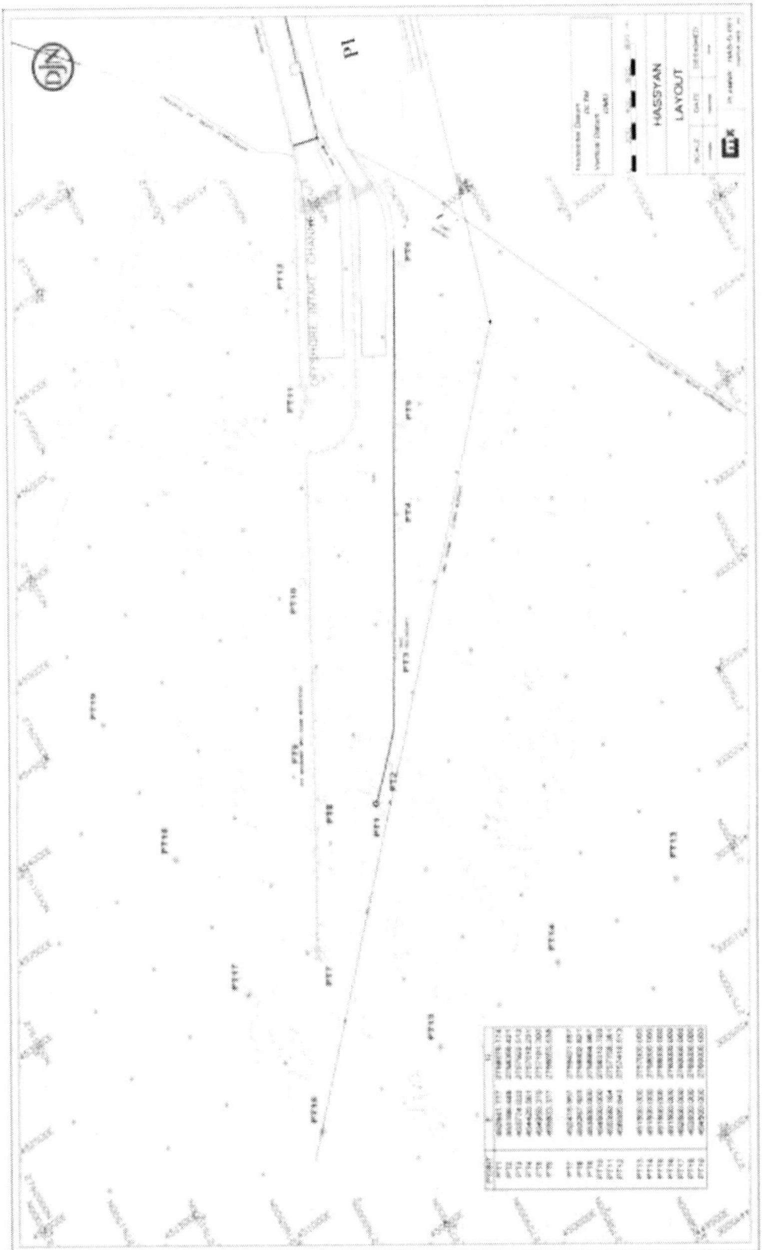

Figure 1. Locations of output points PT1- PT12.

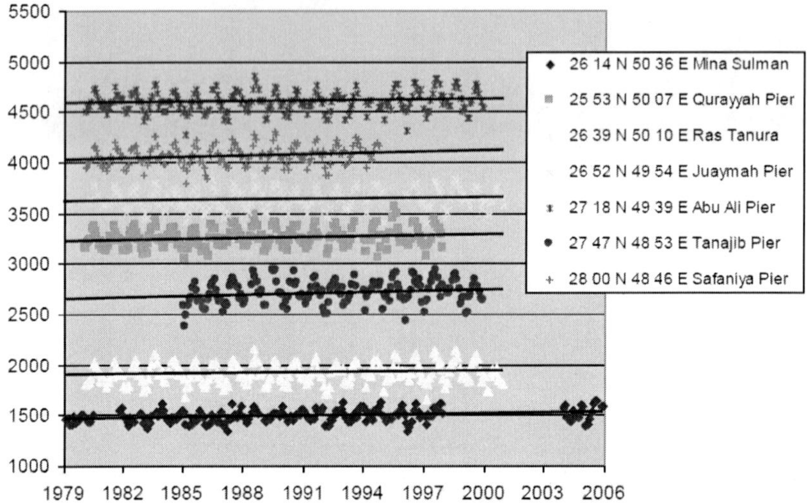

Figure 2. Monthly mean sea levels from several tide gauges in the Gulf

In Figure 2, the vertical scale is in millimetres above an arbitrary datum for each gauge, chosen to separate the data series in the diagram. The trend lines indicate gradual increase in mean sea level, averaging about 2.5 mm/year since 1979 (slightly above the global average rate). The Intergovernmental Panel for Climate Change (IPCC, 2007) established an average global mean sea level rise of about 1.7±0.5 mm/year during the twentieth century and established a slightly higher rate during the period 1961 - 2003. Sultan et al. (1995) noted an upward trend in mean sea level of 2.1 mm/year during the 1980s in the central part Red Sea. Recent studies in southern Gulf, where more joint probability assessments have been carried out, show low levels of dependence between high waves and high sea levels (Sultan et al., 2003). This dependence, quantified as a correlation coefficient (ρ), has always been lesser than 0.4. Figure 3 illustrates the simultaneous occurrence of sea level and wave height within storm data.

2.3 The Simulating WAves Nearshore (SWAN) wave model

Shallow water transformations that the waves undergo during their propagation toward shore, together with the generation of local waves, contribute to the complete description of wave conditions in coastal seas. The wave model SWAN is used to study the wave transformation from the offshore to the nearshore site and the wave generation by local winds. SWAN (Simulating WAves Nearshore) is a 3rd Generation spectral wave model that computes random, short-crested, wind-generated waves in coastal regions and inland waters considering the processes of shoaling, refraction, breaking, reflection, diffraction and nonlinear wave-wave interactions. The offshore wave conditions are derived from ocean weather data and the nearshore wave conditions from the spectral wave model SWAN (2007).

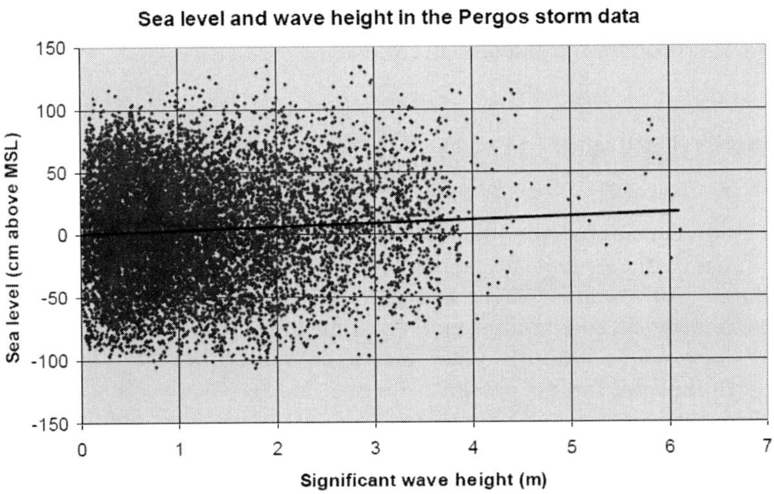

Figure 3. Simultaneous occurrence of sea level and wave height within the storm data.

In this study, a SWAN model is set up covering the area of study. Figure 4 shows the survey depths incoroporated in SWAN model and Figure 5 depicts the SWAN model grids and bathmetry. Lack bathymetric field data are supplemented with information from digital charts. The SWAN model covers an area of approximately 25 km by 25 km in the southern Gulf. It comprises of three nested grids with spatial grid sizes: 1000, 300 and 100 m, respectively. It includes the Dubai waterfront development area and the breakwater layouts. In the SWAN model, the waterfront development boundaries are assigned a reflection coefficient of 0.5 and the breakwaters are assigned a zero reflection coefficient. Standard JONSWAP spectral shape and Cos2(θ) directional spread are assumed for all SWAN model runs.

2.4 Wind and offshore wave conditions

The SWAN model is driven by wind and offshore wave data at 25.125°N, 54.8125°E (water depth: 19.25 m) supplied by ocean weather. This data is based on a computational model hindcast of wave conditions derived from input wind fields and have been calibrated extensively in the region. The storm data set is used to determine extreme wave conditions in the area of study. Continuous time series is employed to derive the nearshore directional extremes (for sectors other than the most severe), and the more frequently occurring extremes are selected. The wind speed and direction, offshore wave height, peak period and direction, and water level are used as inputs to the SWAN model. The storms also include the water level, which includes the surge and tide. Hence, the results took account of the tide, surge, wind conditions and offshore waves.

3. RESULTS AND DISCUSSION

The study considered the joint probability based on peak over threshold combined with Weibul fit to study the extremes Jabel Ali. The joint probability of extreme waves and extreme water levels is derived using three- hourly extreme wind speeds as input into the SWAN model. Table 1 gives the extreme 3-hourly wind speeds for return periods ranging from 0.06 – 200 years. Table 2 presents the extreme significant wave conditions at the output locations. Extreme high sea level predictions obtained in this study are given in Table 3. Figure 6 shows the predicted significant wave heights over the SWAN model for a severe storm (Offshore: Hs=4.4 m, Tp=9.8 s, Direction=304°N). Table 4 presents the joint probability of waves and water levels at points PT1- PT12 for a joint return period of 200 years in each of the five 30o direction sectors (225 - 255°, 255 - 285°, 285 - 315°, 315 - 345° and 345 - 015°).

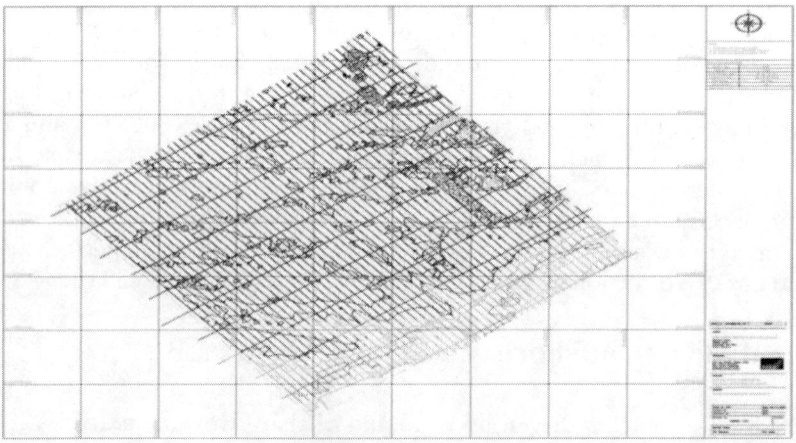

Figure 4. Survey depths incorporated in SWAN model.

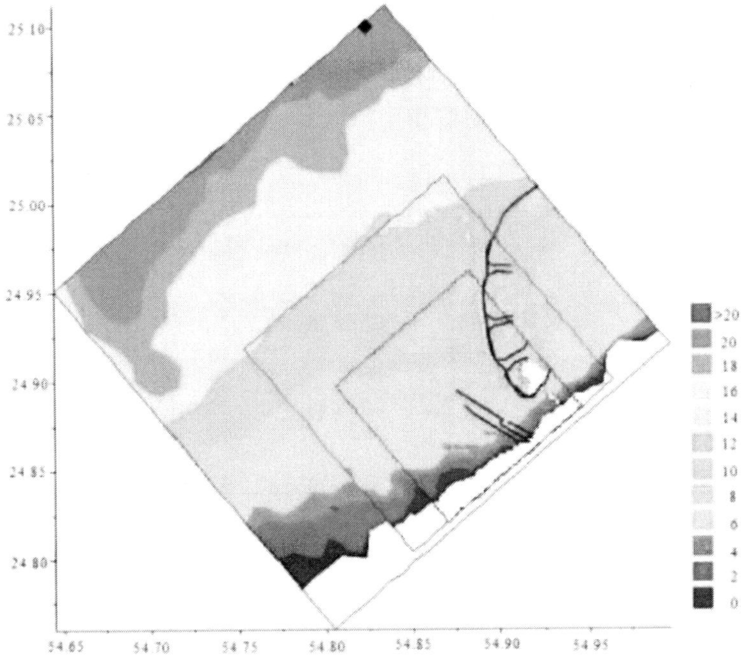

Figure 5. SWAN model grids and bathymetry.

3.1 Nearshore directional extreme waves

For most storms,the most severe offshore wave conditions occur in the direction sectors 285 - 315° and 315 - 345°. As explained above, using 'peaks-over-threshold' method, extreme wave conditions have been derived for a range of return periods. Non-linear shoaling and depth-limited breaking are obtained by extrapolating from the nearshore wave conditions (Goda, 1997). Extreme wave conditions are encountered when the storm approaches from the direction sectors considered. The peaks-over-threshold analysis for narrow direction sectors does not normally give reliable results. Hence, for the directional sector extremes, the three-parameter Weibull method is used to derive the required return period significant wave heights from the nearshore wave climate. 3-h storm duration provides good agreement between the extremes derived for return periods of 1-2 years from climate data and the peaks-over-threshold method. The directional wave spectrum shows similarity to Shadrin's spectrum (Shadrin, 1982).

Table 1. Extreme 3-hourly wind speeds (ms-1).

Return period (Years)	Direction Sector (°N)												
	0	30	60	90	120	150	180	210	240	270	300	330	All
0.06	5.5	5.8	7.3	6.7	7.0	5.4	4.3	4.4	5.1	10.1	10.6	7.2	11.4
0.2	6.6	7.1	8.7	8.2	8.7	7.3	5.8	5.6	6.4	11.5	12.2	8.7	12.8
0.6	7.5	8.2	9.8	9.5	9.9	8.7	7.3	6.8	7.5	12.7	13.6	10.1	14.0
2	8.6	9.3	10.8	10.9	11.0	10.0	9.0	8.1	8.9	13.9	15.0	11.5	15.3
6	9.6	10.3	11.7	12.0	11.9	11.1	10.6	9.3	10.1	14.9	16.3	12.8	16.4
20	10.6	11.4	12.5	13.3	12.9	12.3	12.4	10.7	11.5	16.0	17.6	14.2	17.6
60	11.6	12.3	13.3	14.4	13.6	13.2	14.1	12.0	12.7	16.9	18.7	15.5	18.7
200	12.6	13.2	14.0	15.6	14.4	14.2	16.0	13.4	14.1	17.8	19.9	16.9	19.8

Table 2. Extreme significant wave conditions at Locations PT1-12; Hs (m) and Tp (s).

Location		Return period years					
		1	5	10	50	100	200
PT 1	Hs	1.7	2.5	2.7	3.1	3.3	3.4
	Tp	7.6	9.2	9.6	10.4	10.7	10.9
PT 2	Hs	2	2.8	3.1	3.5	3.6	3.8
	Tp	7.6	9.2	9.6	10.4	10.7	10.9
PT 3	Hs	1.1	1.6	1.8	2	2.1	2.2
	Tp	7.6	9.2	9.6	10.4	10.7	10.9
PT 4	Hs	0.9	1.3	1.4	1.6	1.6	1.7
	Tp	7.6	9.2	9.6	10.4	10.7	10.9
PT 5	Hs	0.8	1.2	1.3	1.5	1.6	1.6
	Tp	7.6	9.2	9.6	10.4	10.7	10.9
PT 6	Hs	0.7	1	1.1	1.3	1.4	1.4
	Tp	7.6	9.2	9.6	10.4	10.7	10.9
PT 7	Hs	2	2.8	3.1	3.5	3.6	3.8
	Tp	7.6	9.2	9.6	10.4	10.7	10.9
PT 8	Hs	0.9	1.3	1.4	1.6	1.6	1.7
	Tp	7.6	9.2	9.6	10.4	10.7	10.9
PT 9	Hs	1.5	2.3	2.5	3	3.1	3.3
	Tp	7.6	9.2	9.6	10.4	10.7	10.9
PT 10	Hs	1.4	2.1	2.3	2.7	2.8	3
	Tp	7.6	9.2	9.6	10.4	10.7	10.9
PT 11	Hs	1.3	1.9	2.1	2.5	2.5	2.7
	Tp	7.6	9.2	9.6	10.4	10.7	10.9
PT 12	tp	1.2	1.7	1.9	2.2	2.3	2.4
	Hs	7.6	9.2	9.6	10.4	10.7	10.9

With SWAN wave model, it is not necessary to run every condition of the offshore time series to generate nearshore wave climate. The offshore wave climate is analysed and 100 wave conditions are selected to represent the climate. SWAN is run with these input conditions and results extracted to provide look-up tables for the wave prediction points 1 to 12 at two water levels: +1.7 mCD (present day MHHW) and +2.8 mCD(1/200 year water level including sea level rise to 2058). The look-up tables are applied to time series to produce corresponding time series of nearshore wave conditions at the output points. The predicted nearshore annual wave climates (for present day MHHW water level) show that waves above 1.0 m occur at PT7 for approximately 7% of the time and waves above 2.0 m occur less than 0.1% of the time. Highest waves occur from the nearshore direction sectors centred on 300°N and 330ºN, and these are the dominant wave direction sectors. Highest predicted wave heights occur at PT2 and PT7. At these points the 200 year conditions are predicted to be H_s=3.8 m and T_p=10.9 s.

Table 3. Extreme high sea level predictions for present and future levels

Return period (years)	Metres above MSL datum		Metres above CD, Jebel Ali		
	Time series data	Storm data	Time series data	Storm data	Future (2058) levels
	2008 levels				
1	1.19	0.89	1.19	2.19	2.49
10	1.26	1.18	1.26	2.26	2.56
50	1.30	1.34	1.34	2.34	2.64
100	1.32	1.41	1.41	2.41	2.71
250	1.35	1.49	1.49	2.49	2.79

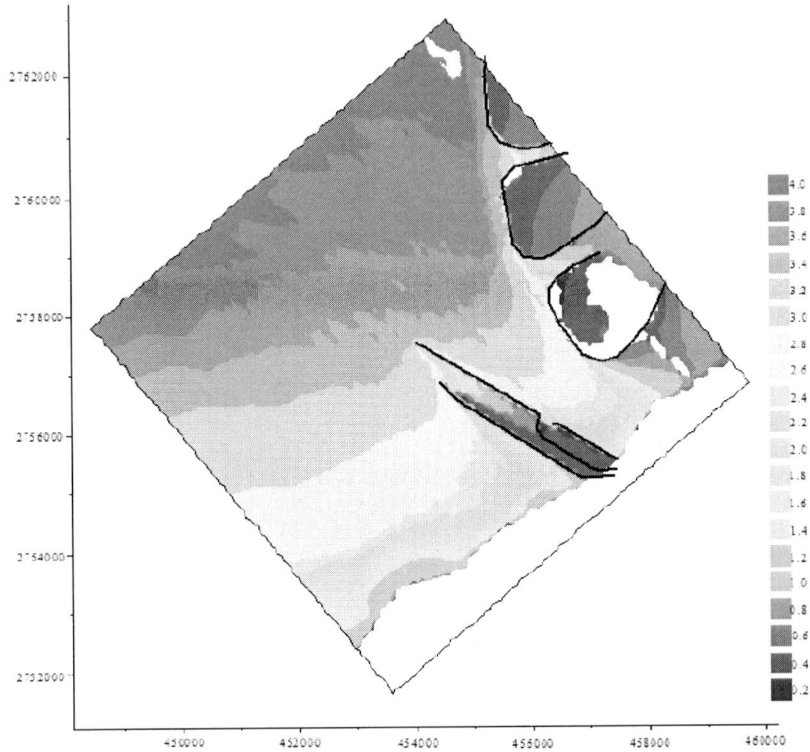

Figure 6. Predicted significant wave heights for a severe storm considered.

3.2 Locally generated wave conditions

The Dubai Waterfront Development provides a degree of shelter to the northern side of the breakwaters (PT 9- 12) while the southern side (PT 2-6) are exposed to locally generated waves. The SWAN model runs predict locally generated wave conditions due to winds from the north and northeast (Table 1). The predicted locally generated significant wave heights, when combined on an energy basis with the transformed offshore waves, give the local wave climate. The nearshore wave climate and extremes are caused by shallow water modifications of offshore waves and interaction with locally generated waves. Local nearshore wave measurements are essential for a validation of the SWAN model.

Table 4. Joint probability of waves and water levels at points PT1 to PT12 for joint return period of 200 years (Chart Datum, Jebel Ali).

Direction	Joint RP	Significant wave height (m)									Level (mCD,Jebel Ali, 2058)
		PT1	PT2	PT3	PT4	PT5	PT6	PT7	PT8	PT9	PT1 TO PT9
225-255	0.06	0.30	0.30	0.30	0.30	0.30	0.20	0.30	0.20	1.30	2.77
	0.20	0.40	0.40	0.40	0.40	0.40	0.30	0.40	0.30	1.50	2.66
	0.60	0.50	0.50	0.40	0.40	0.40	0.40	0.50	0.30	1.80	2.58
	2.00	0.70	0.70	0.70	0.60	0.60	0.50	0.70	0.40	1.90	2.54
	6.00	0.90	0.90	0.80	0.70	0.60	0.50	0.90	0.50	2.40	2.51
	20.00	1.00	1.00	0.90	0.70	0.70	0.60	1.00	0.50	2.70	2.47
	60.00	1.20	1.20	1.00	0.90	0.80	0.70	1.20	0.70	3.00	2.39
	200.00	1.20	1.30	1.10	0.90	0.80	0.70	1.30	0.70	3.30	2.29
255-285	0.06	0.90	0.90	0.90	0.80	0.80	0.60	0.80	0.50	0.50	2.77
	0.20	1.10	1.00	1.10	1.00	1.00	0.70	1.00	0.60	0.60	2.66
	0.60	1.30	1.30	1.20	1.10	1.10	0.80	1.30	0.80	0.70	2.58
	2.00	1.50	1.50	1.40	1.20	1.20	0.90	1.50	0.90	0.90	2.54
	6.00	1.60	1.60	1.40	1.30	1.30	0.90	1.60	0.90	1.00	2.51
	20.00	1.80	1.80	1.60	1.40	1.40	1.10	1.80	1.10	1.20	2.47
	60.00	1.90	1.90	1.60	1.40	1.50	1.10	1.90	1.10	1.30	2.39
	200.00	2.10	2.10	1.70	1.50	1.50	1.20	2.10	1.20	1.50	2.29
285-315	0.06	1.40	1.30	0.90	0.70	0.60	0.50	1.50	0.50	0.30	2.77
	0.20	1.60	1.50	1.00	0.80	0.70	0.60	1.80	0.60	0.40	2.66
	0.60	1.80	1.70	1.10	0.80	0.80	0.70	2.10	0.70	0.50	2.58
	2.00	2.10	2.40	1.40	1.10	1.00	0.90	2.20	1.10	0.50	2.54
	6.00	2.50	2.90	1.70	1.30	1.20	1.10	2.40	1.30	0.60	2.51
	20.00	2.90	3.20	1.90	1.50	1.40	1.20	2.60	1.50	0.70	2.47
	60.00	3.20	3.50	2.10	1.60	1.50	1.30	2.80	1.60	0.80	2.39
	200.00	3.40	3.70	2.20	1.70	1.60	1.50	3.00	1.70	0.80	2.29

3.3 Mean sea level rise in the southern Gulf

Sultan et al. (2003) examined monthly mean sea level variations in the Arabian Gulf over a period of 11 years (1980–1990) by spectral and regression analyses. The mean sea level is higher in summer and lower in winter with the range of 26 cm. About 80% of the monthly mean sea level variance is related to seasonal changes. The annual component of the seasonal variation is 11.3 cm while that of the semi-annual is about 2.5 cm. The results of regression analysis show that atmospheric pressure changes can explain about 75% of the variance of the seasonal signal. The remaining 25% is due to seasonal changes in density. Regression of sea level on time reveals the presence of a rising trend of 2.3 cm over the 11-year study period. The present analysis of water levels indicates 1:50, 1:100 and 1:250 return period water levels of 2.64, 2.71 and 2.79 mCD Jebel Ali, respectively.

4. CONCLUSIONS

An assessment of the joint occurrence of extreme sea levels and high wave heights with extreme wind in the southern Gulf has been carried out. Nearshore wave climate and extremes have been derived at selected points using a combination of the storms and continuous data transformation SWAN model. The SWAN model has also been used to derive locally generated wave (Wind Sea) conditions in Nearshore Sea. The results indicate combinations of water levels with return periods of 0.06, 2 and 200 years with extreme wave conditions having corresponding return periods of 200, 6 and 0.06 years, respectively. Future sea level rise to 2058 indicates 1:50, 1:100 and 1:250 return period water levels of 2.64, 2.71 and 2.79 mCD respectively at Jebel Ali. The Intergovernmental Panel for Climate Change estimated an average global mean sea level rise of about 1.7±0.5 mm/year during the 20th century and indicated slightly higher rates during the period 1961 - 2003. Sultan et al. (1995) also noted an upward trend of 2.1 mm/year in mean sea level during the 1980's in the central part of the Red Sea. The present study strongly corroborates these suggestions of rising trends in mean sea levels. Estimated parameters especially the one year and 10 year water level are matching with the CO-OPS (NOAA Center for Operational Oceanographic Products and Services) estimate.

REFERENCES

1. Abdullah MA (2010). Tide and sea level characteristics at Juaymah, west coast of the Arabian Gulf. King Abdul-Aziz Univ. Mar. Sci. 21:133-149.

2. Defra / Environment Agency (2005a). Joint probability, Dependence mapping and best practice. Technical Report FD2308/TR1.
3. Defra / Environment Agency (2005b). Use of joint probability methods in flood management. A guide to best practice. Technical Report FD2308/TR2.
4. Draper L (1964). Freak ocean waves. Oceanus. 10:13-15.
5. Draper L (1970). The Canadian wave climate study- the formative years. Proc. 12th Coast. Eng. Conf. ASCE. pp. 1-11.
6. Draper L (1973). Extreme wave condition in British and adjacent waters. Proc. 13th Coast. Eng. Conf. ASCE. pp. 157-165.
7. Goda Y (1997). Random wave concept as the tool of engineering practice. Proc. 2nd Indian Nat. Conf. Har. Ocean Eng. 1:1-12.
8. Hawkes PJ, Gouldby BP, Tawn JA, Owen MW (2010). Joint probability of waves and water levels in coastal engineering design. J. Hydraulic Res. Special Issue Marit. Hydraulics. 40:241-251.
9. IPCC (2007). Report of Intergovernmental Panel on Climate Change, Working Group 1: The Physical Science Basis.
10. Irish JL, Song YK, Chang K (2011). Probabilistic hurricane surge forecasting using parameterised surge response functions. Geophys. Res. Lett. 38:L03606.
11. Kurup PG, Rao AD, Muraleedharan G, Mourani S (2007). Design and operational wave height statistics for extreme sea states. Proc. 4th Indian Nat. Conf. Har. Ocean. Eng. pp. 53-57.
12. Li CW, Song Y (2005). Correlation of extreme waves and water levels using a third-generation wave model and a 3D flow model. Ocean Eng. 33:635-653.
13. Mendez FJ, Menendez M, Luce-o A, Losada IJ (2007). Analyzing monthly extreme sea levels with a time-dependent GEV model. J. Atmos. Oceanic Technol. 24:894-911.
14. Muraleedharan G, Claudia L, Guedes SC, Unnikrishnan NN, Kurup PG (2012). Modeling significant wave height distributions with quantile functions for estimation of extreme wave heights. Ocean Eng. 54:119-131.
15. Rashed Chowdhury Md, Pao-Shin C, Xin Z, Thomas AS, John JM (2010). Sea level extremes in the U.S.-Affiliated Pacific Islands—a coastal hazard scenario to aid in decision analyses. J. Coast Conserv. DOI 10.1007/s11852-010-0086-3.
16. Shadrin IF (1982). Deformed wave spectrum in the coastal zone. Oceanology 22:694-696.
17. Sultan SAR, Ahmad F, El-Hassan A (1995). Seasonal variations of the sea level in the central part of the Red Sea. Estuarine. Coast. Shelf Sci. 40:1-8.
18. Sultan SAR, Ahmad F, Elghribi NM, Al-Subhi AM (2003). An analysis of Arabian Gulf monthly mean sea level. Cont. Shelf Res. 15:11-12.
19. SWAN (2007). Delft University of Technology, Environmental Fluid Mechanics Section. Version 40.51 AB.
20. Thom HCS (1971). Asymptotic extreme value distributions of wave heights in the open ocean. J. Mar. Res. 29:19-27.
21. Haigh ID, Wijeratne EMS, MacPherson LR, Pattiaratchi CB, Mason MS, Crompton RP, George S (2013). Estimating present day extreme total water level exceedance probabilities around the coastline of Australia: tides, extra-tropical storm surges and mean sea level. Clim. Dyn. 42(1–2):121-138.
22. Jain I, Rao AD, Jitendra V, Dube SK (2010). Computation of expected total water levels along the east coast of India. J. Coast. Res. 26(4):681-687.
23. Coles S (2001). An Introduction to the Statistical Modelling of Extreme Values. Springer Texts in Statistics, Springer- Verlag: London.
24. Coles S, Simiu E (2003). Estimating uncertainty in the extreme value analysis of data generated by a hurricane simulation model. J. Eng. Mech. 129(11):1288-1294.

CHAPTER 2

Self-Organizing Maps-based ocean currents forecasting system

Ivica Vilibić, Jadranka Šepić, Hrvoje Mihanović, Hrvoje Kalinić, Simone Cosoli, Ivica Janeković, Nedjeljka Žagar, Blaž Jesenko, Martina Tudor, Vlado Dadić & Damir Ivanković

AUTHOR CONTRIBUTIONS

I.V. coordinated the work and wrote the initial version of the text, J.S. and H.M. created the figures, H.K. did the neural network computations, J.S. and H.K. compared SOM and ROMS-based forecasting systems, I.J. ran the ROMS operational model, S.C. reprocessed the HF radar data, M.T. ran the ALADIN/HR operational model, V.D. and D.I. managed the HF radar network, N.Ž. and B.J. contributed to creation of SOM-based forecasting system; all authors actively contributed to the manuscript writing, polishing and improvement.

ABSTRACT

An ocean surface currents forecasting system, based on a Self-Organizing Maps (SOM) neural network algorithm, high-frequency (HF) ocean radar measurements and numerical weather prediction (NWP) products, has been developed for a coastal area of the northern Adriatic and compared with operational ROMS-derived surface currents. The two systems differ significantly in architecture and algorithms, being based on either unsupervised learning techniques or ocean physics. To compare performance of the two methods, their forecasting skills were tested on independent datasets. The SOM-based forecasting system has a slightly better forecasting skill, especially during strong wind conditions, with potential for further improvement when data sets of higher quality and longer duration are used for training.

1. INTRODUCTION

Ocean and environmental forecasting at regional or basin-wide scales[1]can be done in different ways, depending on the processes to be investigated, and the time and spatial scales of interest[2]. The most common approach at regional scales and for short-term forecasts (up to a week), is to use regional ocean models driven by mesoscale atmospheric model outputs. Forcing is usually carried out through one-way coupling, using multiple nesting techniques and with lateral boundary conditions coming from larger or global models[3]. Other approaches, developed in recent decades, include multi-ensemble methodologies, different data assimilation techniques[4], hybrid modeling[5] and neural network modeling[6]. The latter is often based on the Self-Organizing Maps (SOM) algorithm[7].

Since the 2000s the Self-Organizing Maps method has been introduced more extensively in atmospheric and ocean sciences[8], above all for mapping of: climate and atmospheric processes[9], remote sensing data[10], and ocean modeling data[11]. The method performs best on comprehensive and long time series, and it is particularly suitable for mapping surface current fields from moored current-meters and high-frequency (HF) radars in coastal areas[12,13,14]. However, a SOM-based operational forecasting method has not been implemented yet in any operational oceanography network, in spite of the fact that a similar approach, in combination with other techniques, is already used for instance in energy systems[15], and hydrology[16]. This paper documents the architecture of a SOM-based forecasting system, specifically designed for surface currents, and its evaluation vs. more conventional ocean forecasting tools (specifically, the ROMS ocean model). The system is based on numerical weather prediction (NWP) from the atmospheric model ALADIN/HR, on HF radar data and on an unsupervised SOM training. Skill assessment for such system is done and relevant metrics are estimated. System performance is compared to the existing operational ocean forecasting system based on the ROMS ocean model, which is coupled with the same ALADIN/HR NWP model.

2. THE SOM-BASED FORECASTING SYSTEM AND THE OPERATIONAL ROMS SETUP

The SOM-based operational system was developed with the purpose of providing surface current forecasts (up to 72 h) in the northern Adriatic Sea (Fig. 1) as part of the NEURAL project (http://jadran.izor.hr/neural). The system is based on the assumption that local wind is the major driver of near-surface circulation in the region. The temporal extent of the ocean current forecast window is dictated by the available NWP weather forecast, which extends to 72 hours.

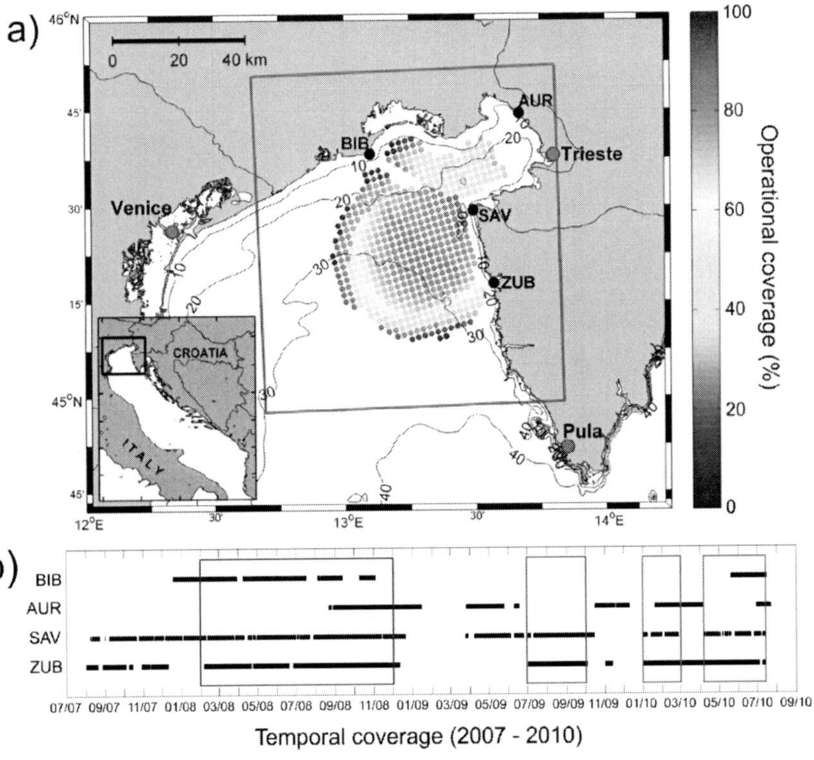

Figure 1 (**a**) The domain of the SOM-based operational forecasting system in the northern Adriatic with marked HF radar stations (BIB – Bibione, AUR – Aurisina, SAV – Savudrija, ZUB – Zub). Operational coverage of hourly surface currents over the predefined Cartesian grid during testing periods is given in percent, while spatial coverage of the Aladin/HR model used for training and forecasting is denoted by the red rectangle. Operational coverage at the same Cartesian grid during testing

period was higher than 60%. (**b**) HF radar operability between 2007 and 2010 with marked training (blue rectangle) and testing (red rectangles) periods. The figure has been created using MATLAB (www. mathworks. com) and CorelDRAW (www.corel.com) software.

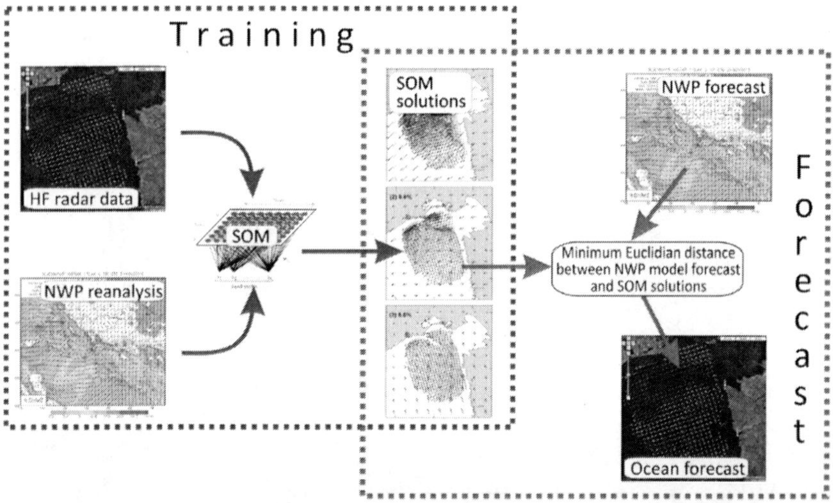

Figure 2: The architecture of the SOM-based operational forecasting system.

The system is based on training and forecast phases (Fig. 2). The main goal of the training phase is to find a robust link between ocean surface currents and surface wind fields. In order to do so, the following steps are undertaken: (1) surface currents in the area are measured using HF radars during a prolonged time interval; (2) sea-surface wind fields are obtained from a high-resolution mesoscale atmospheric model; (3) the characteristic surface current patterns are extracted from HF radar data using the SOM method; (4) joint surface current and wind patterns are extracted from wind and surface currents using the SOM method; (5) then, optimal SOM parameters are estimated and the method applicability is tested on available data. The forecasting phase is based on: (1) forecasting surface winds with a high resolution mesoscale atmospheric model; (2) recognizing characteristic SOM patterns (as determined during the training phase) in the surface wind forecast by searching for the minimum Euclidian distance between characteristic patterns and the NWP-forecasted surface winds; and (3) assigning the corresponding surface current pattern as the surface current forecast.

For this study HF radar data were acquired from four SeaSonde radar systems, located at Bibione (BIB), Aurisina (AUR), Savudrija (SAV) and Zub (ZUB) (Fig. 1), that were operational from 2007 to 2010. Radars worked at 25 MHz, were set up with 512 fft points, 2 Hz sampling rate and a bandwidth of 100 kHz, corresponding to a radial resolution of 1.5 km in range, and 2.3 cm/s in radial velocity. Angular resolution was 5°, and the maximum operating range was 50 km. Hourly surface currents were estimated on a predefined 2 km × 2 km Cartesian grid (Fig. 1) by applying a least-squares algorithm[17]. Radial currents were quality-checked to remove Doppler lines that were either poorly constrained by signal-to-noise (SNR) ratio or had insufficient spectral quality factor[18]. Total vectors were also checked for spikes[19] and spatial gaps were filled using a Gaussian-weighted spatial interpolation. The period with the best temporal coverage (operational coverage from 60% at the distant points up to 100% in the center of the radar domain) and having the highest quality of data was the period between February and November 2008, so this dataset was used for the SOM training phase.

High-resolution surface wind fields (2 km × 2 km) obtained with the ALADIN (Aire Limitée Adaptation Dynamique Développement Inter National) mesoscale NWP model[20], have been used both in training (jointly with surface currents data in the period February-November 2008) and forecast phase. The ALADIN/HR version of the NWP model which has been used for the Adriatic Sea has 37 levels in the vertical, 2 km resolution in horizontal, and a full set of physics parameterizations, including prognostic TKE (turbulent kinetic energy), microphysics and a prognostic convection scheme21. Hourly files were used for the training.

Both surface currents and surface winds were preprocessed with a 33-h low-pass filter, to remove tides and high-frequency atmospheric and ocean processes from the data. Then, current data in the training period (February-November 2008) were introduced to the 4 × 5 SOM of the sheet type and 20 characteristic surface current patterns were mapped. Next, a joint set of surface currents and surface winds was introduced, again to the 4 × 5 SOM of the sheet type and another 20 characteristic joint surface currents surface wind patterns have been obtained. Complex correlations coefficients with values from 0.880 to 0.999 between SOM patterns based on surface current data only and patterns based on the joint surface current and surface wind data were estimated[22]. High correlation between these two sets of SOM patterns is a prerequisite for creating SOM ocean surface currents based on wind forecasts.

Following the testing phase, three distinct periods, 1: July-September 2009, 2: January-February 2010, and 3: April-July 2010, were used for testing reliability of the SOM-based operational forecast (Fig. 1). BIB and

AUR stations were largely out of operation during the three testing periods, therefore operational coverage was as low as 20% at some grid points (Fig. 1), much lower than operational coverage during the training phase (at least 60% at the same Cartesian grid). A forecast of surface currents was verified on measured data and standard metrics (RMSE) were computed. The 4 × 5 SOM matrix was chosen for the forecasting system, resulting in 20 patterns of forecast surface currents, as this choice provides the minimum root-mean-square error (RMSE) between the measured and forecast surface current patterns when compared to other tested sets of SOM matrices (from 2 × 2 to 16 × 12 SOM).

The parallel ocean forecasting system, based on the ROMS ocean model (Regional Ocean Modeling System[23]), is operational in the Adriatic Sea since 2008. The Adriatic ROMS model has a 2 × 2 km horizontal resolution, 20 sigma layers in the vertical; it is forced by the ALADIN/HR NWP fields at the surface. Lateral boundary conditions come from the larger AREG model[24] and river climatology[25]. Further details of the modeling system can be found in this paper[25]. Modeled ROMS surface current data filtered with a 33-h low pass filter were recomputed over the HF radar grid for the three testing periods to allow comparison with the SOM-based surface current forecast.

3. EVALUATION OF THE SOM-BASED FORECAST

Average RMSE from the SOM-based forecast is 9.4 cm/s for period 1, 7.7 cm/s for period 2 and 8.4 cm/s for period 3. The best metrics are estimated during the winter period (period 2), while they are much larger during summer and spring periods (periods 1 and 3, respectively). The respective ROMS-based metrics have a similar level of quality, with slightly higher RMSE values during summer (period 1: 10.0 cm/s) and spring (period 3: 9.8 cm/s) and lower during winter period (period 2: 7.5 cm/s). Averaged over all testing periods, the ROMS-based forecast is 8.7% worse than the SOM-based forecast.

We have further estimated both ROMS and SOM forecast system skill changes over different SOM solutions (Fig. 3). For any given SOM pattern we computed: (1) RMSE between the current field characteristic for that pattern and the surface current field obtained from measurements; (2) RMSE between ROMS currents and the surface current field obtained from measurements; both (1) and (2) were computed for those times for which a particular SOM pattern was forecasted. The pattern BMU16, associated to strong bora wind (Fig. 4), has the largest RMSE values. During wintertime (period 2), this pattern appears more frequently than during other investigated periods (9% in period 2, 1.5% in period 1 and

5% during period 3) and is slightly better forecast by the ROMS than by the SOM forecast system (Fig. 3). On the other hand, it is better forecast by the SOM during periods 1 and 3, but its representation in all patterns is smaller (1.5 and 5%, respectively). Another important extreme wind pattern, BMU20, is associated with the sirocco wind blowing from the south (Fig. 4). This pattern has a much lower frequency (up to 3%) during testing periods (Fig. 3). Its SOM-based forecast also provides larger errors than the ROMS forecast during periods 1 and 2, but smaller errors in period 3 when the pattern BMU20 is slightly more frequent. The overall shape of the RMSE distribution over the BMUs (Fig. 3) shows more resemblance between periods 1 and 3, which are characterized by vertical stratification in the sea, than in period 2, when the sea is usually homogenized[26]. Also, RMSE is much smaller during period 2 than during periods 1 and 3. Therefore, baroclinicity introduces additional error to both SOM- and ROMS-based forecasts.

SOM and ROMS forecast for BMU 16 and BMU 20 are analyzed in more detail. The BMU16 is a result of extreme bora conditions. The bora wind is strongly horizontally variable over a kilometer spatial scale in the northern Adriatic[27], and it might not be properly represented by the 2-km resolution ALADIN/HR mesoscale model. Mesoscale atmospheric models tend to largely underestimate energies at high frequencies[28], thus implying that higher resolution is needed in the model to fully describe the bora wind dynamics.

For that reason surface currents forecast by the SOM-based operational system – which are based on real measurements – are more intense and reliable during extreme events than surface currents forecast by ROMS. As seen in Fig. 4, the SOM-forecasted surface currents are much stronger (30–40%) than ROMS-forecasted values. Also, ROMS-forecasted fields corresponding to BMU16 wind conditions contain a northward current in the southern part of the domain, being a part of the cyclonic circulation known to occur in the area of the northern Adriatic during strong bora forcing[29]. By contrast, the SOM-based forecast has very weak currents in the southern part of the domain associated with the BMU16 pattern, followed by a lower RMSE than the one of the ROMS-based forecast (Fig. 4). Two alternative explanations for the discrepancy are feasible: (i) ALADIN/HR wind forcing during wintertime extreme bora wind creates an artificial strengthening of the cyclonic gyre in the area of the northern Adriatic, frequently reproduced by ocean numerical models which use atmospheric mesoscale models[30], or (ii) HF radar measurements are of lower quality at the southernmost part of the domain, which is covered by two HF radars only.

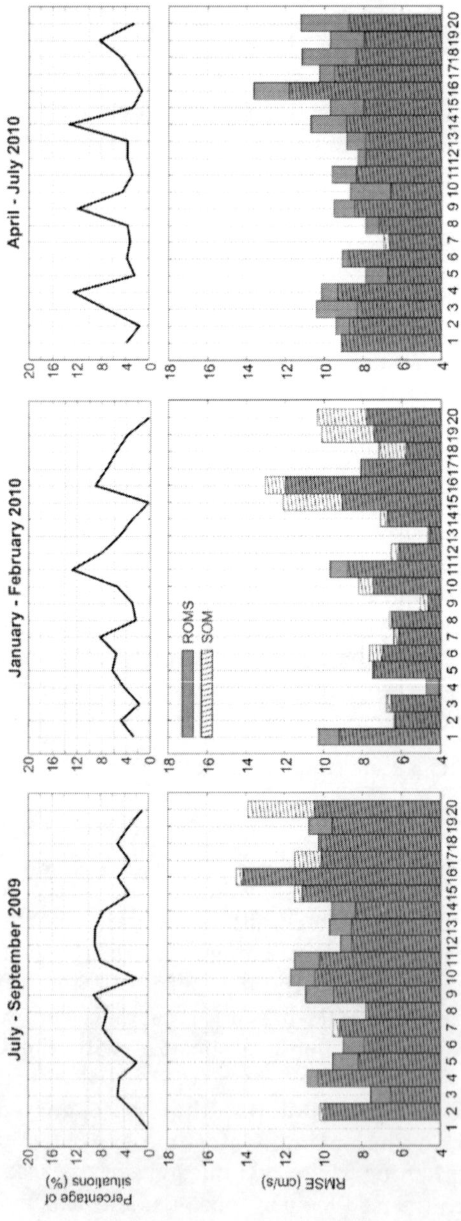

Figure 3: Distribution of root-mean-square error (RMSE) between SOM- and ROMS-derived forecast of surface currents and the respective measurements per BMU and for three testing periods. Percentage of situations ascribed to a particular BMU is also shown. The figure has been created using MATLAB (www.mathworks.com) software.

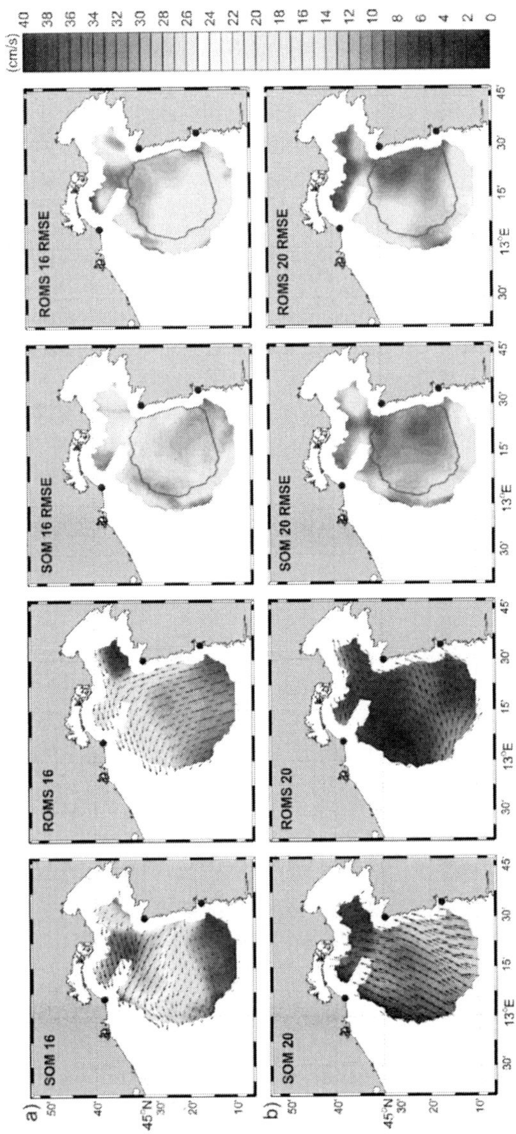

Figure 4 Surface currents obtained by the SOM-based forecasting system, the respective average surface currents obtained by ROMS forecasting system, and the RMSE between SOM-based and ROMS-based forecast of surface currents and the measurements, computed for (**a**) BMU16 (associated with strong bora wind), and (**b**) BMU20 (associated with sirocco wind). Encircled area includes grid points with at least 60% operational coverage during the testing period. The figure has been created using MATLAB (www.mathworks.com) software.

The difference between the SOM- and ROMS-based forecasts is also evident during intervals represented by other patterns, like the BMU20 (Fig. 4), which is representative of the second dominant wind in the northern Adriatic, the sirocco[31]. Although currents are not as strong as during strong bora episodes, the SOM-based forecast still gives 20–40% stronger currents than the ROMS-based forecast. In addition, there is a 10–30% clockwise offset in ROMS-forecasted current direction.

The spatial distribution of RMSE during periods corresponding to BMU16 and the BMU20 patterns (Fig. 4) reveals that both SOM- and ROMS-based forecasting systems reproduce mean current field relatively well (small RMSE in area with temporal coverage higher than 60%). On the other hand, very high RMSE values in areas with relatively low temporal coverage reveal that variability of the measurements is much higher than variability of both SOM and ROMS forecasts and that this variability is not adequately reproduced by either of the two forecast models.

4. DISCUSSION AND PERSPECTIVES

Both SOM- and ROMS-based forecasting systems perform short-term (72 h) forecasting of surface currents in the northern Adriatic with similar skill and quality. However, a limited dataset used for training of the SOM-based forecast, including only 10 months of HF currents in the northern Adriatic, presumably lower the quality of the SOM-based forecast. An increase of the training dataset to several years would likely improve the skill of the forecast, allowing for better and more reliable selection of an optimal number of SOM patterns. Such datasets are now available from HF radar measurements in numerous other regions of the World Ocean[32], so this approach can be easily tested.

There are several advantages when using SOM-based neural networks for forecasting ocean parameters, in our case surface currents. The first is that the system is based on real data, with a much simpler model learning approach than in other techniques based on real data, such as data assimilation[33]. Another advantage is that, after HF radar time series of sufficient length are collected and a SOM-based forecasting system for an area has been created, HF radars can be redeployed in other coastal areas, potentially allowing for a spatial extension of the SOM-based operational forecasting. Finally, the required time for execution of the SOM-based forecast is much smaller, an order of magnitude, than for forecasts provided by ocean numerical models.

On the other hand, a prerequisite for the proposed forecasting system is a high correlation between predictor and predictand, in our case between wind and surface currents. However, such a high correlation is common for coastal areas[34], widening application of a SOM-based forecasting system for surface currents at a number of coastal regions where HF radar measurements are available. Another problem is an under-representation of extreme situations in the forecasting system - and forecasting of extremes has much more practical use than forecasting of mean situations - if not trained by a satisfactorily number of BMU patterns. Such a problem may be overcome by a two-step SOM, i.e. by first training it on all measured current fields and then by re-training it only on those current fields which are representative for extreme conditions (e.g., BMU16 and BMU20 in our case). This would downscale the training process through use of a number of BMU solutions instead of one representative for extreme conditions. However, this should be accompanied with an increasing resolution of atmospheric numerical models - enabling them to recognize more variable wind fields. Also, a use of supervised neural network algorithms[35] with prescribed patterns in predictands might be a solution for improving the skill of the SOM-based forecasting system.

The evaluated SOM-based forecasting system in the northern Adriatic is performed on low-pass (33 h) filtered ocean currents and winds, and therefore it does not account for tides and other processes over hourly timescales, such as inertial oscillations or mesoscale phenomena originating outside of the domain. These processes might be significant in marginal seas (tidal currents are about 10 cm/s in the area[36]). For that reason, we tested both SOM and ROMS-based forecasting systems without preprocessing (low-pass filtering) on the same training and testing dataset, resulting in RMSE increase for about 60% for both forecasting systems compared to the computations with preprocessing. Nevertheless, the RMSE of the SOM forecast is on average about 9.5% smaller than the RMSE of the ROMS forecast. However, the increase of the RMSE in the SOM-based forecasting system might be lowered by removing tides prior to the SOM analyses and re-adding them later to the forecast patterns. Also, a significant increase of the testing dataset might allow for use of much more SOM patterns than used here (20 patterns), which, in the end, might result in more reliable reproduction of all processes appearing in a domain.

The proposed SOM-based forecasting system may be particularly applicable for areas with heavy marine traffic, harbors, touristic areas, etc., where forecasting of surface currents may be essential for preventing and mitigating impacts of marine accidents (e.g., oil spill, rescue at sea, safety of navigation). A similar architecture of the forecasting system could also be appropriate for forecasting different parameters in marine systems (in biogeochemistry, trophic relations, fisheries[37]) which are too complex for "classical" forecasting tools based on numerical models.

REFERENCES

1. Bryan, K. A numerical method for the study of the circulation of the world ocean. *J. Comp. Phys.* **4,** 347–376 (1969).
2. Haidvogel, D. B. & Beckmann, A. *Numerical Ocean Circulation Modeling* 344 p (Imperial College Press, 1999).
3. Oddo, P. *et al.* A nested Atlantic-Mediterranean Sea general circulation model for operational forecasting. *Ocean Sci.* **5,** 461–473 (2009).
4. Krishnamurti, T. N. *et al.* Improved weather and seasonal climate forecasts from multimodel superensemble. *Science* **285,** 1548–1550 (1999).
5. Chassignet, E. P. *et al.* The HYCOM (HYbrid Coordinate Ocean Model) data assimilative system. *J. Mar. Syst.* **65,** 60–83 (2009).
6. Lee, T. L. Neural network prediction of a storm surge. *Ocean Eng.* **33,** 483–494 (2006).
7. Kohonen, T. Self-organized information of topologically correct features maps. *Biol. Cybern.* **43,** 59–69 (1982).
8. Liu, Y. & Weisberg, R. H. In *Self-Organizing Maps: Applications and Novel Algorithm Design* (ed Mwasiagi, J. I.) 253–272 (InTech, 2011).
9. Chattopadhyay, R., Vintzileos, A. & Zhang, C. D. A description of the Madden-Julian Oscillation based on a Self-Organizing Map. *J. Clim.* **26,** 1716–1732 (2013).
10. Richardson, A. J., Risien, C. & Shillington, F. A. Using self-organizing maps to identify patterns in satellite imagery. *Prog. Oceanogr* **59,** 223–239 (2003).
11. Williams, R. N., de Souza, P. A. & Jones, E. M. Analysing coastal ocean model outputs using competitive-learning pattern recognition techniques. *Environ. Modell. Softw.* **57,** 165–176 (2014).
12. Liu, Y. & Weisberg, R. H. Patterns of ocean current variability on the West Florida Shelf using the self-organizing map. *J. Geophys. Res.* **110,** C06003 (2005).
13. Liu, Y., Weisberg, R. H. & Mooers, C. N. K. Performance evaluation of the self-organizing map for feature extraction. *J. Geophys. Res.* **111,** C05018 (2006).
14. Hisaki, Y. Classification of surface current maps. *Deep-Sea Res. I* **73,** 117–126 (2013).
15. Chen, C. S., Duan, S. X., Cai, T. & Liu, B. Y. Online 24-h solar power forecasting based on weather type classification using artificial neural network. *Solar Energy* **85,** 2856–2870 (2010).
16. Chang, L. C., Shen, H. Y. & Chang, F. Y. Regional flood inundation nowcast using hybrid SOM and dynamic neural networks. *J. Hydrol.* **519,** 476–489 (2014).
17. Lipa, B.J. & Barrick, D. Least-squares methods for the extraction of surface currents from Codar crossed-loop data – application at ARSLOE. *IEEE J. Ocean. Engineer.* **8,** 226–253 (1983).
18. Cosoli, S., Bolzon, G. & Mazzoldi, A. A real-time and offline quality control methodology for SeaSonde high-frequency radar currents. *J. Atmos. Oceanic Technol.* **29,** 1313–1328 (2012).
19. Kovačević, V., Gačić, M., Mancero Mosquera, I., Mazzoldi, A. & Marinetti, S. HF radar observations in the northern Adriatic: Surface current field in front of the Venetian Lagoon. *J. Mar. Syst.* **51,** 95–122 (2004).
20. Aladin International Team. The ALADIN project: Mesoscale modelling seen as a basic tool for weather forecasting and atmospheric research. *WMO Bulletin* **46,** 317–324 (1997).
21. Tudor, M. & Ivatek-Šahdan, S. The case study of bura of 1st and 3rd February 2007. *Meteorol. Z.* **19,** 453–466 (2010).

22. Mihanović, H. *et al.* Surface current patterns in the northern Adriatic extracted from High-Frequency radar data using Self-Organizing Map analysis, *J. Geophys. Res.* **116,** C08033 (2011).
23. Shchepetkin, A. F. & McWilliams, J. C. The regional oceanic modeling system (ROMS): A split-explicit, free-surface, topographyfollowing coordinate ocean model. *Ocean Modell.* **9,** 347–404 (2005).
24. Oddo, P., Pinardi, N., Zavatarelli, M. & Coluccelli, A. The Adriatic basin forecasting system. *Acta Adriat.* **47** (Suppl.), 169–184 (2006).
25. Janeković, I., Mihanović, H., Vilibić, I. & Tudor, M. Extreme cooling and dense water formation estimates in open and coastal regions of the Adriatic Sea during the winter of 2012. *J. Geophys. Res.* **119,** 3200–3218 (2014).
26. Buljan, M. & Zore-Armanda, M. Oceanographic properties of the Adriatic Sea. *Oceanogr. Mar. Biol. Ann. Rev.* **14,** 11–98 (1976).
27. Grubišić, V. Bora-driven potential vorticity banners over the Adriatic. *Q. J. R. Met. Soc.* **130,** 2571–2603 (2004).
28. Skaramock, W. C. Evaluating mesoscale NWP models using kinetic energy spectra. *Mon. Wea. Rev.* **132,** 3019–3032 (2004).
29. Beg-Paklar, G., Isakov, V. Koračin, K., Kourafalou, V. & Orlić, M. A case study of bora-driven flow and density changes on the Adriatic shelf (January 1987). *Cont. Shelf Res.* **21,** 1751–1783 (2001).
30. Kuzmić, M., Janeković, I., Book, J. W., Martin, P. J. & Doyle, J. D. Modeling the northern Adriatic double-gyre response to intense bora wind: A revisit. *J. Geophys. Res.* **111,** C03S13 (2006).
31. Jurčec, V., Ivančan-Picek, B., Tutiš, V. & Vukičević, V. Severe Adriatic jugo wind. *Meteorol. Z.* **5,** 67–75 (1996).
32. Harlan, J. *et al.* The integrated ocean observing system high-frequency radar network: Status and local, regional, and national applications. *Mar. Technol. Soc. J.* **44,** 122–132 (2010).
33. Moore, A. M. *et al.* The Regional Ocean Modeling System (ROMS) 4-dimensional variational data assimilation systems Part I - System overview and formulation. *Prog. Oceanogr.* **91,** 34–49 (2011).
34. Paduan, J. D. & Rosenfeld, L. K. Remotely sensed surface currents in Monterey Bay from shore-based HF radar (Coastal Ocean Dynamics Application Radar), *J. Geophys. Res.* **101,** 20669–20686 (1996).
35. Melssen, W., Wehrens, R. & Buydens, L. Supervised Kohonen networks for classification problems. *Chemometrics Intell. Lab. Syst.* **83,** 99–113 (2006).
36. Malačič, V., Viezzoli, D. & Cushman-Roisin, B. Tidal dynamics in the northern Adriatic Sea, *J. Geophys. Res.* **105,** 26265–26280 (2000).
37. Cheng, L., Lek, S., Lek-Ang, S. & Li, Z. J. Predicting fish assemblages and diversity in shallow lakes in the Yangtze River basin. *Limnologica* 42, 127–136 (2012).

CHAPTER 3

Novel Tools for the Evaluation of the Health Status of Coral Reefs Ecosystems and for the Prediction of Their Biodiversity in the Face of Climatic Changes

Stéphane La Barre[1, 2]

[1] *Université Pierre et Marie Curie-Paris 6, UMR 7139 Végétaux marins et Biomolécules, Station Biologique F-29680, Roscoff, France*

[2] *CNRS, UMR 7139 Végétaux marins et Biomolécules, Station Biologique F-29680, Roscoff, France*

1. INTRODUCTION

Coral reefs concentrate between one quarter and one third of the total marine biodiversity, according to different estimates, though they only cover about 0.1% of the global oceanic surface and are confined to tropical and sub-tropical latitudes. The extraordinary diversity in invertebrates and fish species is often compared to that of arthropods and vertebrates of tropical primary rainforests. Half a billion humans depend partly or totally on the goods and services provided by coral reef ecosystems.

However, coral reefs are now recognized as being among the most fragile of all environments in the face of localized anthropic pressures (overfishing and various atmospheric and water pollutions) and of their climatic consequences of planetary dimensions. Within the last 30 years, about 20% of all coral reefs have being totally destroyed while another estimated 60% are damaged to some extent – a few beyond recovery - and only 20% can still be regarded as unharmed.

With a human population growing at faster rates in developing nations, biodiversity concerns are conflicting with the pressures to exploit local mineral resources and to develop agricultural and seafood production. While residents often deplore the gradual changes in the natural habitats in which they were brought up, socio-economic pressures originally due to post-colonial emancipation and now to global trade tend to resist any policy that may be considered restrictive or receding. Almost every aspect of this economic growth relies on fossil fuel energy consumption and to a lesser extent on vegetal biomass burning, which links development with climatic changes, and also with the generation of polluting wastes.

Educational programs and non-government associations propose alternatives to established practices in habitat and resource management and waste disposal by end users. Research scientists continue to explore natural biodiversity in remote pristine environments (especially *biodiversity hotspots*, a very useful conservation-promoting concept) and to observe its losses in degrading habitats. Programs on bioremediation of impacted sites are attracting funds, and classification of natural habitats as protected sites is gaining public support. Yet the scientific community has very little leverage on the decision-making of potentially impacting industrial, commercial and urban development projects, and on the say-so broadcast by their promoters in the media. In-house consultancy is often hampered by the lack of appropriate analytical tools, and reports are often biased in favor of employment and cash return prospects.

In a previous chapter of this series on biodiversity [1] the various ecological consequences of climatic changes, of chemical and microbial pollutions and of overexploitation of natural resources have been reviewed for coral reef ecosystems. Suggestions have been made on the basis of recent publications by experts on various subjects, as to how modern techniques and innovative approaches could be used appropriately to complement the above initiatives.

In this chapter, a holistic concept is proposed that (i) integrates cutting-edge molecular research and standard technologies with field sampling and laboratory simulations of natural habitats (ii) using holobiont-based *sentinel* systems, (iii) into a single tool that "shows evidence" of ongoing degradation rather than aftermath "score loss". Corrective action can then be taken in specific directions before no-return limits have been reached and total ecosystem collapse is on the way.

2. BIODIVERSITY IS OUR RESPONSIBILITY: THE FUTURE OF MANKIND DEPENDS ON IT

2.1. If the origins of life remain controversial, biodiversity is a miracle of sorts

Basic forms of life like bacteria are reputedly capable of withstanding extreme conditions, to the point that scientists of repute such as Sir Francis Crick, Carl Sagan or Stephen Hawking have speculated on an alien origin to terrestrial life, which is now held as a tenet of the modern panspermia theory by some exobiologists. Dormant bacterial spores or alike would have been seeded on our planet, possibly from different sources and at different times. Those that encountered favorable "starting" conditions, supposedly in the chemolithotrophic environments around oceanic ridges, would have initiated the evolutionary scenario we know. Posed under such terms, the true *origin of life* escapes our observation, but the *origin of biodiversity* does not, as it remains intrinsically terrestrial.

Satellite views of the ionized portion of our atmosphere show it as a barely perceptible glow that outlines the shape of our planet against the black outer space background. Just under it, blue expanses of oceanic waters spread as a delicate film less than one-thousand times thinner than the supporting "blue marble", but over two thirds of its surface (Figure1). Life forms occur at the sea-air and soil-air interfaces, just where geoclimatic fluxes and exchanges are the most rapid and subjected to biotic influences. Thus marine biodiversity thrives under specific conditions that are only found between the oxygen-rich lower atmosphere (at cloud canopy level on Figure 1), and the sunlit oceanic surfaces.

One-third of all known marine species is concentrated on confettis occupying one-thousandth of the oceanic surfaces, also thanks to favorable conditions afforded by land influence, i.e. marine volcanoes and continental shelves under tropical-subtropical latitudes. This circumstantial miracle is called coral reefs.

Figure 1. (left) - Composite image of Planet Earth viewed from orbiting satellite Suomi NPP on January 4, 2012 (http://npp.gsfc.nasa.gov/), scaled here at approximately 5 x 10^{-9}; (right) -The biosphere is about one-thousandth the diameter of the planet – just as thin as the surface of a soap bubble that size, a metaphor illustrating its instability and fragility.

2.2. Extinction events are natural over long periods after which biodiversity has to reinvent itself

Any lasting change in the biogeochemistry of any of the three components (atmosphere, seawater and land) will disrupt the interfacial equilibrium that supports the many thousands of life forms that interact constantly within an ecosystem. Mass species extinctions occurred several times (6 or 7 described) in the history of our planet since it became life-supporting, every time followed with new and better adapted life forms and a biodiversity climax attained after long periods of environmental stability. Changes in soil mineral strata indicate the occurrence of biodiversity-modifying events such as occupation by seas or the occurrence of an ice-age. Discrete organic layers may indicate the presence of a tropical rainforest or of a dry land savannah. Datable fossil evidence within these strata, together with paleontological reconstructions, point out the floristic and faunistic peculiarities of the times. Core drills in ice provide datable evidence of biogeoclimatic episodes within the last few millennia, while core drills in massive scleractinian corals give accurate calendar-like records of recurrent or of accidental climatic events affecting their biotope.

Speciation usually goes along with occupation of new territories and new habitats, the first colonizers having acquired the necessary adaptations to cope with evolving external demands – the Cambrian explosion (545 million years ago) being the most dramatic example of such adaptive diversification at all scales.

Along with this, evidences of accidental episodes of massive species extinctions are witnessed by the sudden "disappearance" of terrestrial and of marine life, that are attributable to tectonic, telluric or meteoritic impacts and to their profound and lasting climatologic and geochemical consequences. The most significant mass extinction is undoubtedly the Permian-Triassic Great Dying [2] where a 96% loss of all non-microbial marine life occurred within ten million years. The precise causes of mass extinction events may be in connection with continuous tectonic movements with their telluric and volcanic outbreaks and their climatic consequences, to collisions with meteoritic bodies, and to a lesser extent to the appearance of dominant predators, parasites or microbial diseases, or to combinations thereof. Common to many extinction events, however, is the massive release of greenhouse and of toxic gases (carbon dioxide, methane, hydrogen sulfide etc.). The water solubility of CO_2 being nearly 30 times that of oxygen, water acidification occurs that impacts preferentially all calcifying organisms with low metabolic rates and weak respiratory systems: most coral genera died out during the Great Dying, along with calcareous sponges, calcifying algae, echinoderms, bryozoans etc. [3].

Interestingly, profound taxonomic changes in all major phyla seem to follow extinction events, resulting in a better adapted biodiversity. Nothing is known, however, on the consequence of such changes in microbial life or on the putative role microbial associates had on the *reinvention* (understand: adaptive evolution) of new species. As Falkowski [4] puts it, animals and plants are merely new incarnations of ancient metabolic processes, but the ultimate key to biodiversity may be held by bacteria ferrying the set of core genes that are necessary for life to express itself.

2.3. Brutal human influence may lead to hazardous extinction events

Man is of very recent occurrence in the history of terrestrial life, and until the advent of the industrial age (especially since the middle of the 19th century) his planetary influence was minimal, and the destructive potential of its inventions was purely local. Since then, his population has increased seven fold while consuming or exploiting 42% of the terrestrial net primary production [5]. The single most environment-impacting activity today is the production of energy from fossil fuel, and the single most mechanically destructive invention is that of nuclear weapons. More insidious are the thousands of new chemical species (mostly organic) that are produced and improperly disposed of once used, and the new strains of mutant microorganisms that pop up unpredictably and threaten to create worldwide pandemias affecting both wildlife and humans.

Without reviewing the subject of negative traces of human doings (see [1]), a reflection on the accelerating pace of carbon volatilization in the atmosphere is necessary to appreciate its effect on marine biota and on coral reef ecosystems in particular. In one year, we extract the equivalent of one million years' worth of fossil fuel natural production [4]. In other words, using primitive technology e.g. internal combustion engines, we enrich our atmosphere in carbon dioxide at a pace at which terrestrial and marine autotrophs will not be able to keep up with in their efforts to create new biomass and quench free carbon excess and reduce seawater acidification rates.

3. CORAL REEF MANAGEMENT: BETWEEN THE ECONOMY'S HAMMER AND THE CLIMATE'S ANVIL

Ecosystems at immediate risk are biodiversity hotspots (primary rainforests and land-connected reefs systems), but most ecosystems with endemic species at all latitudes will be profoundly affected by acclimation of competing aliens. Being able to analyze, quantify and predict changes is the first step to avoid losing control... indeed environmental scientists should endorse the role of a general practice physician making an overall check-up on an ailing patient and prescribing a course of medications and exercises for recovery, or that of an investigator using state-of-the-art forensics such as DNA amplification to evidence causes of a crime and provide evidence to lawyers. This will become a necessity, beyond economic pressures worldwide (q.v. [6]).

3.1. Climate changes have profound biogeoclimatic consequences

Climatic changes act globally, and the effects of naturally occurring destructive episodes are now superseded by those due to ever-growing man-induced emissions of various sorts. The latter include carbon, nitrogen or sulfur transfers that are generated essentially through the combustion of fossil organic matter or of agricultural or livestock breeding activities, and manifesting themselves differently according to their biogeochemical state.

1) First, their volatilization as simple molecular species which are responsible for temperature rises through the glasshouse shielding effect, while halocarbon emissions tend to destroy the protective ozone

layer of the upper atmosphere. Elevated surface water temperatures and genotoxic radiations are responsible for the bleaching effect and cellular stresses observed on subtidal corals, and death may follow if the exposures are lasting. Also, strong seasonal evaporation of large volumes of surface waters leads to atmospheric pressure instabilities, resulting in the occurrence of more frequent and more severe hurricanes. The mechanical damage due to wave action and to the sudden input of large volumes of freshwater and alien material (silt and debris) in coastal zones may entirely destroy some reef portions in single episodes;

2) Second, the restitution of most of the above molecules at the air-water interface and their incidence on biomineralization of many marine invertebrates and of their larvae, and of coralline algae; In the 22nd century, seawater acidification may be as large as 0.5 pH units below the pH levels recorded in early 20th century. The extent of seawater acidification due to a shift in bioavailability of HCO_3^- concentration, if still debated in 2012, may shortly become dramatic for the survival of calcifying invertebrates (corals, mollusks, echinoderms, some sponges etc. and their larvae) and of coralline algae.

3.2. Societal changes: think globally, act locally

Socio-economic determinants are linked to the development of industries, commerce and urbanism, their primary effects being direct, localized and traceable.

- industrial activities typically generate important volumes of polluting agents that dissipate during the processing stages. Furthermore, residues are managed according to criteria of productivity and cost-effectiveness i.e. with little concern about recycling alternatives, and about speciation into toxicants that make their way from the discharge areas into waterways and eventually into the sea. Ore excavation or grading of metal-rich top soil cause faunistic and floristic deteriorations, i.e. loss of endemic wildlife, facilitate bioerosion, cause the constant remodeling of river mouths with the finer particles and destroy almost life forms with the accumulation of heavier deposits in basins. In the sea, bioaccumulation of organic pollutants by primary consumers can expectedly reach very high concentrations in top predators, e.g. lipid-bound chlorinated biphenyls. Heavy metals that leach out from anti fouling formulations are accumulated by filter-feeding invertebrates, and take a heavy toll on the reproductive success of many life forms found in intertidal and subtidal life forms around harbor waters;

- intensive agriculture and farming generate various type of pollution, (i) enrichment by fertilizers and farm animals disjecta, (ii) pesticides and systemic herbicides, (iii) veterinary products, hormones that alter the quality of underground water reserves, while excess runoff benefits undesirable primary production in coastal zones, e.g. "green tides" and microalgal blooms;
- global commerce favors dissemination of alien species by shifting large amounts of ballast water by container ships and cleaning of hulls;
- urban development in tropical zones eliminates mangrove forests and seagrass beds that are important coastal component of fringing reefs. Shifting into the sea of very large volumes of untreated sewerage generated on land causes eutrophication in coastal reef systems with the introduction of novel microbial diseases.

3.3. Scientists should play a central role in biodiversity issues

At present, there are no reliable estimates of the rate at which coral reef biodiversity is going to be affected by climate and anthropic forcings by year 2100. The various proxies used by climatologists and reef ecologists predict two types of scenarios: (i) short term collapse of tropical reef biodiversity as early as 2050 if current trends are not curbed, (ii) a good level of resilience if present-day conditions are maintained as an absolute maximum, allowing photosymbiotic systems to adapt gradually. By limiting direct interference (q.v. *overfishing, algal proliferation, alien species*) that introduce phase-shifts in the complex and otherwise self-regulated food chains, acclimatization of most common coral species can be made possible in the face of climatic changes. The second alternative represents the only and narrow line of action, knowing that it will not be realistic to completely stabilize CO_2 emissions to safe levels, i.e. natural quenching capacity of excess by phytoplankton.

In 2012, scientific expertise is mostly directed to the discovery and documentation of new species in remote or in hitherto neglected regions, or to report the loss of biodiversity in areas impacted by human activities. Consulting is sought for the delineation of protected areas and establishing quotas for fishing. Science-based management is necessary, in particular in regions suffering from direct and rapidly growing human influence [7]. Particularly at risk are young developing nations and small islands that are placed under tremendous pressure from major industrial countries to exploit and export their natural resources – a socio-political scenario that tends to divide local populations and sometimes accelerate environmental issues. Thus, care must be taken of the "growth crises" of these culturally fragile ethnic groups undergoing these economic pressures, and a premium should be placed on the management of their natural patrimony, in which their culture is firmly rooted.

What comes out most often in recent discussion groups is: (i) that new monitoring tools, new environmental technologies and new models must be developed within a consensus mode between interested parties (ii) scientific expertise and field experience must be federated into autonomous consortia, (iii) implementation stages must follow a rigorously coordinated and step-wise approach. The present feeling is that scientific expertise is only marginally called upon and often misquoted in the decision-making of matters of economic development and urbanization in developing countries. By following the above recommendations, habilitation of the scientist as an advisor or as a mediator playing a crucial role in decision-making will be greatly facilitating the social dialogue on ecological matters.

4. STATE-OF-THE-ART: NEW CONCEPTS, NOVEL APPROACHES, CUTTING-EDGE TECHNOLOGY ETC...

How to find a suitable management compromise between climate and human forcings without altering the economy and the social development of developing nations? This is everybody's concern: entrepreneurs, merchants, economists, businessmen, politicians, educators, end-users and consumers and of course, scientists.

From the scientist's point of view, each identifiable environmental issue has its specific set of solutions, and should be treated with the same care as an outpatient's condition is in a hospital. The physician makes a global assessment and recommends that a series of analyses be made by specialists before he confirms his diagnosis and prescribes an adapted treatment. In most cases, the patient will recover successfully – sometimes he has to be hospitalized for some time, and on rare occasions he may not leave the hospital alive.

After an overall check-up in which the probable stressor(s) is (are) identified, the environmental investigator will hand the case over to specialists who will each carry out a set of biochemical, molecular, microbiological and imaging tests on specific model organisms, on their tissues, cells, body fluids and associated microbes. Once the diagnosis is confirmed, a preliminary report is made that may contain special recommendations, followed by regular visits to evaluate the resilience of the system and the potential for the recovery of its lost biodiversity.

To be able to adapt this medical approach to an ailing ecosystem, the environmentalist needs to find representative biological study models (i.e. sentinel species that are sensitive to the stressors, but not to the point of immediate eradication at the onset of a mild exposure). He also needs to study characteristic and observable symptoms, their evolution and their succession. Early markers of an organism exposed to a stressor can be detected using functional genomics with suitable molecular tools, and the evolution of the responses can be followed thenceforth. For instance, an organism undergoing abiotic stress is usually more prone to microbial infection than a conspecific control organism, calling for physiological and bacteriological / fungal / viral / parasitic analyses. Unfortunately, very few field investigators are in a position to use these tools routinely, let alone in association - especially molecular tools that have been adapted from the medical world to specific biological models, and only for research.

4.1. Shifting from a consumer-minded to a conservation-minded attitude

Coral reef organisms tend to live in close association in order to gain optimal access to essential resources such as hard substratum, access to light, appropriate food etc., the co-occurrence of which is limited in shallow tropical waters [8], which are naturally efficient at nutrient cycling (oligotrophic regime). These constraints are reflected in the highly sophisticated communal assemblages formed by coralline and fleshy algae, invertebrates, vertebrates, protists etc. which communicate essentially via surface-to-surface or distant interactions of immunological and/or chemical nature. All forms of associations are encountered, ranging from obligate parasitism to symbiosis via commensalism, from prey hunting to filter feeding via surface browsing, etc.

The scientific literature on coral reef biology has traditionally emphasized on competitive aspects between members sharing a same habitat, especially in connection with secondary metabolites emitted by sessile or sedentary invertebrates and having allelopathic or growth inhibiting activities [8,9]. Only recently have the cooperative and functional aspects of interspecies and cross - phyletic communication been explored - stepping from a more "medical" attitude (i.e. pharmacologically-oriented) to a more "ecological" one (i.e. conservation-oriented). Discoveries such as bacterial inter - communication via quorum-sensing signals, biofilm studies, and of course the progresses made in genomics have greatly contributed to this attitude change (e.g. the coral probiotic hypothesis of Reshef et al. [10].

In the eighties, the term *biodiversity* was publicized by [11] to account for the need to evaluate and to manage biological resources in endangered habitats. It gained immediate recognition as a key concept in life sciences, and public adoption followed relayed by media channels, e.g. "citizen science" in which data collection by non-scientists was encouraged to feed statistics on endangered species, as well as public initiation to scientific issues [12]. Ecosystem services and human well-being oriented approaches in general, while often assimilated to biodiversity-conservation strategies, may not ultimately pursue the same goals, calling for trade-offs in difficult issues [13].

The biodiversity concept certainly helped in the funding of sampling expeditions in diversity *hotspots*, another term useful to conservationists and to taxonomists. The Tree of Life (http://tolweb.org/tree/) is a collaborative web project in which over 10,000 web pages provide information about extant groups of organisms and their evolutionary history.

Rapid progress in molecular science, and genomics in particular, has made it theoretically possible to retrieve useful information from crude environmental samples, i.e. doing away with cultivation restrictions and going beyond conventional taxonomy based on morphological traits. However, metagenomic characterization of marine microorganisms from the plankton, from biofilms or associated with macrobiota and their exsudates, generates phenomenal volumes of data from which crucial information extraction is a difficult and costly task for biocomputing specialists. Advocated by some as a new paradigm for biodiversity studies [14], "data-intensive science" taking a naïve (data-driven, i.e. non-theory-based) approach looks for truly novel and surprising patterns that are "born from the data". But because of meta-data sorting problems and costs, and since environmental problems cannot wait for new hypotheses to emerge, others think that "knowledge-based science" can at least confront existing hypotheses against the meta-data background and guide investigators into detecting novel information [15].

4.2. The holobiont as an evolutionary concept, the *extended holobiont* as a conservation concept

Closer to hands-on science, genomists have coined a very useful concept, originaly to account for the functional dynamics associated with bioconstructing organisms such as coral, sponges, or coralline algae in tropical marine ecosystems: that of the *holobiont*, an entity that includes the host (basibiont) and its microbial associates (microbionts) – the *hologenome* being the sum total of the associated genomes [16]. Photosymbiotic systems may function of both autotrophic and heterotrophic modes with the assistance of, respectively, symbiotic microalgae or cyanobacteria during the day, and element-recycling bacteria inside bacteriocytes or associated with the mucus, tissues or skeletal cavities followed by directed uptake at night. This dual mode of operation is conferring an adaptive advantage to short-term disturbances [17].

The hologenome theory was later generalized to terrestrial eukaryotic-prokaryotic systems, including man and its microflora [18], in order to account for the coevolutionay and cross-kingdom aspects of symbiosis [19].

Finally, the holobiont is also a practical concept insofar as it can be conveniently transposed from its natural sites to a microcosm or mesocosm aquarium setup, in order to evaluate stress impacts on both the host and its associated microbiota.

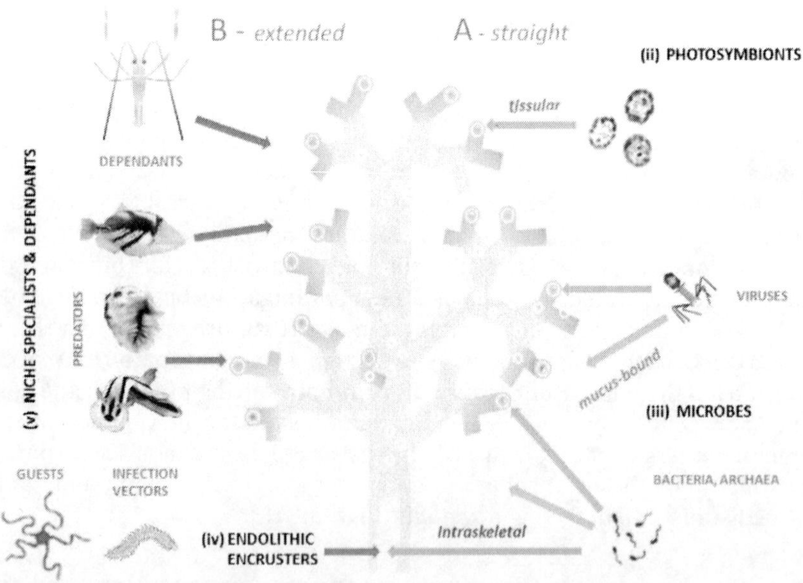

Figure 2. (right) A - Representation of the coral holobiont *sensu stricto* i.e. basically a functional 3-way photosymbiotic system with (i) the sedentary or sessile basibiont (center) and its mucosphere (pale grey halo) with (ii) its photosymbiotic associates and (iii) prokaryotic microbiome; (left) B - The *extended holobiont*, which may include (iv) endolithic associates (fungi, encrusting sponges and algae, protists) and (v) biodiversity that lives in association with the coral and may be critically affected by the loss of the host

The concept of *extended holobiont* is proposed here, as the urge is now felt to develop practical tools to directly address biodiversity issues. In their definition of the coral holobiont, [20] include cryptic associates that are structurally associated with the host (fungi, encrusting sponges and algae, protists). In order to maintain some evolutionary coherence between the holobiont *sensu stricto* and the hologenome concepts of Rosenberg, we may include the dozens of specialized and mobile life forms that live in trophic or parasitic or commensal relationship with the host (e.g. little crabs, shrimps, echinoderms, ascidians, polychaete worms, shelled or naked mollusks, planarians, fish, crustose algae, fungi, foraminifera, etc.) into a wider concept which may account for the host- dependant biodiversity. We propose to refer to it as the *extended holobiont*.

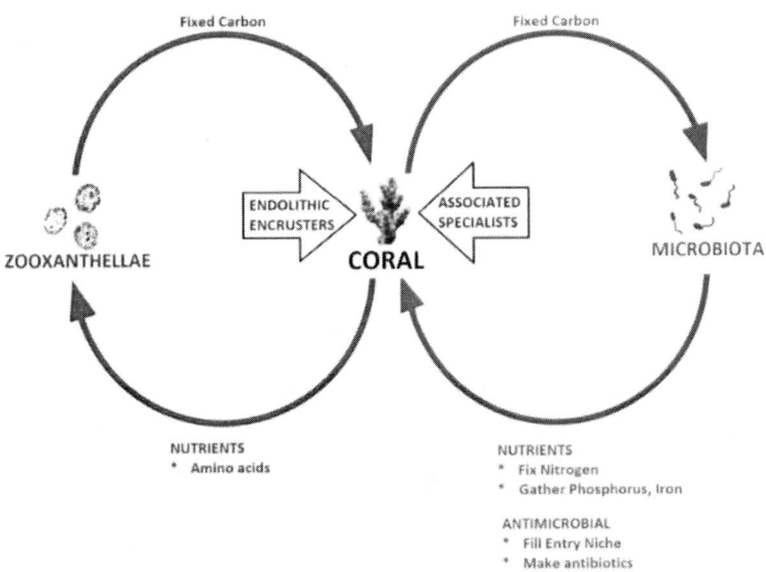

Figure 3. Relationships between the components of the coral holobiont system (developed from [20], with the addition of cryptic organisms and of associates (the *extended holobiont*)

4.3. The "omics" or high-dimensional biology revolution

"Omics" technologies are primarily aimed at the universal detection of genes (*genomics*), mRNA (*transcriptomics*), proteins (*proteomics*) and metabolites (*metabolomics*) in a specific biological sample [21]. Omics can be used in a wide variety of applications, ranging from biomedical to environmental, from biotechnological to ethical, from single-cell to ecosystem-wide, from systems biology to phylogeny. In short, "omics" technologies adopt a holistic view of the molecules that make up a cell, tissue or organism. This integrative approach, together with the "extended holobiont" as a functionally comprehensive biological model, provides a useful conceptual framework when dealing with complex environmental issues.

Transcriptomics - Perhaps the single most informative approach to studying the fitness of a test organism being subjected to an environmental stress is transcriptomics. In essence, the aim is (i) to catalogue all species of gene transcripts, including mRNAs, non-coding RNAs and small RNAs (non-coding bacterial RNAs), (ii) to determine the transcriptional structure of genes in terms of their start sites, 5' or 3' ends, splicing patterns and other post-transcriptional modifications, and (iii) to quantify the changing expression levels of each transcript during development and under different conditions [22]. The set of coding transcripts from activated genes will encode proteins as products of the ribosomal assembly line. Initially, complementary DNA (cDNA) copies are created from the mRNA templates, from which two analytical techniques have been developed: (i) hybridization-based and (ii) sequence-based. In hybridization transcriptomics, DNA microarrays composed of collections of a collection of microscopic DNA probes are used to hybridize cDNA targets - here a gene is activated when a spot is highlighted with a visual label, allowing quantification as well as characterization. Yet, the method relies on existing knowledge about the genome sequence. Modern sequence-based methods (RNA-seq) determine cDNA sequences directly, the latter (30-400 bp.) being attached terminally to adaptors. Each short sequence is aligned with the reference genome or transcriptome and classifies as exon read, junction reads and poly(A) end reads, from which a base-resolution expression profile is generated. A critical comparison between the different transcriptomic techniques is proposed in [22], with an emphasis on RNA-seq.

Stress reactions have been studied on of components of the coral holobiont. In [23] De Salvo and collaborators performed a medium-scale micro transcriptomics experiment on the Carribean coral *Montastrea flaveolata* undergoing thermal stress and bleaching vs. unexposed conspecifics. Their complementary microarray experiment containing 1310 genes revealed that thermal stress and bleaching in this species affected the following processes: oxidative stress, Ca^{2+} homeostasis, cytoskeletal organization, cell death, calcification, metabolisms, protein synthesis, heat-shock protein activity and transposon activity. A dedicated transcriptome database was recently produced on *Pocillopora damicornis*, a ubiquitous and environmentally sensitive scleractinian [24], [http://cnidarians. bu.edu/PocilloporaBase]

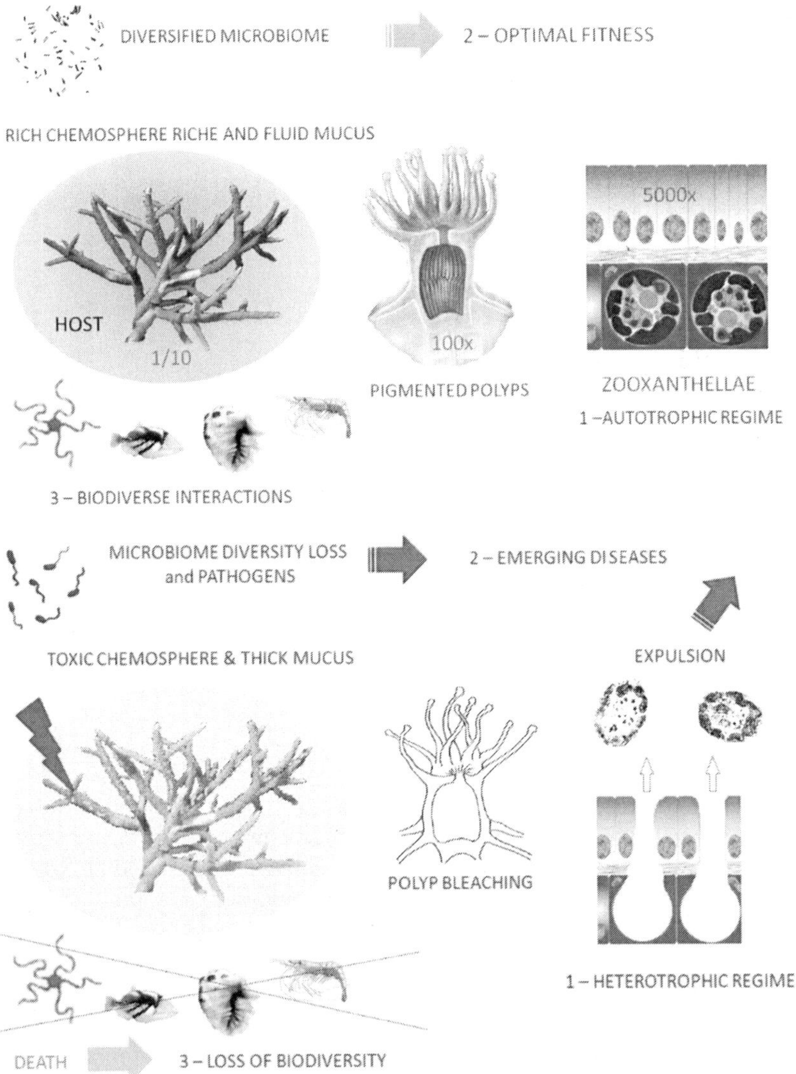

Figure 4. The extended holobiont of a scleractinian sentinel system under contrasting conditions: (1) top: under optimal conditions, the fit colony produces a fluid mucus which hosts a well-diversified and functional microbial flora; the polyps are colourful thanks to a dense population of symbiotic zooxanthellae that colonize the endodermal layer and provide the substantial autrotrophic part of the energy budget of the holobiont. Eukaryotic associates are numerous, engaging into a network of interactions centered on the host. (2) Bottom: under lasting stress, e.g. sea water temperature above 32°C, the colony may produce a thicker mucus, in

which a growing population of pathogenic bacteria tends to displace less competitive strains. Zooxanthellae leave the polyp tissues as a result of loss of immune recognition, and no longer provide photosynthates to the holobiont system which survives on heterotrophic/direct diffusion regime. Rate of associate biodiversity loss depends on the type of relationship with the host, from obligate to occasional

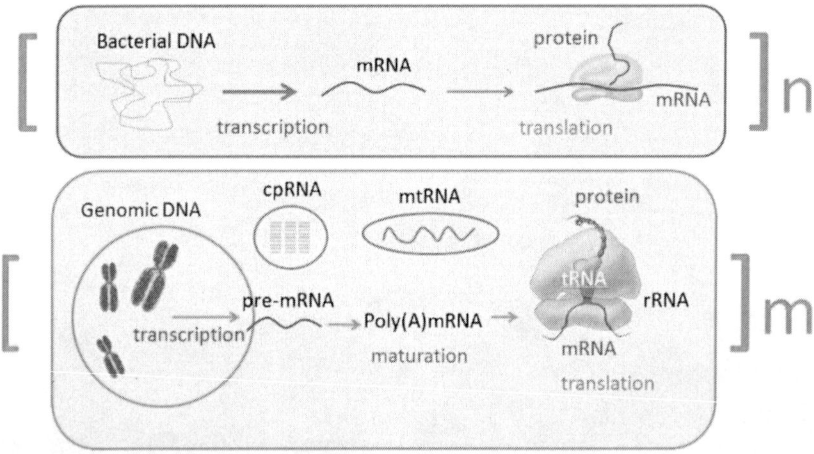

Figure 5. The total transcriptome of a holobiont system (1) top: the total set of bacterial transcripts from a population of *n* cells (2)bottom: the total set of eukaryotic transcripts from a population of *m* cells. Separating the coding RNAs from the non coding ones is made on the basis of the presence or absence of a poly-adenosine monophosphate tail, but only the final mRNAs are usually selected in transcriptome analyses

Proteomics - Physical or chemical stress (i) inactivates or down-regulates many genes, including many housekeeping genes, (ii) while up-regulating stress genes that perform orchestrated induction of key proteins necessary for cellular protein repair. Stress expressions of organisms, or comparative analysis of ecotypes of a species naturally found in contrasting environments, can highlight the production of stress-related enzymes, such as heat-shock proteins. Heat stress does denaturate proteins, causing weakening of polar bonds, unfolding and exposure of core hydrophobic groups. Beyond the cell's tolerance, heat stress will cause its death. Thus the cellular stress response (heat-shock response) protects organisms from damage resulting from critical fluctuations of e.g. heat, UV light, trace metals, and xenobiotics. Stress genes are activated to rapidly synthesize stress proteins, a universal and highly conserved response (from bacteria to humans in which similar roles are played).

Indeed, candidate genes that are directly implicated (e.g. corals' response to stress) or functionally interconnected (e.g. to genes related to immunity) with cnidarians-dinoflagellate symbiosis point out the major role of key proteins [25], and their fast-evolving adaptation in the face of environmental challenges. Oxidative stress in zooxanthellae produces reactive oxygen species (ROS) with hydrogen peroxide diffusion into the host cell which activates a cellular cascade resulting into the photosymbiont expulsion and polyp bleaching. It may be that recognition of a suitable zooxanthellae clade by the coral host is a selective process during which other strains are actively expelled through immunity and apoptosis [26], the photosymbiont being more susceptible than the host to e.g. elevated temperatures and possibly UV levels [27]. Massive and laminar species are supposedly more resistant to environmental fluctuations than shorter-lived branched or encrusting species. Investigators in [28] showed that laminar corals (using *Turbinaria reniformis*) undergo transient Hsp60 heat-shock protein induction under either light or thermal overexposure, and prolonged induction of Hsp60 if the two treatments were combined – whereas branched *Stylophora pistillata* were relatively unresponsive, highlitghting differences in potential for resilience between different growth forms and nutritional regime.

Metabolomics - Notwithstanding their dedicated specialist predators and parasites, scleractinian corals in general are poor producers of secondary (i.e. defence) metabolites since they enjoy protection from a biomineralized skeleton into which polyps can retract – a situation quite different to that of soft corals (alcyonarians) which produce a whole range of cyclized terpenoids against predators, for space occupation and larval protection. However, the presence of several classes of compounds, naturally produced or stress-elicited, has been described for massive coral species (e.g. *Porites* and *Montipora*), and in a number of branched forms. In the former, thermal stress followed by pathogen-associated molecular patterns (PAMPS) triggered defensive responses againt alien threat as well as protective responses, e.g. phenoloxidase activity (production of melanin pigment [29], antibacterial activity, peroxidase and ROS scavenging (oxidative stress) and fluorescent protein [30]. Branched forms are regarded as more sensitive to thermal stress than massive forms, and produce natural antibiotics. Ectodermal cell of *Pocillopora damicornis* release damicornin, a 40-(residue antimicrobial peptide) in response to non-pathogenic challenges, but its expression is repressed by pathogen *V. corallilyticus* [31]. Various *Pseudoalteromonas* strains are known to produced antibiotics, e.g. diketopiperazines, that help them control the bacterial profile sharing the same host-associated resources [32], especially during changes in bacteriome profiling [33]. Mucus-associated *Pseudoalteromonas* spp. are generally considered efficient at protecting the coral holobiont's defense against potential Gram-positive pathogens [34], and are already known to participate in the antibiotic defense of green algae, against surface fouling organisms [35] and fungal epiphytes [36]. Each of these molecules is a potential chemomarker of stress.

Genomics - Marine eukaryotes span 35 phyla, 14 of which are exclusively marine. Genomics is useful in complementing taxonomy beyond classical morphological keys, in exploring evolutionary traits and similarities of small are large taxa, and in investigating the therapeutic potential of target taxa. Furthermore, molecular barcoding allows accurate assessments of environmental samples (especially protists an prokaryotes) – a very useful tool in estimating biodiversity. Corals genomes include 23677 genes (fewer than human genome with 36036 but also than paramecium genome with 39581 genes). The genomes of several common species of scleractinian corals have been characterized at least partially. *Acropora millepora* [37], *Acropora digitifera* [38, 39], genomes have revealed a more complex than hitherto suspected complexity, especially regarding calcification and innate immune repertoire, though scleractinians seems to have lost the ability to carry out *de novo* synthesis of photoprotective mycosporin-like amino acids (MAA) family of compounds. Innate immune responses provide valuable information as predictors of thermal stress susceptibility and disease. A recent paper by [40] usefully reviews key features in coral immunology (recognition, signalling pathways and effector responses) in the general context of invertebrate immunology. Gene expression markers have been recently developed to monitor early responses to acute heat-light stress in *Porites* species, on the basis of differential expression of Hsp16 and actin genes [41]. Genes involved in the immune response against bacterial pathogens present clear differences in their expression patterns between *Vibrio corallilyticus* – exposed *Pocillopora damicornis* vs. unexposed control conspecifics [42]. Such predictors of stress responses can prove very useful if integrated in monitoring tools.

A comprehensive expressed sequence tags (EST) transcriptomics dataset on the symbiotic zooxanthellae has recently indicated some unique regulatory characteristics not found in free dinoflagellates – once completed and annotated, the complete *Symbiodinium* genome will represent a considerable asset in interpreting the coordinated responses of the coral holobiont under stress [43], beyond completely elucidating the molecular bases of the host-photosymbiont association. A recent worldwide survey spanning 20 years of data on coral-*Symbiodinium* photosystems revealed that the transmission mode correlates positively with photosymbiont specificity, not with coral specificity [44]. The authors found fewer generalist species than specialist species. The former group include the common Pocilloporid scleractinians and several Acroporids that can accommodate more than one zooxanthellae clade, and are more environment-sensitive than e.g. common massive corals which may be more resilient to contemporary stressors, especially those genera that are vertical transmitters such as *Porites* and *Montipora.* These findings point out the necessity to select one sentinel species of each group (branched vs. massive) in order to obtain a less biased assessment of the health status and of the resilience potential of a given reef locality under stress.

The third major functional component of the coral holobiont is the bacterial and archaeal microbiome, also susceptible to large composition shifts during e.g. heat stress [45], as revealed by metagenomic studies. Bacterial consortia are described as host-specific, each profile having its specific and its generalist strains, the relative composition of which being affected by environmental conditions.

Ideally, holobiont-wide analysis will benefit from the combined knowledge of the molecular biology of coral, zooxanthellae and prokaryotes that are necessary to define a fully functional system used as a control. The fitness descriptors will need to take into account (i) geographic variations, e.g. chemotypes for widely dispersed taxa), as well as (ii) short natural environmental fluctuations that fall into the natural physiological tolerance of a given population.

The introduction of a stressor (significant in intensity and/or duration) will allow investigators to model the precise interactions of molecular events that affect the three components of this holobiont, and place the physiological responses of the different parties into a system-wide sequence, from early responses to total collapse. Using a holistic and clinical-like approach [46], issue - specific network models can be created by confronting data collected on "stressed" holobionts against homeostatically regulated "no-stress" conspecific controls.

4.4. Merging "omics" with imaging and physiological / ecotoxicological approaches

Creating a multi-approach and comprehensive tool to evaluate the health status of corals under climatic or direct anthropic threat provides a more robust assessment than when using a single analytical method. The omics revolution is coming of age, and large scale data collection (metadata) are more easily tractable using modern bioinformatics algorithms (next generation sequencing) than before. The number of research papers dealing with molecular aspects of photosymbiosis in corals, with stress responses leading to the host-symbiont rupture, with the detection of early markers of stress (before actual symptoms are physiologically or visually expressed), with the resilience potential of massive vs. branched growth forms, with innate immunity, with the onset and development of bacterial pathogenicity following stress, has increased tremendously within the last few years. Each paper brings its unique and useful light into one of the most fascinating biological phenomenon. As of today, however, scientists are little more than spectators of a dramatic acceleration of the destructive impacts of civilization on the most fragile of all marine ecosystems. Politicians and the media are only implementing or relaying preventive anti-pollution policies, and what high-tech research has to offer to monitor what is actually occurring goes well beyond the understanding of the layman.

What is proposed here is to select the most informative analytical strategy as the core component, molecular biology (omics) being the choice alternative to detect early stress responses, and to complement it with physiological measures that are privileged for evidencing (i) adaptability and (ii) loss of function. Physiological monitoring is important in comparing the tolerance range of corals various growth forms, to short lived or limited stress exposure that do not cause changes in the composition of zooxanthellae [47], and allow for gradual acclimation. This laboratory experiment on *Acropora millepora*, a favourite model, may explain why in some localized coral populations, holobionts were capable of resisting temperatures well above the known tolerance limit for the species. Whatever the analytical method used for physiological measurements on the coral host or on its photosymbionts, the instrumentation must be adapted and the measurements repeatable in order to have strong and scalable metrics both in measuring the level and duration of the stressor, and the responses on behalf of the holobiont component under investigation. Respiration and photosynthesis, oxygen production, calcification rates, mucus rheology, pigmentation etc. can be very useful as complementary analyses along with omics.

Imaging tools have a greater impact to non-scientists, and their can offer excellent visual "proof" of an ongoing stress response, in the field (e.g. time-lapse photography of entire holobiont), or inside the component under examination. For example high-resolution imaging mass spectrometry or NanoSIMS can make isotope tracing at single cell level. When linked to molecular visualization methods, such as *in situ* hybridization and antibody labeling, these techniques enable in situ function to be linked to microbial identity and gene expression [48]. Fluorescent pigments, e.g. multi-use pocilloporins, that have a photoprotective and antioxidant roles as well as helping dissipate excess light energy [49] can be detected by classical UV-VIS spectrometry. In fact green fluorescent protein-like (GFP-like) coral fluorescent pigments are routinely used as biomarkers in medicine and can be used to detect coral growth and tissue proliferation, as well as monitoring reactive oxygen species scavenging during stress. Hyperspectral pigment imaging in combination with oxygen profiling provide useful information on competitive interactions between benthic reef organisms, and demonstrate that some turf and fleshy macroalgae can be a constant source of stress for corals, while crustose coralline algae are not [50].

In combination, -omics, physiological/ecotoxicological and imaging tools provide a potentially formidable combination for measuring stress responses in coral holobionts and their separate components.

5. WHAT MODERN TECHNOLOGIES CAN DO FOR THE ENVIRONMENT

Ideally, we need a multi approach diagnosis tool focused on the sentinel species undergoing stress, on the profile of its microbial associates, and to be able to estimate the loss of the epibiotic and encrusting macrofauna and flora which lives in association with the living host, i.e. the evolution of the overall biodiversity from the earliest stress symptoms detectable on the host, to its death.

The basic requirements for a diagnostic tool are:

- to consider different components of the holobiont model (i.e. host, photosymbionts and microbiota) and then integrate the results of the different analyses into a single comprehensive "holistic" diagnosis;
- for each analysis and each approach (molecular, microscopic, ecotoxicologic), to be able to define dose/exposure limits of the stressor that correspond to threshold responses along a continuum such as: normal tolerance/acclimation/resilience/no-return/rapid death;
- to propose corrective measures wherever some critical point is reached in one of the above limits.
- *The basic requirements for a biodiversity-assessment tool are:*
- to list the algal and animal species that are usually associated with the sentinel holobiont;
- to categorize each species with respect to its location within the holobiont system (encrusting, epibiotic, mucus-bound, free-living) and to its degree of dependence to the host for each type (predator, commensal, parasite, symbiont).

5.1. Measuring stress responses, diagnosing overall fitness and proposing corrective measures

On the basis of the above requirements for a diagnostic tool, we propose an 8-step procedure to achieve a compromise between experimental robustness and implementation simplicity. The sequence is shown in Fig. 6, and each step is detailed in the text.

STEP ONE (ISSUE) - Identification of the problem (in the field)

Each biodiversity issue is different, and the first task is to identify the source of the problem. Abiotic stresses should not include first-degree biological interference due to competition, predation or outbreaks of invasive or of alien opportunists, including benthic and pelagic macro and microbiota. On the other hand, we shall consider the departures from "normal" or standard macro and microbiodiversity components resulting from the host species being under abiotic stress, e.g. expulsion or "bail out" of photosymbionts and loss of useful bacterial strains, or disappearance of vegetal and animal associates, which are at the very heart of the biodiversity loss issue. Most common abiotic stressors include: thermal stress, dessication, irradiation, hypo-hypersalinities, silting, heavy metal accumulation, organic compounds, mechanical damage due to wave or wind action and to anchoring. Thus, abiotic stressors can be climatic or pollution bound, chronic or accidental and have their effects combined. It may be necessary to combine field and laboratory (aquarium) studies in which stressors can be analyzed and modulated individually.

STEP TWO (METHODS) - The choice of the adequate holobiont (field)

The choice of a biological taxon may depend on the type of stressor. Corals and coralline algae might be more suitable to evaluate thermal, ultra-violet and acidity stresses, whereas bioaccumulators such as sponges or bivalve mollusks might be more suitable for silting and heavy metal stresses. Furthermore, a good model is one which is sensitive enough to the stressor, at the same time displaying a range of responses that can be calibrated usefully against different concentrations / exposure times. Finally, the model host-species must be (i) representative of the area under investigation, (ii) common enough for sampling at statistically significant scale and (iii) amenable to aquarium studies.

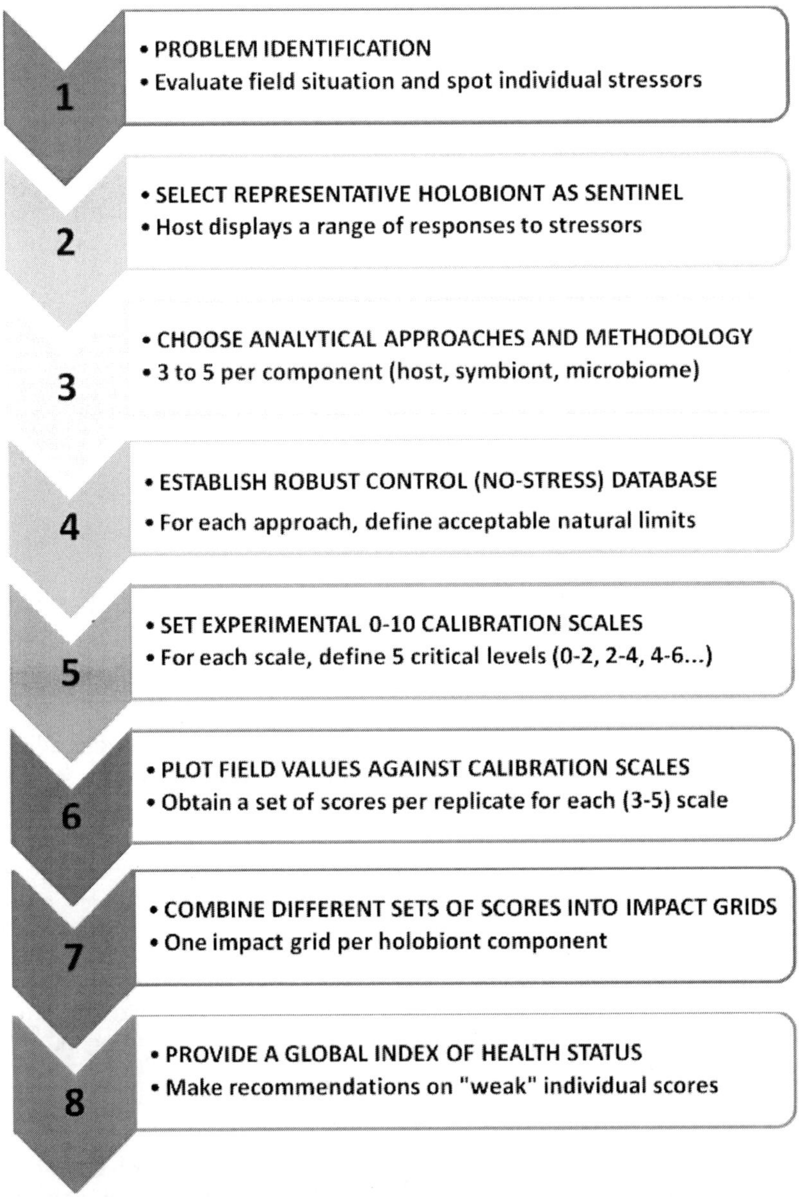

Figure 6. The 8-step logical sequence to create a customized tool (the INDICORAL procedure), each step being described in the text

STEP THREE (METHODS) - How to measure stress in a model holobiont system

Abiotic stress will affect all components of the holobiont diversely and in a network connection manner, with gradual loss of function and accompanying morbidity symptoms. Stress studies will therefore be dealing with the host organism, its photosymbionts, and also with its associated microbiome. This will be achieved through a combination of cutting-edge and classical approaches, e.g. (i) *in aquario*: transcriptomics on the basibiont, microscopy/cytometry and molecular biology on the photosymbiont, (ii) *in the field*: bacterial and viral metagenomic sampling on the associated microbiomes, and ecotoxicology/physiology on the holobiont.

In aquario studies will use a significant number of the selected sentinel species from undisturbed environments, or cloned fragments of same where applicable.

STEP FOUR (DATABASE) - Making a robust set of control data for each analysis

The sentinel holobiont must be sampled in reputedly undisturbed areas, whatever the type of analysis: physiological, metagenomic, taxonomic (associate biodiversity) etc. and natural variations from a statistically significant number of replicates must be recorded.

When dealing with a photosystem holobiont, aquarium studies should consider the three components separately: (i) the holobiont e.g. coral, sponge..., (ii) the photosymbionts, e.g. zooxanthelae, cyanobacteria, (ii) the associated microbes e.g. mucus and tissue-bound bacteria. In each case, a minimum of three (and up to five) independent analyses using different analytical approaches must be undertaken, e.g. one or two –omics, one or two physiological or ecotoxicological, one using imaging etc.

For each analysis, the average values or estimates will set the 0 mark or control score of a future 0-10 scale. This step is crucial and the natural variability must not be too large with scores never exceeding those of responses to mild-severe stresses.

STEP FIVE (DATABASE) – Set experimental calibration scales (Fig. 7)

A useful way to calibrate a response scale *in aquario* is to expose the holobiont or its separate components to different stress regimes (concentration and duration) and establish a lethal value (10 on the 0-10 scale) and in between define threshold values (first symptoms, loss of photosymbionts, loss of microbial biodiversity, irreversible loss of function, death < 12h). Each analytical method will highlight a sequence of events along this "control to immediate death" gradient.

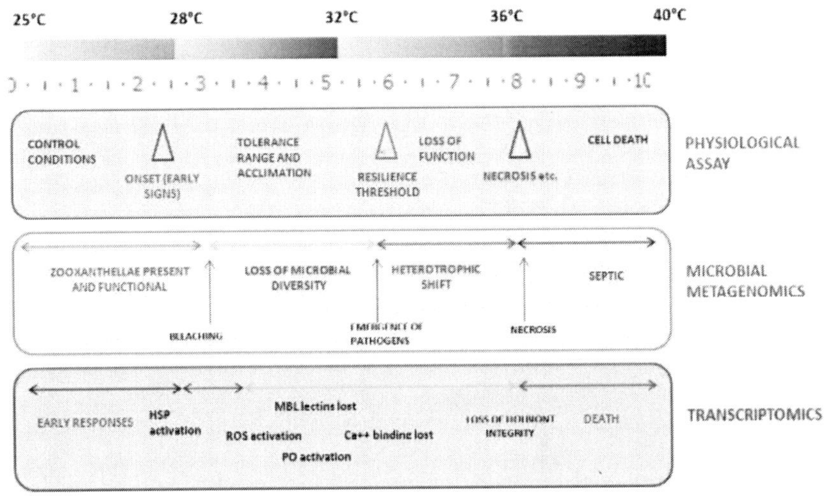

Figure 7. Studying responses of a coral holobiont during experiment heat/light stress using different analytical methods, and establishing a 0-10 scale of responses with critical values. This diagram is indicative only (values and sequences are fictional)

STEP SIX (ANALYSIS) - Compare stress vs. standard profiles

Once each experimental scale is established per analysis, environmental samples can be rated by attributing an average performance score based on *n* replicates.

STEP SEVEN (TREATMENT) - The single grid of impact or radar chart (Fig. 8 top and middle)

Then scores for each of the 3-5 analyses on each of the biological components will be reported to a single grid of impact, e.g. on a coral (5), its photosymbionts (4) and its microbiome (3).

STEP EIGHT (REPORT) - Health status of sentinel species and recommendations for amendment (Fig. 8 bottom)

Let us say we have a 7.4 average score on the coral host, 6.6 on the microbiome and 4.2 on the photosymbiont.. This indicates (i) what biological component is most affected by the environmental stress (ii) where appropriate action is to be taken. Here, holobiont looks normal but early signs of stress are detected on the zooxanthellae (ROS and oxidative stress products), without signs of microbial dysfunction.

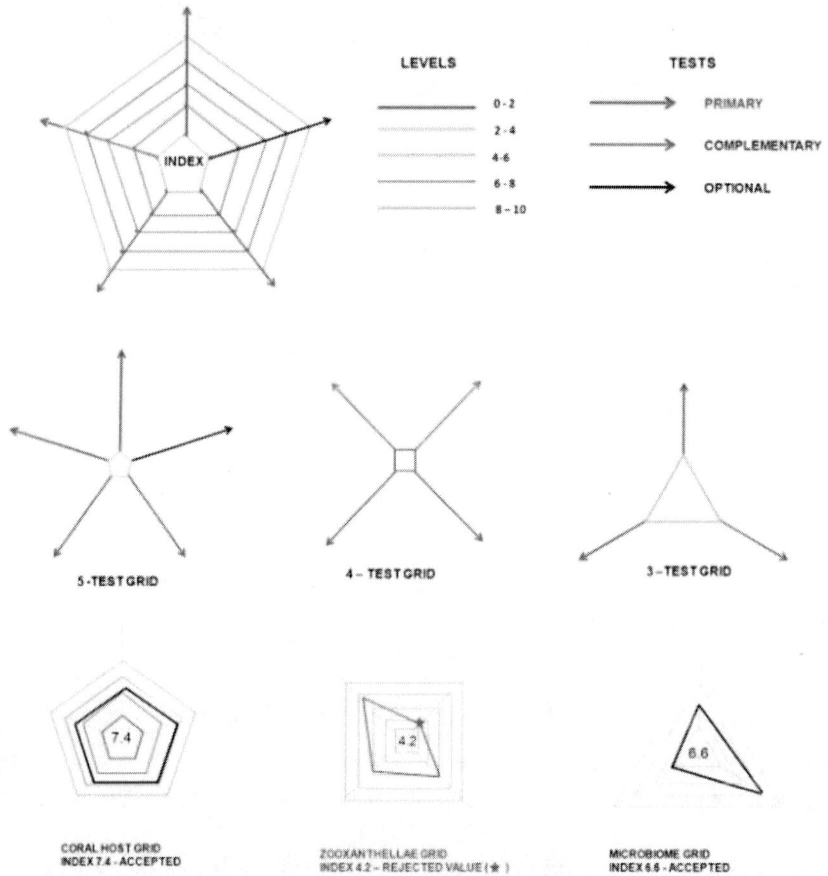

Figure 8. Radar charts – *Top:* typical radar chart - each apex of this polygon is a calibrated scale. *Middle*: types of radar charts with 3, 4 or 5 scales (red: important tests, blue: complementary tests, black: optional test). *Bottom:* Example of a sentinel coral host under heat/UV stress, using a 5-test grid on coral host, a 4-test grid on zooxanthellae photosymbionts and a 3-test grid on bacterial microbiome. Here, the coral holobiont undergoes preliminary signs of stress on photosymbiotic component (red star = oxidative stress on zooxanthellae).

On the basis of these visual diagrams and numerical scores that are explicit enough to be understood and followed by non-scientists. Useful recommendations can be made to customers - if the principle behind this nicknamed "INDICORAL" environmental tool can be validated as a standard, legal enforcement can follow.

5.2. Strengths, limitations and future of environmental diagnosis tools

What is described here is a custom-designed diagnostic tool that integrates the critical snapshot information from different analytical approaches into a single easy-to-read layout, with a consensus fitness index. Using the multivariate "radar chart" model in conjunction with the fitness index then allows us to point out the weaknesses of the holobiont health at a given time, after which corrective measure can be proposed. Such a chart is made up of radiating spokes each representing a performance scale, for example rated from 0 to 10 in a given test. The spokes or performance scales correspond to a single test type, either molecular or visual or physiological. A robust radar chart typically integrates at least one test of each sort, in order to miss as little useful information as possible when establishing the final diagnosis. For example, the holobiont may "look normal" (as compared to control conspecifics) using a visual scale, perform close to optimal using a physiological assay, yet present strong signs of stress using an omics approach that detects early molecular responses. Or else, the holobiont may be suffering from a pollutant that acutely undermines its respiration, with no apparent effect on its microbiome nor on its appearance, and so on.

The objection that immediately comes to mind is that such an endeavour is time and money consuming, knowing how difficult it is to perform discriminating omics tests alone, especially when dealing with meta-data and the mathematical treatment that follows. The answer has to be optimistic: giant strides are being made in analyzing environmental meta-data more efficiently and cost-effectively, a good thing since some environmental issues are becoming rapidly critical. The other objection is that you need a whole team of specialists to devise and run a single "radar experiment".

The answer is two-fold: (i) once a suitable sentinel holobiont is chosen, the tedious part is to establish a reliable control database for each experiment type, mostly in the laboratory and using precise instrumentation. This database may initially require a panel of experts to set it up, but it does not have to be repeated for subsequent investigations. The other aspect is that we are creating a tool, not a thorough investigation into fundamentals, i.e. the experts must make sure that only essential information is retained, e.g. use identified molecular or microbial markers instead of profiles, set critical or threshold values within the 0-10 scale of responses (e.g. onset of symptoms appearance, loss of pigmentation, resilience limits, and so on), 0-2 representing the natural variability observed in controls, and 8-10 representing loss (no-return point to immediate death). This simplified representation relies on strong metrics, not on checking hypotheses which is the job of researchers. The custom association of 3 to 5 different tests types on the coral host and/or the photosymbionts and/or the mucus-bound bacterial flora spans a whole range of dysfunction possibilities.

5.3. Predicting biodiversity loss using the extended holobiont concept

The health status of the coral holobiont directly affects all associated life forms according to the degree of dependence on the host. Ultimately, the loss of the host determines the biodiversity loss of all dependant flora fauna and microbes. Changes in fish biodiversity and pressures on feeding niches have been reported in relation to coral bleaching and loss [51].

Pocillopora damicornis has been used as a choice model in biodiversity issues, not only because of its pan-tropical distribution in the Indo-Pacific region and because of its sensitivity to climatic stress, but also because it provides shelter to a number of commensals that are dependant to varying degrees on the fitness of the coral host [52] and on the state of its bacterial flora. Metagenomic barcoding of the whole extended holobiont system, coupled with appropriate bioinformatics, will be a powerful asset in determining biodiversity loss associated with environmental stressors, and provide a useful link with the analysis of detritic biota after the host's death.

6. MEASURING, PREDICTING AND HOPEFULLY MENDING

As pointed out earlier, over half a billion human live off goods and services of coral reef ecosystems at large, mostly in Asia. But not only: research on some 20,000 chemicals extracted from reef invertebrates has inspired the design of novel anticancer agents, antibiotics, anti-inflammatory, painkillers etc. and the exploration of the complex biosynthetic pathways leading to the production of these "miracle" molecules is only starting. Knowing the paramount importance of host-microbial symbioses in the making of these molecules, biodiversity loss will inevitably lead to chemodiversity loss, and opportunities will no longer exist to investigate the full bio-inspiration potential of what our holobiont systems can produce better than we can.

Marine ecology is a young science, and its developments are accelerating in response (i) to the urge of measuring the reactions of organisms facing the fluctuations of their environment, (ii) to the understanding of how they form complex interaction networks around key molecules acting as mediators of antibioses and symbioses, feeding hierarchies and occupational strategies for essential resources. Holobiont-wide systems biology is coming of age and will allow us to understand how various components of a holobiont system respond to stress in a coordinated manner, in the face of sudden and brutal environmental disasters, or of steadily increasing climatic or anthropogenic forcings, against the background of naturally fluctuating levels of stress. Ecosystem-wide resilience to environmental challenges can no longer be hoped for, and total biodiversity wipe-out of coral reefs is highly unlikely. The most probable scenario is that some reef systems might face near-total biodiversity collapse (e.g. in tropical zones that are within direct influence of urban expansion), while coral species from subtropical and remote localities will be constantly trying to acclimatize, each ecosystem striving to reach an overall "resilience equilibrium" at the cost of some of its biological diversity.

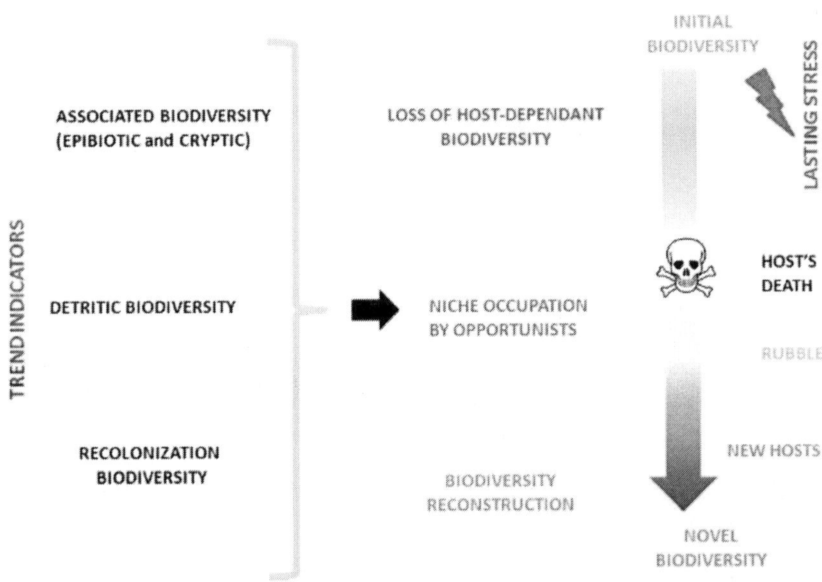

Figure 9. Loss of host-associated biodiversity (i) of original holobiont during lasting stress episode until host's death, (ii) detritic biodiversity finding refuge in coral rubble (iii) replacement biodiversity as substrate is recolonized with new hosts.

In 2013, most biological phenomena can be measured at all scales, from single cells to whole ecosystems, directly or using proxies and appropriate metrics. In combination, -omics, physiological/ecotoxicological and imaging tools provide a potentially formidable combination for measuring stress responses in coral holobionts and their separate components. Designing environmental tools such as proposed here might soon or later become a necessity, not only for coral reefs, but for all endangered ecosystems, marine and terrestrial.

ACKNOWLEDGEMENTS

I wish to dedicate this environmental tool concept to those who guided and encouraged my early endeavours as an enthusiastic observer of nature: to the great naturalist René Catala (1901 - 1987), and to my Australian and New Zealander academic mentors, from Sydney to Townsville via Auckland.

REFERENCES

1. La Barre S. Coral reef biodiversity in the face of climatic changes. Chapter 4 in: Biodiversity Loss in a Changing Planet, vol. 4 of Biodiversity series, Intech (ISBN 979-953-307-252-3), 77-112, http://www.intechopen.com/subject/biological-sciences/ biodiversity (open access, Nov. 2011).
2. Benton MJ. When life nearly died: the greatest mass extinction of all time. London: Thames & Hudson 2005 ISBN 0-500-28573-X.
3. Knoll AH, Bambach RK, Canfield DE, Grotzinger JP (1996). Comparative Earth history and Late Permian mass extinction. Science 1996;273(5274): 452–457, ISSN 0036-8075
4. Falkowski P. Tenth Annual Roger Revelle Commemorative Lecture: The once and future ocean. Oceanography 2009;22(2) 246-251, doi:10.5670/oceang.2009.57
5. Vitousek PM, Mooney HA, Lubchenko J, Mellilo JM. Human domination of Earth's ecosystems. Science 1997;277(5325): 494-499. DOI: 10.1126/science.277.5325.494
6. Connell JH. Diversity in tropical rain forests and coral reef high diversity of trees and corals is maintained only in a nonequilibrium state. Science 1978;199(4335): 1302-1310, ISSN: 0036-8075
7. Baker B. New ocean policy depends on biological research. BioScience 2012;62(5) 524, doi: 10.1525/bio.2012.62.5.19
8. Jackson JBC, Buss L. Allelopathy and spatial competition among coral reef invertebrates. PNAS 1975;72(12) 5160-5163, *ISSN: 0027-8424*
9. Pawlik JR, Steindler L, Helkel TP, Beer S, Ilan M. Chemical warfare on coral reefs: sponge metabolites differentially affect coral symbiosis in situ. Limnol. Oceanogr. 2007;52 (2) 907-911, ISSN: 0024-3590
10. Reshef I, Koren O, Loya Y, Zilber-Rosenberg I, Rosenberg E. The coral probiotic hypothesis. Environmental Microbiology 2006;8(2) 2067-2073, ISSN: 1462-2920
11. Wilson EO, editor, Peter FM, associate editor, Biodiversity, National Academy Press, March 1988, 521 pages ISBN 0-309-03783-2 ; ISBN 0-309-03739-5
12. Bonney R, Cooper CB, Dickinson J, Kelling S, Phillips T, Rosenberg KV, Shirk J. Citizen Science: a developing tool for expanding science knowledge and scientific literacy. BioScience 2009;59(11), 977-984, doi: 10.1525/bio.2009.59.11.9
13. Reyers B, Polasky S, Tallis H, Mooney HA, Larigauderie A. Finding common ground for biodiversity and ecosystem services. BioScience 2012;62(5) 503-507, doi:10.1525/bio.2012.62.5.1
14. Kelling S, Hochachka WM, Fink D, Riedewald M, Caruana R, Ballard G, Hooker G. Data-intensive science: a new paradigm for biodiversity studies. BioScience 2009;59(7) 613-620, doi: 10.1525/bio.2009.59.7.12

15. Nichols JD, Cooch EG, Nichols JM, Sauer JR. Studying biodiversity: is a new paradigm really needed? BioScience 2012;62(5) 497-502, doi: 10.1525/bio.2012.62.5.11
16. Rosenberg E, Koren O, Reshef L, Efrony R, Zilber-Rosenberg I. The role of microorganisms in coral health, disease and evolution. Nature Rev. Microbiol. 2007;5 355–362, doi:10.1038/nrmicro1635
17. Grottoli AG, Rodrigues LJ, Palardy JE. Heterotrophic plasticity and resilience in bleached corals. Nature 2006;440, 1186-1189, doi:10.1038/nature04565
18. Rosenberg E, Kushmaro A, Kramarsky-Winter E, Banin E, Loya Y.The role of microorganisms in coral bleaching.The ISME Journal 2009;3, 139-146, doi:10.1038/ismej.2008.104
19. Rosenberg E, Zilber-Rosenberg I. Symbiosis and development: the hologenome concept. Birth Defects Research, part C 2011;93, 56-66, DOI: 10.1002/bdrc.20196
20. Rohwer F, Seguritan V, Azam F, Knowlton N. Diversity and distribution of coral-associated bacteria. Mar. Ecol. Prog. Ser. 2002;243, 1-10, ISSN: 0171-8630
21. Horgan RP, Kenny LC. Omic technologies: genomics, transcriptomics, proteomics and metabolomics. SAC review, The Obstetrician and Genaecologist 2011;13, 189-195, DOI: 10.1576/toag.13.3.189.27672
22. Wang Z, Gerstein M, Snyder M. RNA-seq: a revolutionary tool for transcriptomics. Nature Reviews Genetics 2009;10, 57-63, doi:10.1038/nrg2484
23. De Salvo MK et al. Differential gene expression during thermal stress and bleaching in Montastrea flaveolota. Molecular Ecology 2008;17(17), 3952-3971, DOI: 10.1111/j.1365-294X.2008.03879.x
24. Traylor-Knowles N et al. Production of a reference transcriptome and transcriptomic database (PocilloporaBase) for thecauliflower coral, Pocillopora damicornis. BMC Genomics 2011;12:585, doi:10.1186/1471-2164-12-585
25. Voolstra CR et al. Rapid evolution of coral proteins responsible for interaction with the environment. PLoS ONE 2011;6(5) e20392, doi: 10.1371/journal.pone.0020392
26. Strychar KB, Coates MC, Sammarco PW, Piva TJ. Bleaching as a pathogenic response in scleractinian corals, evidenced by high concentrations of apoptotic and necrotic zooxanthellae. J. Exp. Mar. Biol. Ecol. 2004;304, pp. 99-121, http://dx.doi.org/ 10.1016/j.jembe.2003.11.023
27. Strychar KB, Sammarco PW. Effets of heat stress on photopigments of zooxanthellae (Symbiodinium spp.) symbiotic with the corals Acropora hyacinthus, Porites solida and Favites complanata. International Journal of Biology, 2012;4(1) 3-19. DOI: 10.5539/ijb.v4n1p3
28. Chow AM, Ferrier-Pagès C, Khalouei S, Reynaud S, Brown IA. Increased light intensity induces heat shock protein Hsp60 in coral species. Cell stress and Chaperones 2009;4, 469-476, doi: 10.1007/s12192-009-0100-6
29. Mydlarz LD, Palmer CV. The presence of multiple phenoloxidases in Carribean reef-building corals. Comparative Biochemistry and Physiology, Part A. 2011;159, 372-378, http://dx.doi.org/10.1016/j.cbpa.2011.03.029
30. Palmer CV, McGinty ES, Cummings DJ, Smith SM, Bartels E, Mydlarz LD. Patterns of ecological immunology: variation in the responses of Carribean corals to elevated temperature and a pathogen elicitor. The Journal of Experimental Biology 2011;214(24), 4220-4249, doi:10.1242/jeb.057349
31. Vidal-Dupiol J et al. Innate immune responses of a scleractinian coral to vibriosis. Journal of Biological Chemistry 2011;286(25) 22688-22698, doi:10.1242/jeb.057349
32. Martinez-Luis S, Ballesteros J, Gutiérrez M. Antibacterial constituents from Pseudoalteromonas sp. Rev. Latinoamer. Quim. 2011;39(1-2).75-83, ISSN: 0370-5943
33. Kvennefors CE et al. Regulation of bacterial communities through antimicrobial activity by the coral holobiont. Microb. Ecol. 2012;63, 605-618, 10.1007/s00248-011-9946-0
34. Shnit-Orland M, Sivan A, Kushmaro A. Antibacterial activity of Pseudoalteromonas in the coral holobiont. Microbiol. Ecol. 2012;64, 851-859, 10.1007/s00248-012-0086-y
35. Rao D, Webb JS, Kjelleberg S. Microbial colonization and competition on the marine alga Ulva australis. Appl. Environ. Microbiol. 2006;72(8) 5547-5555, doi:10.1128/AEM.00449-06

36. Franks A, Egan SH, Holmström C, James S, Lappin-Scott H, Kjelleberg S. Inhibition of fungal colonization by Pseudoalteromonas tunicata provides a competitive advantage during surface colonization. Appl. Environ. Microbiol. 2006;72(9),6079-6087, doi:10.1128/AEM.00559-06
37. Miller DJ et al. The innate immune repertoire in Cnidaria – ancestral complexity and stochastic gene loss. Genome Biology 2007;8, R59, doi: 1186/gb-2007-8-4-r59
38. Shinzato C et al. Using the Acropora digitifera genome to understand coral responses to environmental change. Nature 2011;476, 320-324, doi:10.1038/nature10249
39. Shinzato C et al.The repertoire of chemical defense genes in the coral Acropora digitifera genome. Zoological Science 2012;29, 510-517, doi: http://dx.doi.org/10.2108/zsj.29.510
40. Palmer CV, Traylor-Knowles N.Towards an integrated network of coral immune mechanisms. Proc. Roy. Soc. B – Biological Sciences 2012;279(4), 4106-4114, doi: 10.1098/rspb.2012.1477
41. Kenkel CD et al. Development of gene expression markers of acute heat-light stress in reef-building corals of the genus Porites. PLoS ONE 2011;6(10), e26914, doi: 10.1371/journal.pone.0026914
42. Vidal-Dupiol J, Ladrière O, Meistertzheim A-L, Fouré L, Adjeroud M, Mitta G. Physiological responses of the scleractinian coral Pocillopora damicornis to bacterial stress from Vibrio corallilyticus.The Journal of Experimental Biology 2011;214(9),1533-1545, doi:10.1242/jeb.053165
43. Bayer T et al. Symbiodinium transcriptomes: genome insights into the dinoflagellate symbionts of reef-building corals. PLoS ONE 2012;7(4), e35269, doi: 10.1371/journal.pone.0035269
44. Fabina NS, Putnam HM, Franklin EC, Stat M, Gates R. Transmission mode predicts specificity and interaction patterns in coral-Symbiodinium networks. PLoS ONE 2012;7(9), e44970, doi:10.1371/journal.pone.0044970
45. Mouchka ME, Hewson I, Harvell D. Coral-associated bacterial assemblages: current knowledge and the potential for climate-driven impacts. Integrative and Comparative Biology 2010;50(4), 662-674, doi:10.1093/icb/icq061
46. Barabábasi A-L, Oltvai ZN. Network biology: understanding the cell's functional organization. Nature Reviews – Genetics 2004(5), 101-113, doi:10.1038/nrg1272
47. Bellantuono et al. Resistance to thermal stress in corals without changes in symbiont composition. Proc. Roy. Soc. - series B 2012;279, 1100-1107, doi: 10.1098/rspb.2011.1780
48. Pett-Ridge J, Weber PK. NanoSIP: NanoSIMS Applications for Microbial Ecology. In: Microbial Systems Biology: Methods and Protocols. Methods in Molecular Biology, 2012;881, 375-408, DOI 10.1007/978-1-61779-827-6_13
49. Palmer CV, Modi CK, Mydlarz LD. Coral Fluorescent Proteins as Antioxidants. PLoS ONE 2009;4(10): e7298, doi:10.1371/journal.pone.0007298
50. Barott K, Smith J, Dinsdale E, Hatay M, Sandin S et al. Hyperspectral and physiological analyses of coral-algal interactions. PLoS ONE 2009;4(11): e8043, doi:10.1371/journal.pone.0008043
51. Prachett MS, Hoey AS, Wilson SK, Messmer V, Graham AJ. Changes in biodiversity and functioning of reef fish assemblages following coral bleaching and coral loss. Diversity 2011;3, 424-452, ISSN : 1424-2818
52. Plaisance L, Knowlton N, Paulay G, Meyer C. Reef-associated crustacean fauna: biodiversity estimates using semi-quantitative sampling and DNA barcoding. Coral Reefs 2009;28, 977-986, doi:10.1007/s00338-009-0543-3

CHAPTER 4

Remote-Sensing-Based Estimation of Surface Nitrate and Its Variability in the Southern Peninsular Indian Waters

R. K. Sarangi

Marine Geo and Planetary Sciences Group, Space Applications Centre (ISRO), Ahmedabad 380 015, India

ABSTRACT

A relationship between sea surface temperature (SST) and surface nitrate concentrations has been obtained for the first time based on in situ datasets retrieved from U.S. JGOFS (1991–96) and Indian cruises (2000–2006) in the Arabian Sea, Bay of Bengal and Indian Ocean region around the southern Indian tip. The dataset includes 1537 points. A sigmoid relationship obtained with value 0.912. NOAA-AVHRR pathfinder satellite monthly averaged SST data retrieved from the PODAAC/JPL/NASA archive during July 1999–June 2004. The datasets imported in the ERDAS-Imagine software and SST images generated on monthly and seasonal scales, for latitudes 5–12°N and longitudes 75–85°E. The ocean surface nitrate images retrieved based on the established sigmoid relationship with SST. The nitrate concentrations ranged between 0.01–3.0 μM and categorized into five ranges. The significant seasonal upwelling zone around the southwest coast of India (Kerala coast, Latitude 80.10–9.30°N and Longitude 75.60–

76.20°E) was identified during July–September 1999–2004 with very high nitrate concentration (~1.00 μM). Low nitrate and nitrate-depleted zones observed during summer (March–May). In the Arabian Sea and northern Indian Ocean, high nitrate concentration (~0.50 μM) observed during the southwest monsoon (SWM), whereas the Bay of Bengal was marked with high nitrate (~0.50 μM) during the northeast monsoon (NEM). SST was high (~29°C) in the Bay of Bengal and low (~26°C) in the Arabian Sea and northern Indian Ocean during SWM and vice versa during the NEM. There is a clear inverse relationship between nitrate and SST in the study area during July 1999–June 2004.

1. INTRODUCTION

In the upper several hundred meters of the World Ocean, plant nutrient (nitrate, phosphate, and silicic acid) concentrations tend to increase with depth and temperature tends to decrease with depth [1]. This inverse relationship between temperature and plant nutrient concentration also occurs seasonally in surface water to varying degrees depending on geographic location, except at high latitudes where plant nutrients at the surface annually range from a minimal nonzero concentration (as determined by colorimetric methods [2]) to a seasonal maximum concentration related to large-scale geochemical processes [3, 4]. The spatial and temporal characteristics of the depth distribution of temperature and plant nutrients combine to yield latitudinal-specific relationships in plant nutrients and temperature (PN-T) scatter plots [1]. Incident solar radiation is the primary cause of inverse relationship between water temperature and plant nutrient concentration [5], as a major source of light for phytoplankton photosynthesis and plant nutrient uptake [6, 7]. Nutrient limitation is much less straightforward than light limitation because of the multiple interactions between nutrients, these interactions being different depending on the scale that is being considered. Until recently, nitrogen was assumed to be the universal limiting nutrient in marine waters. Nitrate is only supplied by diffusion through the thermocline. In the early phase of stratification, chlorophyll concentration is uniform throughout the mixed layer and then a maximum develops at depth near the nitracline.

More specifically, the processes affecting the character of the temperature-nutrient relationship include solar heating, advection and vertical mixing, the rate of nutrient uptake by phytoplankton at the surface, and the rate at which biologically compound organic compounds are regenerated subsurface. Zentara and Kamykowski [8] analyzed temperature-nutrient relationships along the west coast of north America and showed that the slope steepened and the temperature intercept increased at higher latitudes, presumably due to latitudinal variation in solar heating. They also noted that variability was greatest in boundary regions where a mixture of water masses is expected. H. J. Minas and M. Minas [9] suggested that the PN-T plots can be used to indicate the intensity of nutrient uptake, by phytoplankton in upwelling areas, with steeper slopes indicating faster nitrate uptake. Under conditions of low phytoplankton uptake the PN-T relationship can be used to study physical processes. The simple model leads to the key concept of new and regenerated production as based on nitrate and ammonia [10]. There has been clear evidence of nitrate utilization with a deepening of nutrient depletion with time. The vertical distribution of chlorophyll concentration also changed with time. Nitrate was the dominant nitrogen source for phytoplankton. The uptake rates measured decreased with declination of nitrate concentration. The current study is based to observe/monitor the distribution of nitrate-nitrogen in the ocean surface. More research is needed to understand the source of nitrate and its flux and know the distribution pattern in regional, and seasonal and intraseasonal timescales. The nitrate-temperature relationships from the drogue depth occasionally suggested and enhanced the heating per unit of nitrate uptake compared to the date from the hydrographic sections. The most direct approach towards a more robust description of the relationship between temperature and plant nutrients in the upper ocean may result from biological additions and relatively subtle physical adjustments to more complete, existing models [11] of the upper ocean mixed layer.

Nitrate transport into the euphotic zone in the southwest, northeast, and intermonsoon phases has been estimated using the satellite-derived SST images, and this agrees with the measured nitrate uptake rates and phytoplankton productivity. It is therefore suggested that the satellite-derived nitrate images can be useful tool for assessing nitrate-based productivity over large geographical areas in short time intervals. Surface temperature, nutrients, and primary productivity relate to each other [1], and prediction of new production from transformation of satellite-derived temperature images to nitrate distribution is a distinct possibility [12, 13]. The regularity of the phenomenon unlike temporally variable upwelling and the fact that the phytoplankton growth was stimulated after each spring tide lead one to use SST measurements from satellite infrared radiance to predict nitrate input to the euphotic zone and thereby compute resultant production [14].

A major fraction of the carbon transported into the oceans interior is associated with allochthonous nitrogen inputs, primary nitrate (N), into the euphotic zone [15, 16]. Hence, understanding the spatial and temporal variations of N over basin and global scales in the euphotic layer is an important requirement for ocean biogeochemical and climate studies [16].

Although N in seawater lacks a distinct electromagnetic signature that can be utilized for its estimation from space, satellites have for long been suggested as a promising measurement tool, because N correlates well with certain remotely measurable seawater properties [17]. The relationship between N and seawater temperature (T), for instance, has consistently been found to be negative as nutrient-rich waters brought up into the euphotic zone from the deep by physical processes such as upwelling or winter convective overturning are colder than the surface waters which are generally nutrient poor. Although empirical algorithms based on PN-T relationships have been existent for a while [14, 18–20], Goes et al. [21] have shown that if biologically modified changes in the character of PN-T relationships are taken into account, it is possible to generate high-resolution maps of surface nitrate over much larger spatial and temporal scales than previously possible.

The Labrador Sea Monitoring Program (LSMP) annual spring-summer missions measure seasonal nutrient concentrations but not winter maximums nor summer minimums for the most part, which are critical for P_{new}estimation [22]. The wintertime nutrients in the North Atlantic are new approaches and implications for new production estimates [22]. Another recent approach, worldwide applicable, is that proposed by Kamykowski et al. [23] and Switzer et al. [24]. They generated a scaled index of nitrate availability by computing the differences between SST (derived from AVHRR data) and "nitrate depletion temperatures" (NDT), inferred from in situdata and geographically variable. Satellites have been suggested as a useful alternative. Although they can provide frequent, large-scale, near-surface views of several variables relevant to phytoplankton ecology such as incident radiation, chlorophyll, and sea surface temperature (SST), nutrient measurements have remained elusive as most nutrients including nitrate lack an electromagnetic signal that can be exploited from space [25]. Henson et al. [26] derived their empirical relationships between surface nitrate (or silicate) and temperature and chlorophyll using broad spatial distributions of these variables at 5 m depth collected on four missions between November 2001 and December 2002 in the Irminger Basin.

Silió-Calzada, et al. [27] has proposed a new approach to estimate NO_3 specifically developed for areas influenced by upwelling processes. It relates the nitrate concentration to the difference between SST and the estimated temperature of the upwelled water (variable with latitude and season), which is an indicator of the time elapsed since upwelling. Algorithm has been developed using multiple linear regression (MLR) from nitrate, SST, and mixed layer depth (MLD) data in the North Atlantic Ocean [28]. The N-T relationships, however, are strongly influenced by the regional biogeochemical characteristics and hydrodynamic conditions and therefore reveal a high spatial and temporal variability. The current study is the first attempt to show those relationships in Indian waters with the collection of 1537 data points retrieved from U.S. JGOFS (1991–1996) and Indian cruises (1997–2002) database. The purpose of this work is to formulate a preliminary integrated model of water temperature and nitrate concentration, to examine the influence of the various terms in the model, and to identify future directions for improved models.

Rationale for this study is to find out the role of nitrate in the ocean biogeochemical cycling, mostly its role in new primary production. The new production is bestowed upon the carbon cycle in specific areas in sea like upwelling, eddies, convection, detrainment, and even in algal bloom conditions. The nitrate has a significant role on the growth, metabolism, and in euphotic-zone-based photosynthesis/productivity of the phytoplankton so it is recognized as a major limiting factor, even most of the blooms occur due to the enrichment of nitrate, the cyanobacteria Trichodesmium sp. fixes the atmospheric nitrogen and blooms. The normal inverse trend of ocean surface nitrate is related to the temperature of water. The bottom nitrate or the nitrate from water column being carried up with column water and hence the higher nitrate and low temperature has been observed in the surface. Here in the current study the relationship has been accurately represented finding its sigmoid curve. The results over Indian southern peninsular water have been exclusively discussed for five-year data and generated nitrate images.

This study is important from the point of view of climate change, such that shipboard datasets from several years have to be combined with generate maps of nitrate at global scale and this would enhance the daily, monthly, seasonal, and interannual scale variations of marine nitrate. In previous and on-going studies, the immense promise and potential of satellite-based nitrate maps for carbon cycling and climate change studies have been demonstrated [29]. Climate forcing greatly impacts processes controlling the distribution of nutrients and hence the productivity. The nutrient distribution also gives the indication about the physical processes interaction in the region of study.

2. STUDY AREA

The present study has been carried out in the southern peninsular Indian waters, which include a part of the Arabian Sea, Bay of Bengal, and Indian Ocean. This is an interesting area, because it consists of three different water types with their own productivity potential and other environmental features (Figure 1).

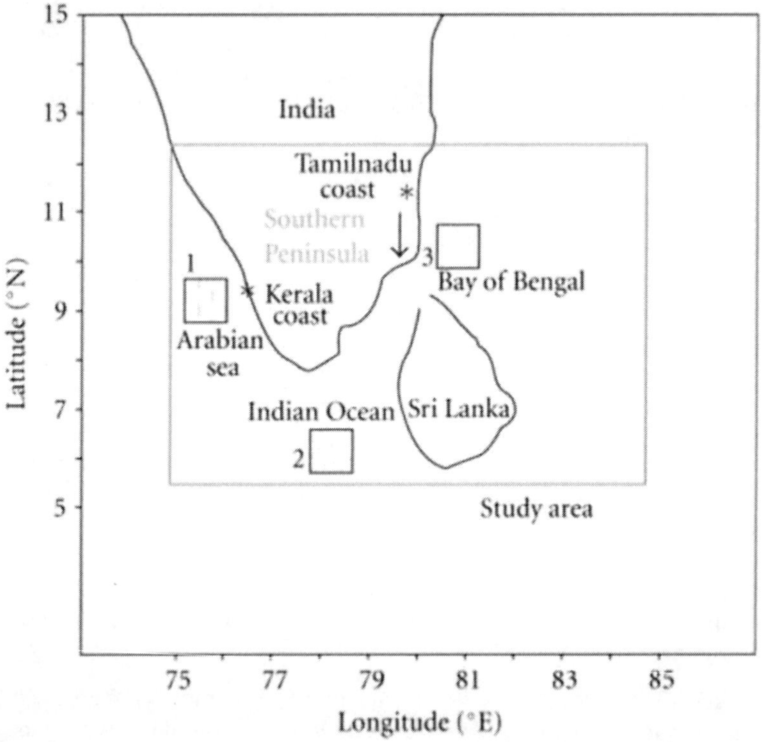

Figure 1: Study area (latitude: 5.5°–12°N and longitude: 75°–84°E) indicating the Arabian Sea, Indian Ocean, and the Bay of Bengal with the boxes drawn for analysis (1, 2, 3).

The physicochemical and biological properties of the Arabian Sea, are affected by monsoon reversal wind systems. The monsoon wind affects the chlorophyll distribution pattern besides other properties [30]. Several phenomena like the high concentration of algae occur during winter as a consequence of vertical convection. Several other interesting phenomena like upwelling and its season-specific increase in algal biomass are also noticed as common events in this sea [31]. The southern tip of India is associated with Indian Ocean. Mixing and convergence takes place between the waters of the Arabian Sea and the Bay of Bengal [32]. The shelf flow also determines the productivity pattern of the Indian Ocean. The Bay of Bengal is activated by the cyclonic winds, storms, gyres and sea surface is rough compared to the Arabian Sea [33–38].

Unlike the northern Bay of Bengal, the southern part happens to be generally nutrient limited largely due to limitations in external river inputs.

The southern rivers are often seasonal and bring external nutrients only during shorter monsoon/postmonsoon periods. The Bay of Bengal, in general, is a data-poor region compared to the rest of the Indian Ocean. Despite its importance, shipboard nitrate measurements fall far short of the space and time scales required for climate research. The nitrate has important role to demonstrate the high productivity and new production zones in different seasons in Indian water. There are limited numbers of studies for Bay of Bengal waters with reference to ocean remote sensing. A detailed study is required to understand the surface nitrate dynamics on time-series basis.

3. MATERIALS AND METHODOLOGY

3.1. Generation of SST Composite Images

The NOAA-AVHRR Pathfinder-5 based monthly averaged SST ASCII data with latitude, longitude, and SST columns were retrieved from the http://www.nasa.gov archive for the period July 1999–June 2004. The datasets were imported in the ERDAS Imagine software to generate the SST images in monthly scales during the period July 1999–June 2004. The monthly images were averaged to seasonal composite images covering the latitude 5°–12°N and longitude 75°–85°E. The images were gridded and colour coded using ERDAS Imagine and Envisoftware. The clouds and unavailable data have been displayed blank in the images. The Pathfinder-5 sensor ascending/morning pass data of resolution 4 kilometers has been used and provides SST with accuracy of 0.3°C [39, 40].

3.2. Nitrate Retrieval Algorithm

In this study an attempt was made to find the relationship between nitrate and temperature from the U.S. JGOFS dataset for the Arabian Sea (1994–1997) and Indian cruise datasets (2000–2006) covering the Arabian Sea and Bay of Bengal. A linear relationship has been established in the European waters [14]. The ocean surface nitrate concentration has been retrieved based on the established relationship as follows.

The regression best fit gives the following equation:

$$\text{Nitrate } (NO_3) = \frac{23.924}{[1 + e^{\{(SST-23.793)/0.662\}}]}. \tag{1}$$

R^2=0.912 Standard Error of Estimate = 2.2755.

Linear, polynomial (2nd and 3rd order), and sigmoid relationship between in situ retrieved temperature and nitrate data has been established for the Arabian Sea data collected from JGOFS and Indian cruises. The respective values have been observed to be 0.765, 0.834, 0.872, and 0.912. The respective standard error of estimate (SEE) has been observed 3.731, 3.136, 2.752, and 2.275 for the 1537 data points. So, the sigmoid relationship has been found with the best curve fitting (Figures 2(a)–2(d)). Figure 3 displays a regression plot indicating the relationship of in situ nitrate data from Indian cruises with pathfinder and MODIS-Aqua-derived SST. The nitrate-algorithm-derived nitrate image and data points have been validated with the 13 in situ data retrieved during 2003 February and March cruises in the Arabian Sea on board cruises of ORV "Sagar Kanya" and FORV "Sagar Sampada." Validation result with the available 13 in situ points from Indian cruises has shown the value 0.843 (Figure 3(a)). Validation result of in situ nitrate with SST from the available cloud-free satellite passes of Pathfinder (February-March 2003) has shown the value 0.815 (Figure 3(b)). Even, the relationship of nitrate with other corollary parameters like chlorophyll and wind speed has been taken attempt besides SST (Figures 4(a), 4(b), 4(c)).

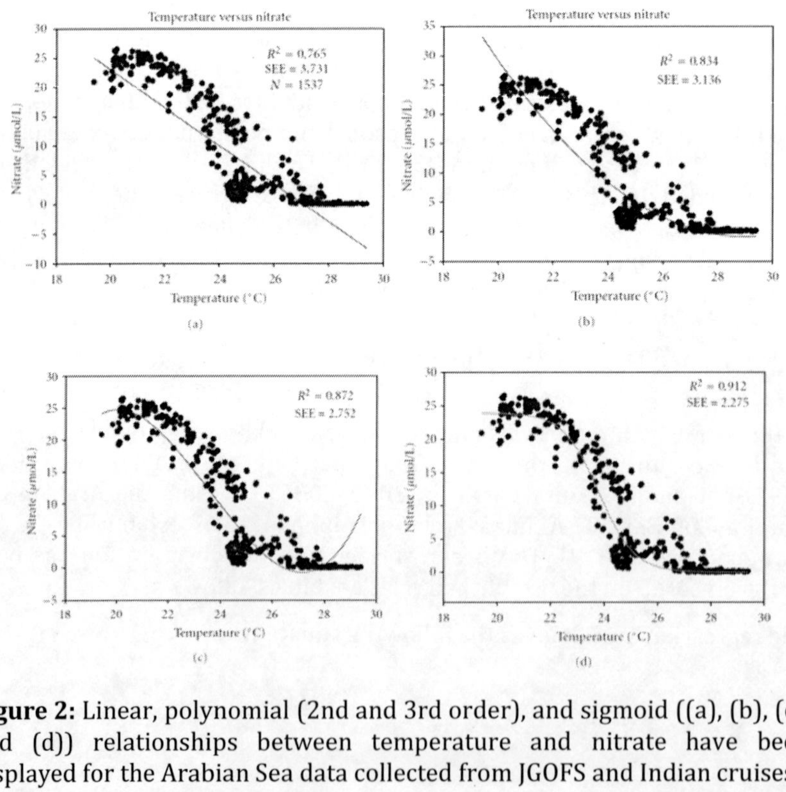

Figure 2: Linear, polynomial (2nd and 3rd order), and sigmoid ((a), (b), (c), and (d)) relationships between temperature and nitrate have been displayed for the Arabian Sea data collected from JGOFS and Indian cruises.

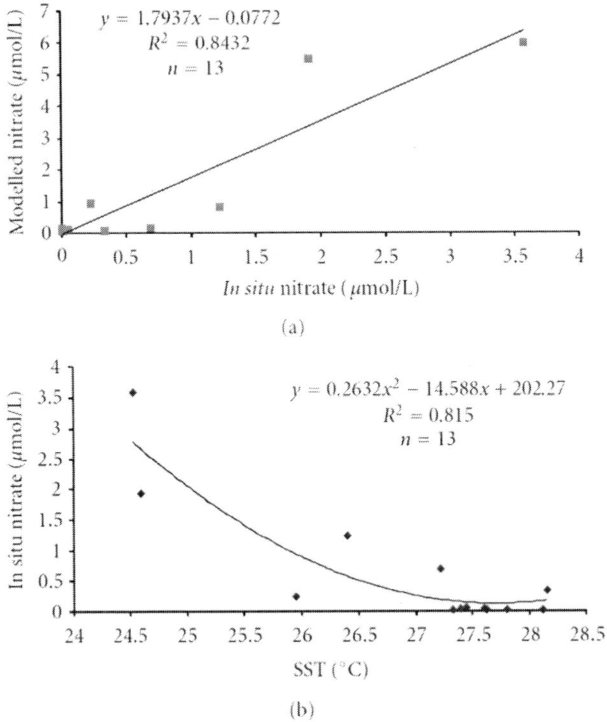

Figure 3: ((a), (b)) Regression plots displaying the relationship of in situ nitrate from Indian cruises with modeled nitrate data and with pathfinder (February-March 2003) derived SST data.

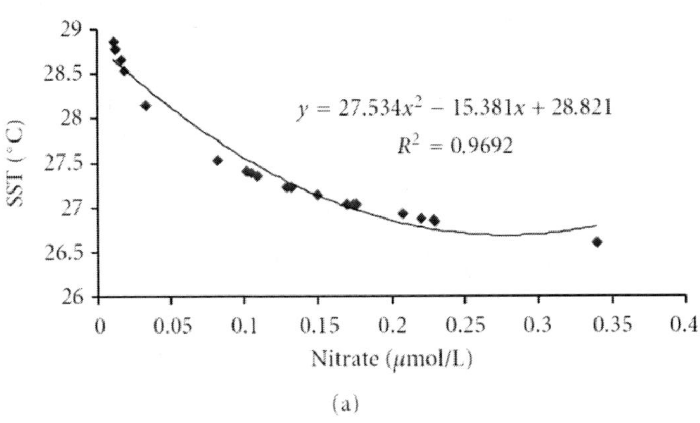

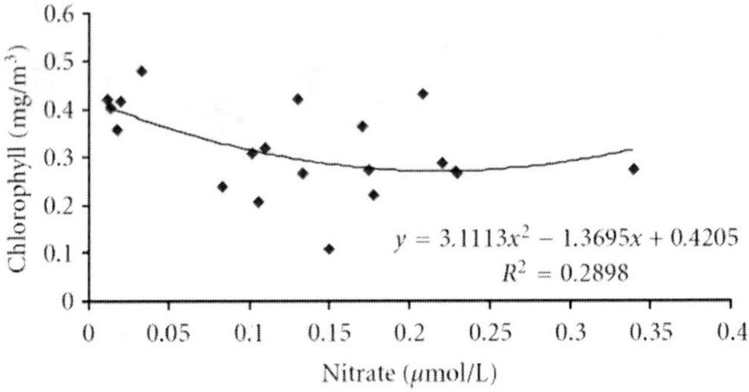

(b)

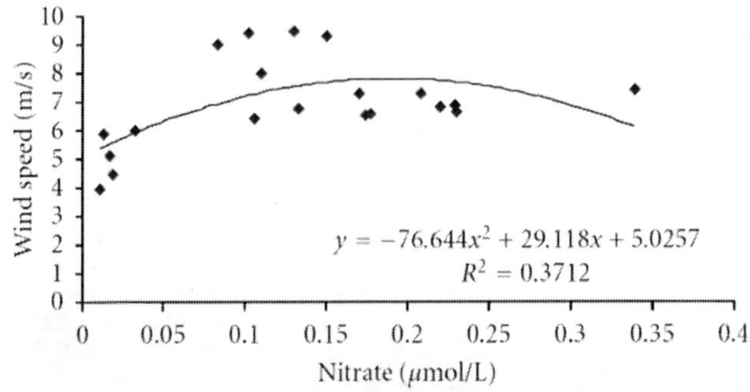

(c)

Figure 4: Regressions plots are displayed indicating the relationship of nitrate with SST, chlorophyll, and wind speed, retrieved from modeled and satellite datasets for April 2000 covering southern peninsular Indian water.

3.3. Generation of Nitrate Images

The monthly, seasonally averaged nitrate composite images have been generated for the study area. The monthly images were averaged to seasonal composite images with the category of seasons; (1) southwest monsoon (SWM)—June–September, (2) fall intermonsoon (FIM)—October-November, (3) northeast monsoon (NEM)—December–March, and the (4) spring intermonsoon (SIM)—April-May [41] (Figures 5, 6, 7, 8, 9, 10 and 11).

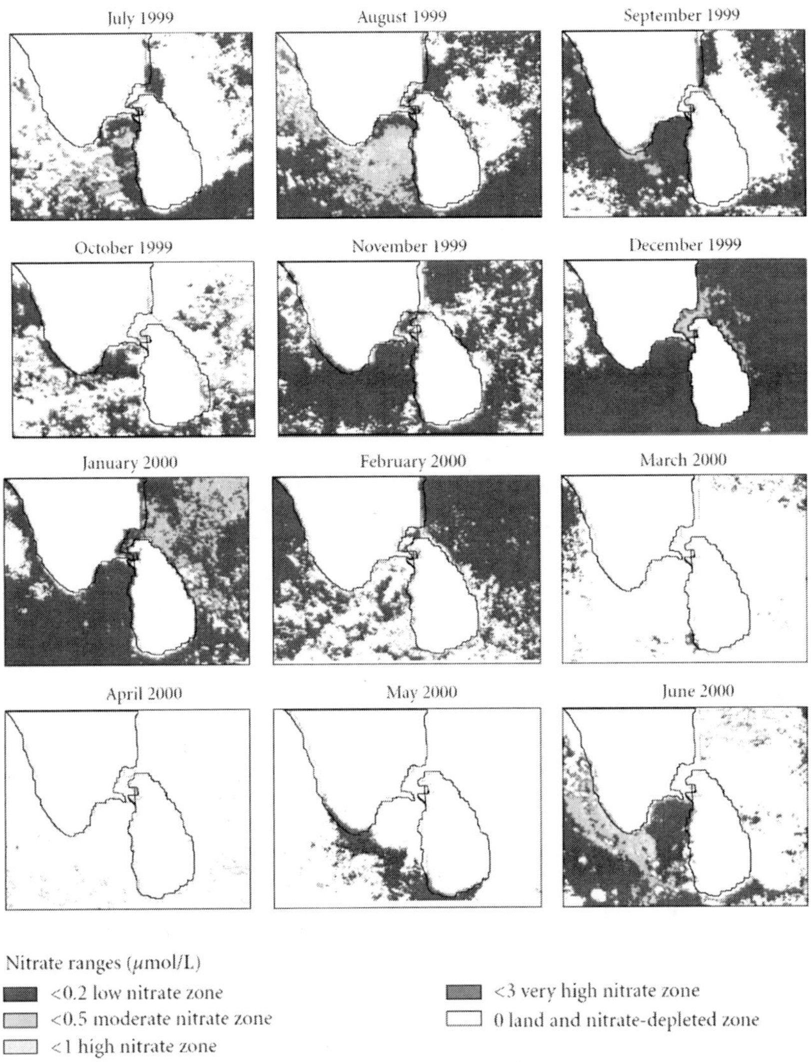

Figure 5: In situ nitrate- and temperature-based modeled monthly composite nitrate images during July 1999–June 2000 retrieved from NOAA-AVHRR pathfinder data.

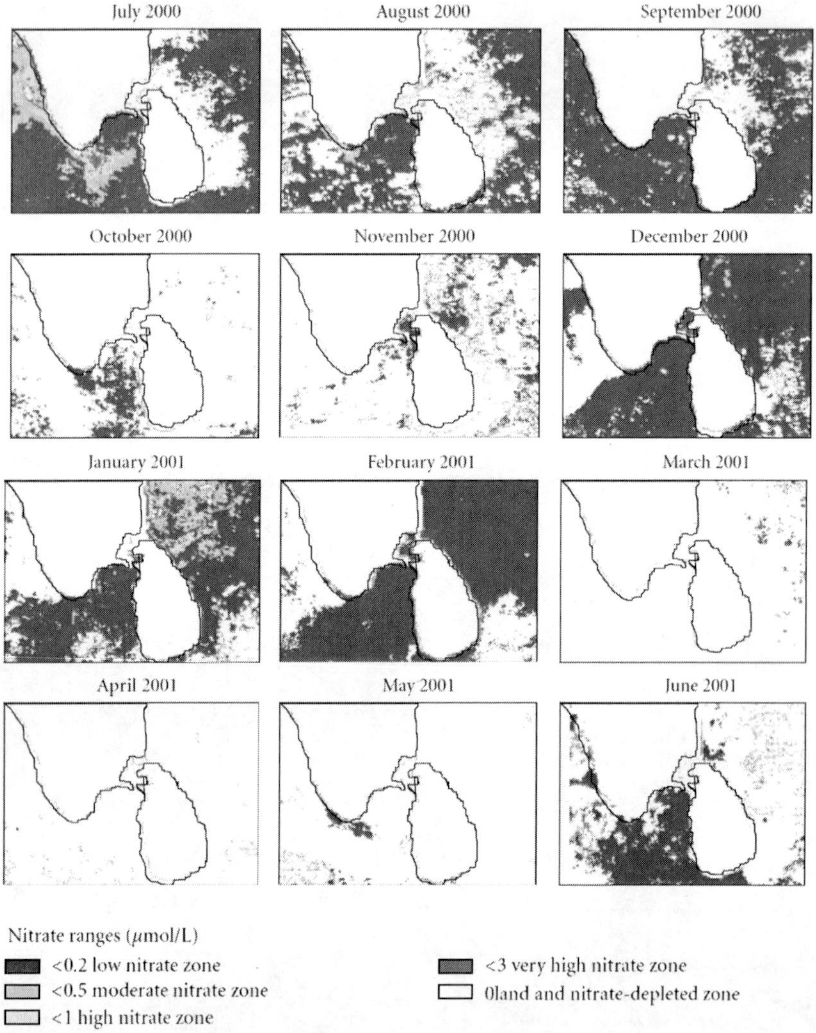

Figure 6: In situ nitrate- and temperature-based modeled monthly composite nitrate images during July 2000–June 2001 retrieved from NOAA-AVHRR pathfinder data.

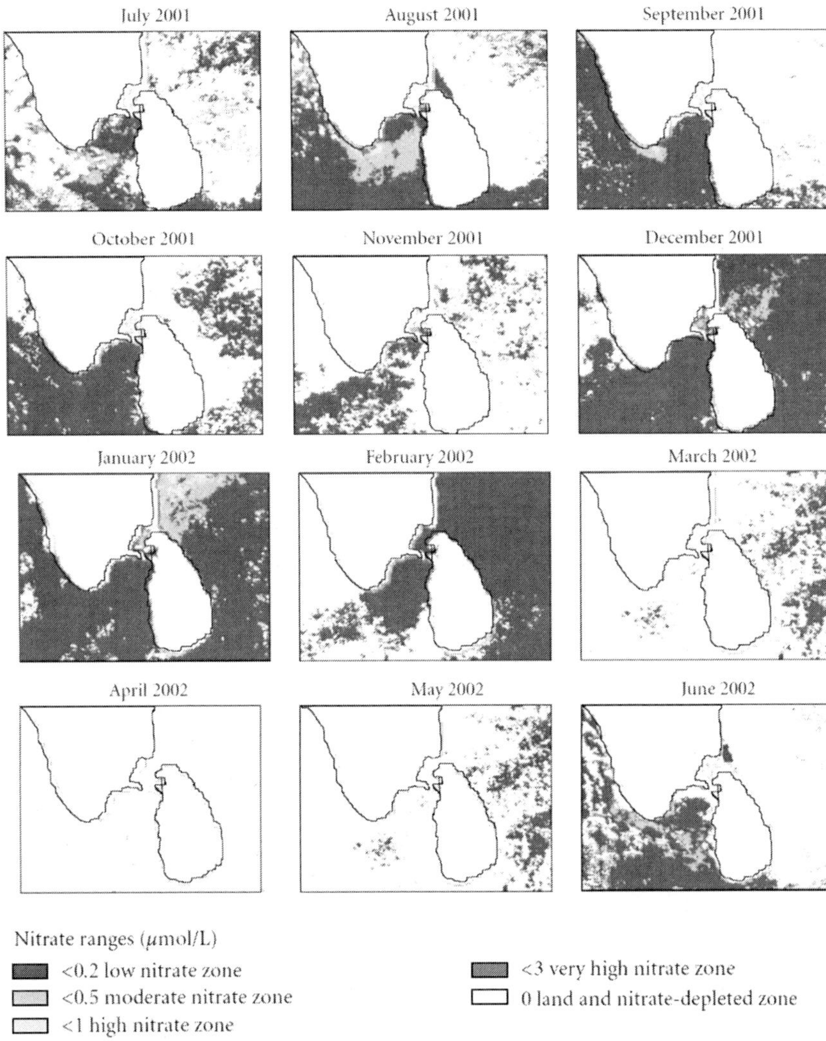

Figure 7: In situ nitrate and temperature based modeled monthly composite nitrate images during July 2001–June 2002 retrieved from NOAA-AVHRR pathfinder data.

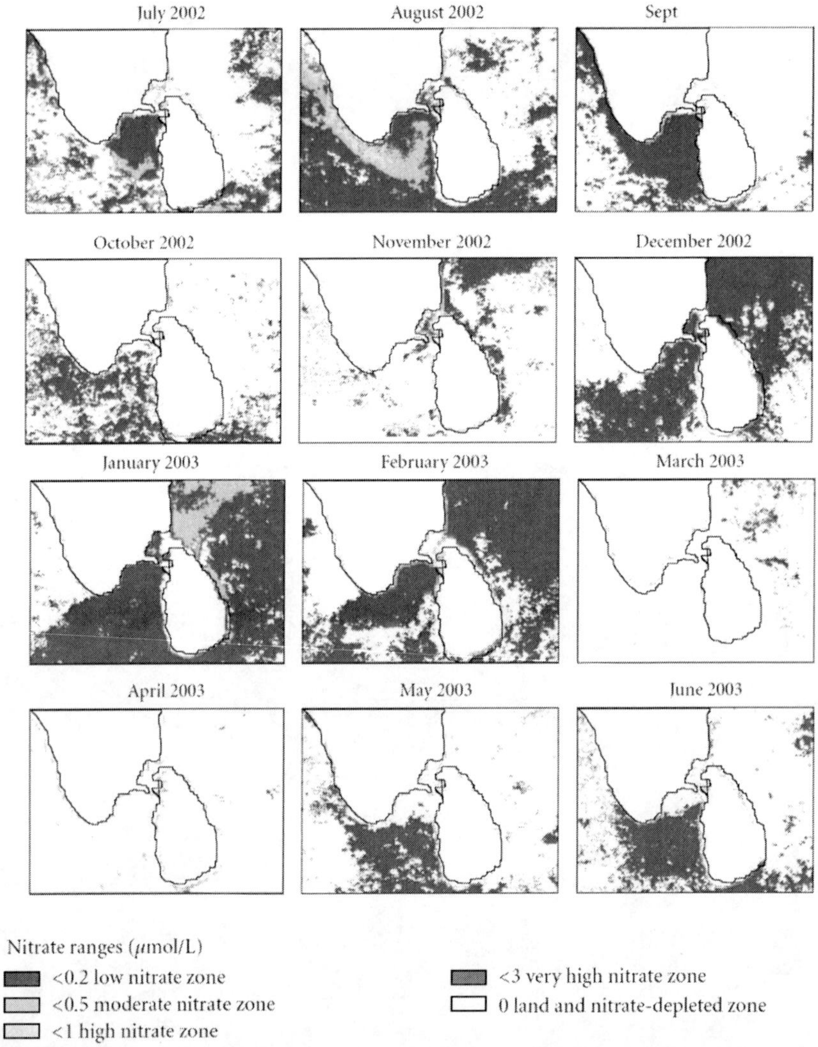

Figure 8: In situ nitrate- and temperature-based modelled monthly composite nitrate images during July 2002–June 2003 retrieved from NOAA-AVHRR pathfinder data.

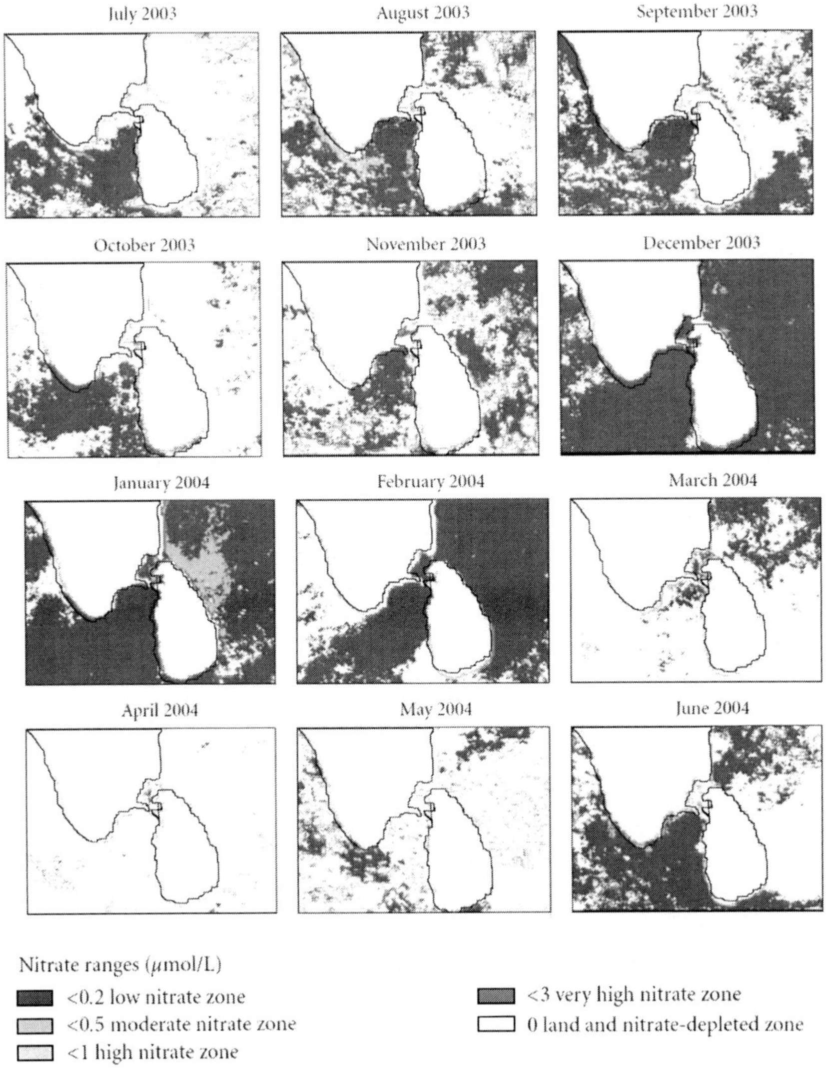

Figure 9: In situ nitrate- and temperature-based modelled monthly composite nitrate images during July 2003–June 2004 retrieved from NOAA-AVHRR pathfinder data.

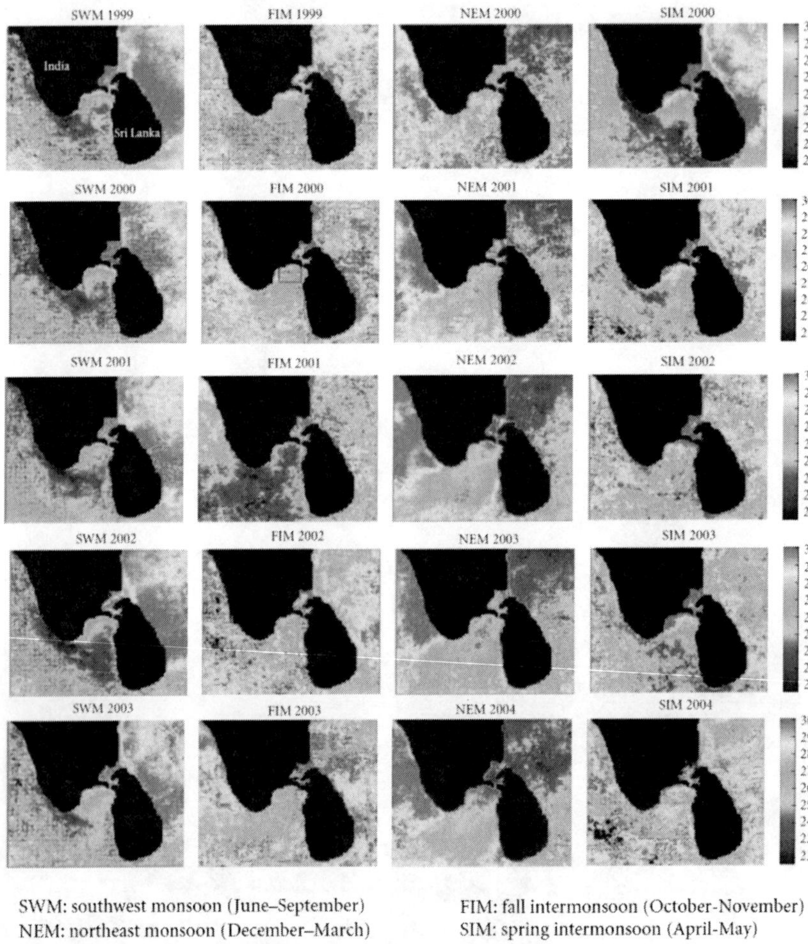

Figure 10: NOAA-AVHRR derived seasonally averaged SST images during 1999–2004 in the southern peninsular Indian water.

3.4. Subset and Transect Data Retrieval

We focus on the three different seas to reflect their properties in terms of satellite-derived SST and nitrate concentrations and their interrelationship. This region is very significant from the point of view of its oceanography, regulated by all the four monsoons, related coastal/oceanic features, circulations, and currents.

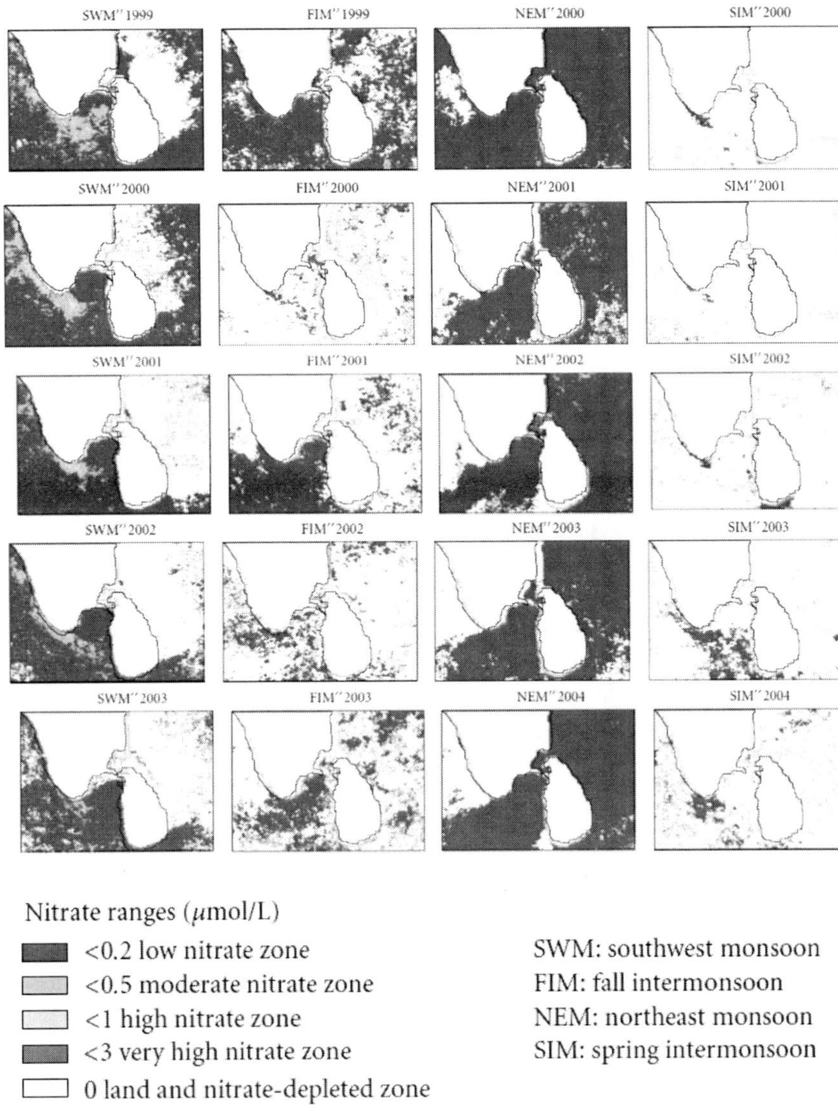

Figure 11: In situ nitrate- and temperature-based five years seasonal composite modelled nitrate images during July 1999–June 2004 retrieved from NOAA-AVHRR pathfinder data.

The pockets/boxes were taken from the three different oceans: (1) Arabian Sea (latitude: 8.95–9.95°N and longitude: 74.6°–75.6°E), (2) Indian Ocean (latitude: 6.24°–7.24°N and longitude: 77.1°–78.1°E), and (3) Bay of Bengal (latitude: 10.2°–11.2°N and longitude: 80.4°–81.4°E) (Figure 1). The three pockets are off the coast and covering the water in the depth ranges to 1000–3000 meter (http://www.ngdc.noaa.gov/mgg/bathymetry/ relief. html/).

The pixels/data points falling in the pockets (1, 2, 3) were respectively averaged to generate subset images for the parameters SST and nitrate. The mean and standard deviation of the respective months and seasonal composites have been calculated and plotted. Monthly mean and standard deviation data for each pocket has been plotted as line plots, and the seasonal, annual, and four-year composite has been plotted as histogram/bar charts to understand the variability in respective time scale and seasons. The colour-coded images for each parameter SST and nitrate concentration have been displayed for different time scales with respective colour scale ranges, namely, SST (22°–30°C) and nitrate (0.01–3 μM) (Figures 10, 11).

4. RESULTS AND DISCUSSION

4.1. Variability of Nitrate (NO_3)

The nitrate concentration varied depending on the conditions like the four monsoon seasons and respective months and on the regional scale. The nitrate distribution has been classified into five categories depending on the nitrate ranges (<0.2 = low NO_3 zone, <0.50 = medium zone, <1.0 = high NO_3 zone, <3.0 = very high NO_3zone, 0 = land mass and NO_3-depleted zone).

4.2. Monthly Scale Nitrate Images

During this period the medium, high, and very high nitrate zones have been observed around the Gulf of Mannar and in the southern tip of Indian region. The stretches of high nitrate have been observed along the Kerala coast in the Arabian Sea, which indicates and supports the foresaid and earlier studied upwelling events with our nitrate maps (Figure 5). The upwelling has been reflected in the nitrate maps of July and August 1999. Even during the June month, the event has been seen along the Kerala coast (latitude 80.10–9.30°N and longitude 75.60–76.20°E). During September and October, the moderate nitrate patches have been observed at the bottom of Indian southern peninsula. Large nitrate-depleted areas can be observed in the Arabian Sea and in the Bay of Bengal during the months July–October.

In December, around the Gulf of Mannar and in the east coast of Sri Lanka, moderate NO_3 range patches have been observed and, during January, it has become intense in the Bay of Bengal. During February, low nitrate patches are seen in the Bay of Bengal and Arabian Sea, and nitrate-depleted zones have been observed in the northern Indian Ocean. March–May indicated the nitrate-depleted zones throughout, and, in the month of May, low NO_3 stripe has been seen from the bottom of India moving towards the Sri Lankan bottom. During June, the Arabian Sea indicated low nitrate zone with moderate, high, and very high NO_3 stripes along Kerala coast and offshore. The Bay of Bengal indicated NO_3-depleted water during June (Figure 5).

During July 2000, the Kerala coast/southeast coast of India has reflected high upwelling zone with high NO_3concentration (Figure 6). rest of the study area showed low NO_3 concentration zones, and similar patterns are seen. The upwelling has been observed during the period June–August (SWM) around Kerala coast and off southeast Arabian Sea with medium, high, and very high NO_3 concentration bands. High NO_3 concentration patches have been seen during January (NEM) in the Bay of Bengal water with moderate, high, and very high NO_3 concentration zones (Figures 5–9), (Table 1).

Table 1: Depiction of high nitrate distribution months in the Indian water.

Study Area	High nitrate/months
Arabian Sea	June–September
Indian Ocean	June–September and December–February
Bay of Bengal	December–February

*April-May showed fully NO_3-depleted zones throughout the study area.

4.3. Interpretation of Seasonal SST Images

Figure 10 indicates the seasonal variability of SST in the southern peninsular Indian water. There has been observation of high SST (~29°C) in the Bay of Bengal during the southwest monsoon and low SST (~25°C) in the Arabian Sea and around southern Indian tip region. The SST was exceptionally low (~24°C) during the fall intermonsoon 2001 around the southern tip of India in the Indian Ocean. During northeast monsoon, SST was low in the Bay of Bengal (~25°C) and high in the Arabian Sea (~29°C) and medium (~27°C) in the southern Indian tip in the Indian Ocean. During spring intermonsoon, low SST (~25°C) patches were seen around the southern Indian tip in the Indian Ocean.

4.4. Interpretation of Seasonal Nitrate Images

The seasonal NO_3 maps indicated the high and very high NO_3 concentration stripe along Kerala coast/southeast coast of India during the SWM season. Few high concentration patches were seen during the NEM, and rest of the months show low NO_3 in the Bay of Bengal and NO_3-depleted zone in the Arabian Sea (Figure 11). During the season spring intermonsoon (SIM), completely NO_3-depleted zones have been seen in the study area. Fall inter monsoon (FIM) showed mostly NO_3 depleted zones throughout with few low NO_3patches in the north Indian Ocean during 2000–2003. But, during the year 1999, the FIM (October-November) indicated the distribution of low NO_3 throughout the study area, probably enhanced with the super cyclone that had hit the northeast coast of India/Orissa during October 1999 with very high wind speed (>200 m/sec) and storm surges with wave heights (~20 meter) [42]. The super cyclone might have mixed the surface and bottom water, making the water turbulent, churning nutrients, and bringing up to surface. So, the exceptional widespread of NO_3 has been observed during FIM 1999. The sources of nitrate can be from terrestrial origin as observed during the NEM and FIM in the Indian Ocean and the Bay of Bengal. The sources can be from the bottom and deep water of ocean as the case of upwelling in the southwest coast of India during SWM (Figure 11).

4.5. Monthly Scale Nitrate Variability/Line Plots

The monthly mean nitrate peaks have been observed to be highest with two major peaks during the months August 1999 and January 2000 (~0.13 μM), and the monthly mean is lowest with two low dips during October and March ($NO_3 \sim$ a0.01 μM) (Figure 12).

4.5.1. Subset 1

For the subset 1 the monthly mean nitrate concentration has been seen highest during June–August (SWM) months (0.09–0.71 μM)) and similar in all the years. The rest of the months indicated less variability and equal trend during all the five-year period (Figure 12).

4.5.2. Subset 2

For the subset 2 of the study area, the northern Indian Ocean, the mean NO_3 peak has been obtained during the SWM and FIM seasons. The peak is seen in the month of October (0.6 μM) and during May–August, with the effect of onset and activation of SWM in the northern Indian Ocean region. January month (NEM) shows a small peak in the NO_3 concentration (0.09 μM) (Figure 12). The rest of the months do not indicate any peak and more variability. So, the Indian Ocean is mostly affected by high NO_3 during SWM and NEM months.

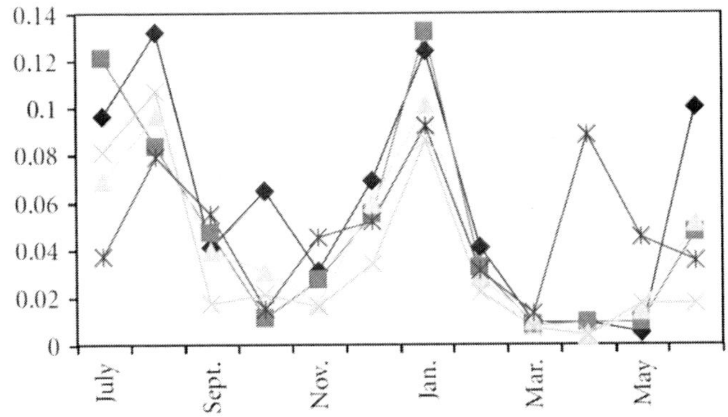

(a)

(b)

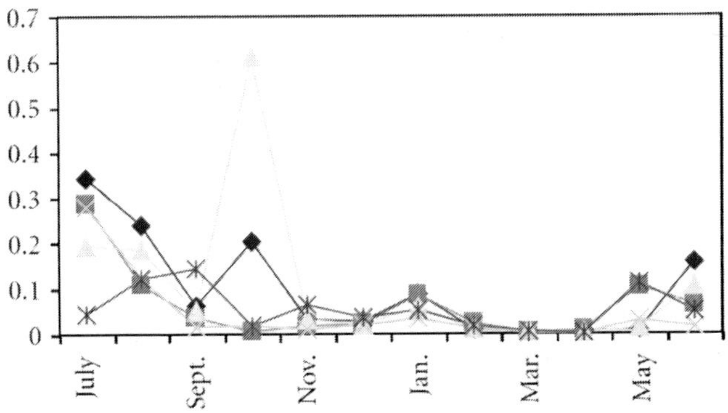

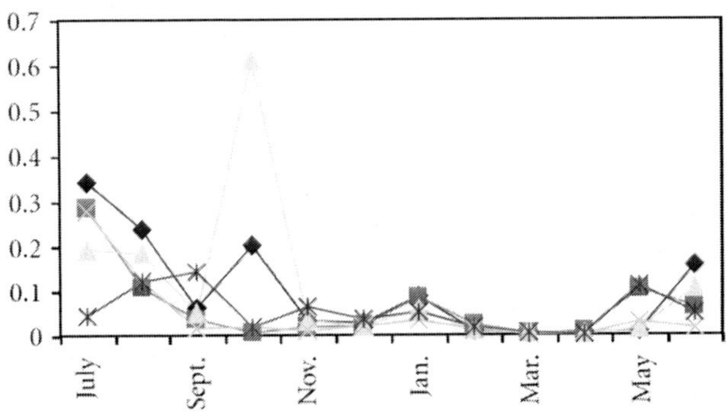

(c)

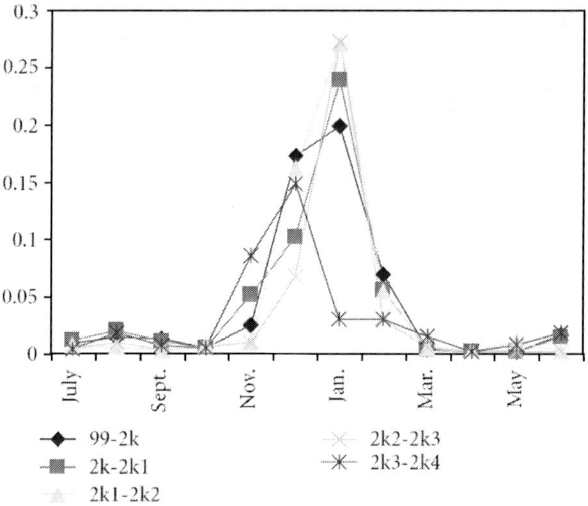

(d)

Figure 12: (a) Full image, (b) subset 1 (Arabian Sea), (c) subset 2 (Indian Ocean), and (d) subset 3 (Bay of Bengal) indicating surface nitrate mean (μM) in monthly scale retrieved from temperature-based in situ modeled data (the annotations on figures 2k–2k4 are mentioned addressing years spanning 2000–2004, resp.).

4.5.3. Subset 3

The case is different for the Bay of Bengal. The NO_3 concentration has been highest during the month of January (Figure 12). The peak of nitrate has been evidenced by other months between November and February for all the years. There is no impact of SWM and enhancement of NO_3 in the Bay of Bengal, like it was in the case of Arabian Sea and to some extent in the Indian Ocean.

4.6. Variability of Surface Nitrate (NO_3)/Histogram Plot

4.6.1. Full Image

The seasonal nitrate (NO_3) trend has been observed significantly. The NO_3 range was highest during SWM (0.05–0.098 μM)) and with the year-wise trend: 1999 > 2000 > 2001 > 2002 > 2003.

The trend was lowest during SIM (0.008–0.021 μM) (Figure 13). The seasonal nitrate trend is as follows: SWM > NEM > FIM > SIM.

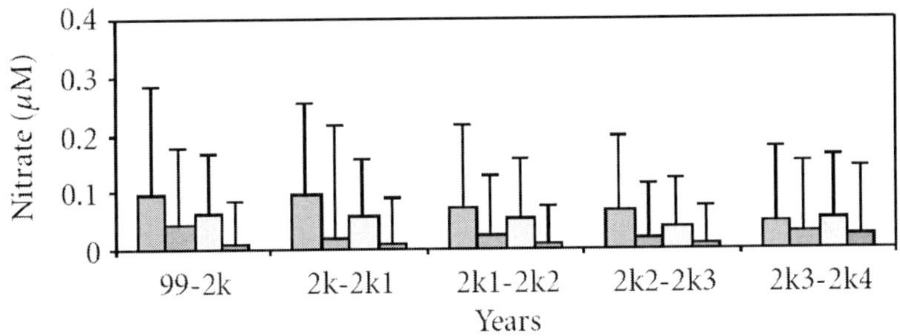
0.4
0.3
0.2
0.1
0
Nitrate (μM)
99-2k
2k-2k1
2k1-2k2
2k2-2k3
2k3-2k4
Years

(a)

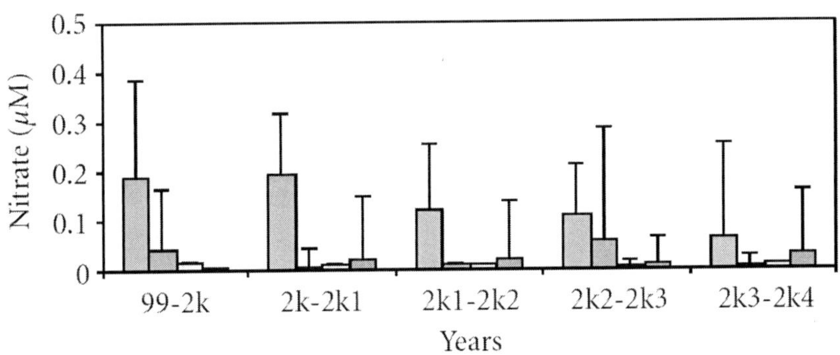
0.5
0.4
0.3
0.2
0.1
0
Nitrate (μM)
99-2k
2k-2k1
2k1-2k2
2k2-2k3
2k3-2k4
Years

(b)

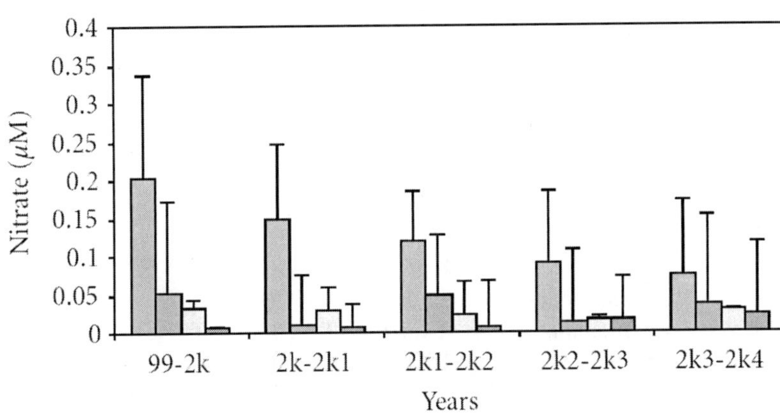
0.4
0.35
0.3
0.25
0.2
0.15
0.1
0.05
0
Nitrate (μM)
99-2k
2k-2k1
2k1-2k2
2k2-2k3
2k3-2k4
Years

(c)

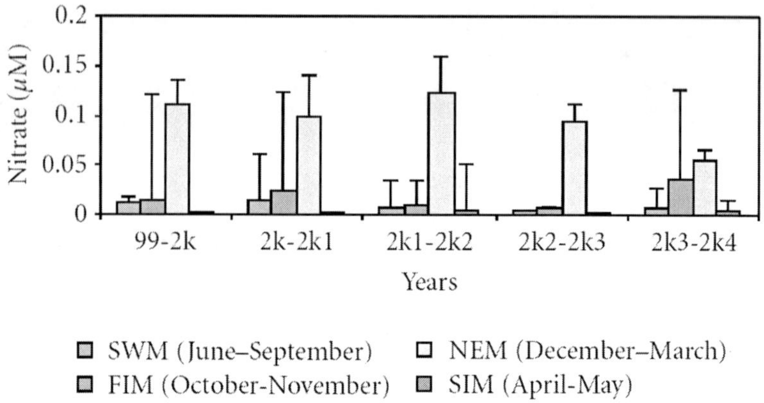

(d)

Figure 13: (a) Full image, (b) subset 1 (Arabian Sea), (c) subset 2 (Indian Ocean), and (d) subset 3 (Bay of Bengal), indicating surface nitrate mean (μM) with standar d deviation on seasonal scale retrieved from temperature-based in situ modeled data (SWM: southwest monsoon (June–September), FIM: fall intermonsoon (October-November), NEM: northeast monsoon (December–March), and SIM: summer intermonsoon (April-May)). The annotations on figures 2k-2k4 are mentioned addressing years spanning 2000–2004.

4.6.2. Subset 1 (Arabian Sea)

The seasonal nitrate (NO_3) trend has been observed significantly for the Arabian Sea surface water. The nitrate mean range was highest during SWM (0.06–0.198 uM) with the year-wise trend 2000 > 1999 > 2001 > 2002 > 2003. The nitrate mean range is lowest during NEM (0.005–0.015 uM) (Figure 13). The seasonal trend is as follows: SWM > FIM > SIM > NEM.

4.6.3. Subset 2 (Indian Ocean)

The seasonal mean NO_3 range has been highest during the SWM (0.07–0.205 uM) with the year-wise trend 1999 > 2000 > 2001 > 2002 > 2003. The nitrate mean range is lowest during SIM (0.005–0.02 uM) (Figure 13). The seasonal trend is as follows: SWM > FIM > NEM > SIM.

4.6.4. Subset 3 (Bay of Bengal)

The seasonal mean NO_3 range has been highest during the NEM (0.058–0.122 uM) with the year-wise trend 2002 > 2000 > 2001 > 2003 > 2004.

The seasonal nitrate mean range is lowest during SIM (0.001–0.004 uM) (Figure 13). The seasonal trend is as follows: NEM > FIM > SWM > SIM.

4.7. Standard Deviation (SD) of Nitrate

4.7.1. Full Image

This indicated the higher variability in the season FIM (Oct.-Nov.) and highest during the year 2000, and variability was lowest during the season SIM (April-May) during the year 2003 (Figure 13). The SD trend of the seasonal nitrate is as follows: FIM > SWM > NEM > SIM.

4.7.2. Subset 1

This indicated the higher variability of nitrate in the season FIM (Oct.-Nov.) and highest during the year 2002, and variability was lowest during the season SIM (April-May) during the year 2002 (Figure 13). The SD trend of the seasonal nitrate is as follows: SWM > FIM > SIM > NEM.

4.7.3. Subset 2

This indicated the higher variability of nitrate in the season SWM (Oct.-Nov.) and highest during the year 1999, and variability was lowest during the season SIM (April-May) during the year 2000 (Figure 13). The SD trend of the seasonal nitrate is as follows: SWM > FIM > SIM > NEM.

4.7.4. Subset 3

This indicated the higher variability of nitrate in the season FIM (Oct.-Nov.) and highest during the year 1999, and variability was lowest during the season SIM (April-May) during the year 1999 (Figure 13). The SD trend of the seasonal nitrate is as follows: FIM > SWM > NEM > SIM.

The chlorophyll in the Indian Ocean is observed to be primarily dependent on nitrate with some influences of SST and wind speed. But the Arabian Sea and Bay of Bengal chlorophyll is observed to be more dependent on SST followed by nitrate and then wind speed, obtained from regression analysis of five-year data sets [43]. So the nitrate has been a limiting factor in the ocean productivity around southern peninsular Indian water.

4.8. Annual Variability of Nitrates

The nitrate range has been observed between 0.01 and 0.31 μM in the study area. Moderate range >2.0 μM has been observed around the southern and southwest tip of Indian peninsula, off Kerala coast, and in the southern Tamilnadu coast in the annual averaged nitrate plot (Figure 13).

The peculiar feature has been observed off the southwest coast of India in the Arabian Sea. The high nitrate distribution pattern has been observed, which spreads and moves up to the Sri Lankan coast. This is most probably with the impact of seasonal upwelling off Kerala coast. The coastal water around Kerala is known to be the region of upwelling resulting in a cyclic injection of nutrients into the euphotic zone and pulses of phytoplankton development [14].

During the southwest monsoon, in the Bay of Bengal very few high nitrate concentration patches (around 0.05 μM) are seen to be spread and the rest of the region in the study area indicates low and very low concentration of nitrate (0.01–0.2 μM). Even in the monthly maps nitrate depletion has been observed significantly, which has been recorded during several cruises in the Bay of Bengal. At present models and algorithms have been developed and applied to know the nitrate depletion temperature (NDT) in the global oceans [44].

The relationship between temperature and nitrate (NO_3) has been established with significant correlation (= 0.912) with the in situ data from U.S. JGOFS and Indian cruises in the Arabian Sea and Bay of Bengal water. The surface nitrate concentration was highest in the Arabian Sea during June–September (SWM) period, in the Bay of Bengal the nitrate was highest during December–February (NEM), and the Indian Ocean exhibited high nitrate during both the seasons SWM and NEM. The nitrate-depleted zones were seen during April-May period. Upwelling during the SWM has the good link to the high nitrate concentration, enhancing the chlorophyll concentration in the Arabian Sea and around southern tip of India (Indian Ocean).

4.9. Sources of Nitrate and Nitrogen Fixation in Indian Water

The nutrients distribution in the Bay of Bengal is hence influenced by the external inputs which makes the waters of the upper layer less saline and highly stratified [45, 46], while the nutrient supply in the Arabian sea is largely due to the seasonal upwelling processes. Additionally the semiannual reversal of the upper-layer circulation also plays a role in transporting the sediments and nutrients through long shore currents in the Bay of Bengal [47–49].

In the northern Indian Ocean, cyanobacteria play a major role in the pelagic food web especially when the upper euphotic zone is stratified and provided with very low concentrations of nitrate (<0.01 μM, [50]). The Bay of Bengal, in the north eastern region of the Indian Ocean, is known for nitrogen limitation in the surface layers due to strong thermohaline stratification. Stratification prevents the advection of nitrate from the subsurface layer resulting in low biological production in the surface waters [32, 51]. Although upwelling and subsurface eddies were reported in some regions of the BoB during some seasons, its manifestations on primary production is much less compared to the Arabian Sea [32, 52]. Even during the summer monsoon period, except along the very coastal and upwelling regions of the Bay of Bengal, nitrate remains depleted in the surface layers. Nitrate input from the rivers is restricted to the coastal regions of the Bay of Bengal, and most of the shelf and offshore regions are devoid of any riverine nitrate in the surface layers [32]. The current study depicts high nitrate concentration (~3.0 μM) during the southwest monsoon upwelling period during August-September. In the summer time in April-May, there is observation of nitrate-depleted zones in the BoB; this is similar to the analysis in [32, 52].

Analyses of seawater collected during two cruises in coastal waters at 18–22°N latitude revealed that nitrogen was low in comparison with phosphorus and could be limiting primary production in the surface waters of the Bay [53]. N_2-fixation has been observed in all ocean basins, with maximum rates in the Indian Ocean, intermediate rates in the Atlantic, and minor rates in the Pacific [54]. Nitrogen nutrients are considered to be the major limiting factor of oceanic primary production in many regions [55]. Primary production is partitioned on the basis of nitrogen source: "new" production supported by N-sources, principally nitrate brought into the euphotic zone from the deep and secondarily, N_2-fixation, riverine, and atmospheric inputs; "regenerated" production [10] supported by ammonium and urea, derived from biological processes occurring within the euphotic zone [56]. So, the role of nitrate and its uptake in the marine carbon cycle is vital from the point of ocean biogeochemistry in local, regional, and global scales. Detailed analysis on the nitrate sources from upwelling, convective processes, terrestrial fluxes, atmospheric sources, and its characterization would help to understand the nitrogen cycle along with carbon cycle and ocean productivity in Indian water.

5. CONCLUSION

Algorithm for generating surface nitrate concentration in the Indian water covering the Arabian Sea, Bay of Bengal, and adjacent north Indian Ocean has been developed and implemented. The nitrate algorithm which models the satellite-derived SST with sigmoid curve has been successfully applied over the Indian water, and the nitrate concentration during the upwelling, southwest, and northeast monsoons has been observed high (0.50–1.00 μM). Low-concentration nitrate (~0.05) and nitrate depleting zones have been observed during the summer (March–May) in the Indian water. The nitrate-algorithm-derived surface nitrate maps can be useful for productivity assessment, algal bloom study, and even in the fishery forecasting approaches. The important role of the nitrate maps will be useful in studying nitrate uptake rates by the phytoplankton and the new production zones identification in different water types throughout the year, induced by the physical and biological processes.

ACKNOWLEDGMENTS

The author is thankful to Dr. R. M. Dwivedi, Head, MBD, Dr. Ajai, Group Director, MPSG, Dr. J. S. Parihar, Deputy Director, EPSA, and Dr. R. R. Navalgund, Director, SAC for providing necessary guidance and facilities for carrying out the work. Data archived from U.S. JGOFS archive, NODC/NIO, and Indian cruises have been acknowledged. The author mentions thanks to two anonymous reviewers for their critical overview and suggestions to improve the paper and to make the study more relevant to the regional and global context.

REFERENCES

1. D. Kamykowski, "A preliminary biophysical model of the relationship between temperature and plant nutrients in the upper ocean," Deep Sea Research A, vol. 34, no. 7, pp. 1067–1079, 1987.
2. J. D. H. Strickland and T. R. Parsons, "Inorganic carbon assimilation by marine biota," Bulletin/Fisheries Research Board of Canada, vol. 167, article 311, 1968.
3. H. U. Sverdrup, M. W. Johnson, and R. H. Fleming, The Oceans, Prentice Hall, Englewood Cliffs, NJ, USA, 1942.
4. W. S Broecker and T.-H. Peng, Tracers in the Sea, Lamont-Doherty Geological Observatory, Palisades, NY, USA, 1982.
5. J. D. H. Strickland, "Research on the marine planktonic food web at the institute of marine resources: a review of the past seven years of work," Oceanography and Marine Biology. An Annual Review, vol. 10, pp. 349–414, 1972.
6. J. T. O. Kirk, Light and Photosynthesis in Aquatic Ecosystems, Cambridge University Press, Cambridge, UK, 1983.
7. J. A. Raven, Energetics and Transport in Aquatic Plants, AR Liss, New York, NY, USA, 1984.
8. S. J. Zentara and D. Kamykowski, "Latitudinal relationships among temperature and selected plant nutrients along the west coast of North and South America," Journal of Marine Research, vol. 35, pp. 321–336, 1977.
9. H. J. Minas and M. Minas, "Net community production in "high nutrient-low chlorophyll" waters of the tropical and Antarctic oceans: Grazing vs iron hypothesis," Oceanologica Acta, vol. 15, no. 2, pp. 145–162, 1992.
10. R. C. Dugdale and J. J. Goering, "Uptake of new and regenerated forms of nitrogen in primary productivity," Limnology & Oceanography, vol. 12, pp. 196–206, 1967.
11. P. Klein and B. Coste, "Effects of wind-stress variability on nutrient transport into the mixed layer," Deep Sea Research A, vol. 31, no. 1, pp. 21–37, 1984.

12. E. D. Traganza, V. M. Silva, D. M. Austin, W. L. Hanson, and S. H. Bonsink, "Nutrient mapping and recurrence of coastal upwelling centers by satellite remote sensing: it's implication to primary production and the sediment record," in Coastal Upwelling, Part A, E. Suess and J. Thiede, Eds., pp. 61–83, Plenum, New York, NY, USA, 1983.
13. R. C. Dugdale, A. Morel, A. Bricaud, and F. P. Wilkerson, "Modeling new production in upwelling centers: a case study of modeling new production from remotely sensed temperature and color," Journal of Geophysical Research, vol. 94, pp. 18119–18132, 1989.
14. P. Morin, M. V. M. Wafar, and P. Le Corre, "Estimation of nitrate flux in a tidal front from satellite-derived temperature data," Journal of Geophysical Research, vol. 98, no. 3, pp. 4689–4695, 1993.
15. R. W. Eppley and B. J. Peterson, "Particulate organic matter flux and planktonic new production in the deep ocean," Nature, vol. 282, no. 5740, pp. 677–680, 1979. View at Publisher·
16. T. Platt, W. G. Harrison, M. R. Lewis et al., "Estimating new production in the sea—a case for a consensus," Marine Ecology Progress Series, vol. 52, pp. 77–88, 1989.
17. C. Garside and J. C. Garside, "Euphotic-zone nutrient algorithms for the NABE and EqPac study sites," Deep-Sea Research II, vol. 42, pp. 335–342, 1995.
18. D. Kamykowski and S. J. Zentara, "Predicting plant nutrient concentrations from temperature and sigma-t in the upper kilometer of the world ocean," Deep Sea Research A, vol. 33, no. 1, pp. 89–105, 1986.
19. S. Sathyendranath, T. Platt, E. P. W. Horne et al., "Estimation of new production in the ocean by compound remote sensing," Nature, vol. 353, no. 6340, pp. 129–133, 1991.
20. F. P. Chavez and S. K. Service, "Temperature-nitrate relationships in the central and eastern tropical Pacific," Journal of Geophysical Research, vol. 101, no. 9, pp. 20553–20563, 1996.
21. J. I. Goes, T. Saino, H. Oaku, and D. L. Jiang, "Method for estimating sea surface nitrate concentrations from remotely sensed SST and chlorophyll a—a case study for the north Pacific Ocean using OCTS/ADEOS data," IEEE Transactions on Geoscience and Remote Sensing, vol. 37, no. 3, pp. 1633–1644, 1999.
22. W. Koeve, "Wintertime nutrients in the north atlantic—new approaches and implications for new production estimates," Marine Chemistry, vol. 74, no. 4, pp. 245–260, 2001.
23. D. Kamykowski, S. J. Zentara, J. M. Morrison, and A. C. Switzer, "Dynamic global patterns of nitrate, phosphate, silicate, and iron availability and phytoplankton community composition from remote sensing data," Global Biogeochemical Cycles, vol. 16, Article ID 1077, p. 29, 2002.
24. C. Switzer, D. Kamykowski, and S. J. Zentara, "Mapping nitrate in the global ocean using remotely sensed sea surface temperature," Journal of Geophysical Research C, vol. 108, no. 8, Article ID 3280, 12 pages, 2003.
25. J. I. Goes, H. D. R. Gomes, T. Saino, C. S. Wong, and C. W. Mordy, "Exploiting MODIS data for estimating sea surface nitrate from space," EOS, Transactions, vol. 85, no. 44, pp. 449–454, 2004.
26. S. A. Henson, R. Sanders, C. Holeton, and J. T. Allen, "Timing of nutrient depletion, diatom dominance and a lower-boundary estimate of export production for Irminger Basin, North Atlantic," Marine Ecology Progress Series, vol. 313, pp. 73–84, 2006.
27. Silió-Calzada, A. Bricaud, and B. Gentili, "Estimates of sea surface nitrate concentrations from sea surface temperature and chlorophyll concentration in upwelling areas: a case study for the Benguela system," Remote Sensing of Environment, vol. 112, no. 6, pp. 3173–3180, 2008.
28. T. Steinhoff, T. Friedrich, S. E. Hartman, A. Oschlies, D. W. R. Wallace, and A. Körtzinger, "Estimating mixed layer nitrate in the North Atlantic Ocean," Biogeosciences, vol. 7, no. 3, pp. 795–807, 2010.
29. J. I. Goes, H. D. R. Gomes, A. Limsakul, and T. Saino, "The influence of large-scale environmental changes on carbon export in the North Pacific Ocean using satellite and shipboard data," Deep Sea Research II, vol. 51, no. 1–3, pp. 247–279, 2004.

30. D. M. Bartolacci and M. E. Luther, "Patterns of co-variability between physical and biological parameters in the Arabian Sea," Deep-Sea Research II, vol. 46, no. 8-9, pp. 1933–1964, 1999.
31. J. S. Sarupria and R. M. S. Bhargava, "Seasonal distribution of chlorophyll-a in the exclusive economic zone (EEZ) of India," Indian Journal of Marine Sciences, vol. 27, pp. 292–297, 1998.
32. S. Prasanna Kumar, P. M. Muraleedharan, T. G. Prasad et al., "Why is the Bay of Bengal less productive during summer monsoon compared to the Arabian Sea?" Geophysical Research Letters, vol. 29, Article ID 2235, 4 pages, 2002.
33. P. H. Burkill, R. F. C. Mantoura, and N. J. P. Owens, "Biogeochemical cycling in the northwestern Indian Ocean: a brief overview," Deep-Sea Research II, vol. 40, no. 3, pp. 643–649, 1993.
34. S. R. Shetye, A. D. Gouveia, and S. S. C. Shenoi, "Circulation and water masses of the Arabian Sea," Proceedings of the Indian Academy of Sciences : Earth and Planetary Sciences, vol. 103, no. 2, pp. 107–123, 1994.
35. M. Madhupratap, T. C. Gopalakrishnan, P. Haridas et al., "Lack of seasonal and geographic variation in mesozooplankton biomass in the Arabian Sea and its structure in the mixed layer," Current Science, vol. 71, no. 11, pp. 863–868, 1996.
36. M. Madhupratap, S. P. Kumar, P. M. A. Bhattathiri et al., "Mechanism of the biological response to winter cooling in the northeastern Arabian Sea," Nature, vol. 384, no. 6609, pp. 549–552, 1996.
37. M. Madhupratap, M. Gauns, N. Ramaiah et al., "Biogeochemistry of the Bay of Bengal: physical, chemical and primary productivity characteristics of the central and western Bay of Bengal during summer monsoon 2001," Deep Sea Research II, vol. 50, no. 5, pp. 881–896, 2003.
38. J. M. Morrison, L. A. Codispoti, S. Gaurin, B. Jones, V. Manghnani, and Z. Zheng, "Seasonal variation of hydrographic and nutrient fields during the US JGOFS Arabian Sea Process Study," Deep-Sea Research II, vol. 45, no. 10-11, pp. 2053–2101, 1998
39. March 2008, http://www.nodc.noaa.gov/sog/pathfinder4km/userguide.html.
40. R. K. Sarangi, S. Nayak, and R. C. Panigraphy, "Monthly variability of chlorophyll and associated physical parameters in the southwest Bay of Bengal water using remote sensing data," Indian Journal of Marine Sciences, vol. 37, no. 3, pp. 256–266, 2008.
41. JGOFS, "Protocols for the Joint Global Ocean Flux Study (JGOFS) core measurements," Tech. Rep. 19, Intergovernmental Oceanographic Commission, Bergen, Norway, 1996.
42. S. R. Nayak, R. K. Sarangi, and A. S. Rajawat, "Application of IRS-P4 OCM data to study the impact of cyclone on coastal environment of Orissa," Current Science, vol. 80, no. 10, pp. 101–106, 2001.
43. R. K. Sarangi, Impact of surface feature changes on the phytoplankton variability in Seas off the southern parts of the Indian Peninsula: a remote sensing approach, Ph.D. thesis, Berhampur University, Orissa, India, 2007.
44. D. Kamykowski and S. J. Zentara, "Can phytoplankton community structure be inferred from satellite-derived sea surface temperature anomalies calculated relative to nitrate depletion temperatures?" Remote Sensing of Environment, vol. 86, no. 4, pp. 444–457, 2003.
45. V. Subramanian, "Sediment load of Indian rivers," Current Science, vol. 64, pp. 928–930, 1993.
46. P. N. Vinayachandran and S. Mathew, "Phytoplankton bloom in the Bay of Bengal during the northeast monsoon and its intensification by cyclones," Geophysical Research Letters, vol. 30, Article ID 1572, 4 pages, 2003.
47. J. P. McCreary, W. Han, D. Shankar, and S. R. Shetye, "Dynamics of the East India coastal current, numerical solutions," Journal of Geophysical Research C, vol. 101, no. 6, pp. 13993–14010, 1996.
48. P. N. Vinayachandran, S. R. Shetye, D. Sengupta, and S. Gadgil, "Forcing mechanisms of the Bay of Bengal circulation," Current Science, vol. 71, no. 10, pp. 753–763, 1996.

49. D. Shankar, J. P. McCreary, W. Han, and S. R. Shetye, "Dynamics of the East India Coastal Current 1. Analytic solutions forced by interior Ekman pumping and local alongshore winds," Journal of Geophysical Research C, vol. 101, pp. 13975–13991, 1996.
50. P. H. Burkill, R. J. G. Leaky, N. J. P. Owens, and R. F. C. Mantoura, "Synechococcus and its importance in the microbial food web of the northern Indian Ocean," Deep Sea Research II, vol. 40, pp. 773–782, 1992.
51. H. R. Gomes, J. I. Goes, and T. Saino, "Influence of physical processes and freshwater discharge on the seasonality of phytoplankton regime in the Bay of Bengal," Continental Shelf Research, vol. 20, no. 3, pp. 313–330, 2000.
52. S. Prasanna Kumar, M. Nuncio, J. Narvekar et al., "Are eddies nature's trigger to enhance biological productivity in the Bay of Bengal?" Geophysical Research Letters, vol. 31, no. 7, Article ID L07309, 5 pages, 2004.
53. C. P. Valsaraj and V. N. R. Rao, "Nitrogen limitation in the tropical waters of the Bay of Bengal," Hydrobiologia, vol. 286, no. 3, pp. 139–148, 1994.
54. D. G. Capone and E. J. Carpenter, "Nitrogen fixation in the marine environment," Science, vol. 217, no. 4565, pp. 1140–1142, 1982.
55. W. G. Harrison, T. Platt, and M. R. Lewis, "f-ratio and its relationship to ambient nitrate concentration in coastal waters," Journal of Plankton Research, vol. 9, no. 1, pp. 235–248, 1987.
56. N. Gandhi, S. Prakash, R. Ramesh, and S. Kumar, "Nitrogen uptake rates and new production in the Northern Indian Ocean," Indian Journal of Marine Sciences, vol. 39, no. 3, pp. 362–368, 2010.

CHAPTER 5

Stressors and Disturbance Regimes on Back Reef Systems: Scale and Scope from Natural and Anthropogenic Sources

Sullivan Sealey KM[1], Valerie P[2], Smith GW[3]* and Slingsby S[4]

[1]*Department of Biology, University of Miami, FL 33124, USA*

[2]*Smithsonian Marine Station at Fort Pierce, Fort Pierce, FL 34949 USA*

[3]*Biology Department, University of South Carolina Aiken, SC 29801, USA*

[4]*National Marine Fisheries Service, Coral Reef Protection Program, Silver Springs, MD 20910, USA*

ABSTRACT

Back reef environments present a particular challenge in terms of characterization and management. Shallow near shore and lagoonal habitats associated with reef systems are the first areas to be impacted by land-based sources of pollutants and disturbances, defined as "stressors". Stressors are a link in the chain of events that lead to environmental change or "phase shifts" in natural systems. Stressors as "damaging stimuli" need to be defined in terms of thresholds for both populations of organisms, and ecological communities. Stressors alter the abundance and dynamics of individual populations of organisms, and thus impact the community structure and function. Microbial communities are key to processing nutrients and pollutants from land-based sources, but poorly characterized. The effects of changes in microbial communities usually go unnoticed until visual signs of change occur in macro-organisms (death, disease or a shift in predominate species). Because microbial communities have a major influence on the transformation of nutrients as well as a protective influence, short-term stressors often can result in latent changes in larger back reef communities. The most notable changes in back reef environments from stressors include: 1) Changes in coastal species abundance and diversity (including local extirpation), 2) Changes in natural community structure, 3) Changes in coastal water quality (or the dynamic" habitat), and 4) Changes associated with exotic species invasion. Advances in technologies and research have facilitated the detection of stressors and stress responses. Recent researches in stable isotopes signatures as indicators of nutrient flux across the land-sea interface have helped identify eutrophication sources. Detection of pathogens, advances in microbial ecology (genetic probes), and disease ecology, bleaching and pollutant studies (e.g. pesticides, fungicides, pharmaceuticals) have helped identify chronic stressors to back reef systems. Historical ecology and studies of long-term environmental change remain key to understanding the nature of change and chronic degradation of complex coastal systems. Although we understand the basic ecology of tropical marine systems in a general sense, researchers need to initiate programs to look at integrated processes across the shoreline, particularly in terms of ecosystem function in back reef environments.

Keywords: Lagoonal habitats; Ecological; Stressors; Pathogens; Fungicides; Isotopes

1. INTRODUCTION

Back reef environments present a particular challenge in terms of characterization and management. Shallow near shore and lagoonal habitats associated with reef systems are the first areas to be impacted by land-based sources of pollutants and disturbances. Stressors can be described as a component of the chain of events that lead to environmental change or "phase shifts" in natural systems [1,2]. Stressors are defined as stimuli that produce damage, and can be abiotic and biotic in nature. It is relatively easy to identify stressors that are catastrophic in nature, regardless of the source (a human-induced oil spill or natural hurricane event). However, it is often the chronic and long-term stressors that are more relevant to management to understand both the nature of variability in back reef systems, and to understand how stress can undermine the resistance and recovery of both individual species as well as natural communities.

Researchers have focused on the importance of a holistic approach to coral reef and fisheries management, with a call for "ecosystem management" to quantify and project "system services" from back reef and coastal environments [3]. This management philosophy seeks to place economic value on reefs for their roles in coast line stability, fisheries production and pollutant processing. The health of human communities is to a large part dependent on the health and services provided by natural ecological systems [4]. Ultimately, the impact of pollutants and physical alteration on any coastal systems will be the loss of production and diversity of large reef systems, but these changes are not easily prevented by current management paradigms. There are few real and measurable ecological criteria for judging "sustainability" or "compatibility" of land-based activities with adjacent marine environments [5]. At issue is the determination of threshold levels of disturbance or stress that impact a population of organisms in a given system for a given region.

"Stressors" have both long-term and short-term impacts on communities, but degraded or compromised natural systems can change irreversibly; thus, "how much of what type of stressors can natural systems take?" When does the natural variability of an abiotic or biotic factor move beyond a normal range, and constitute a "stress"? Stressors are produced by both natural and anthropogenic sources. Intact ecological systems are thought to have resistance and resilience to withstand natural stressors such as hurricanes and severe weather events. Communities such as coral reefs and rain forests are thought to benefit from intermediate levels of disturbance, and thus the community overall benefits from the stress on some populations [6]. Human-produced stressors range in scale from local alterations to the coastal environment to over-fishing to global climate change.

This paper will provide an overall characterization of human-produced stressors, with an overview of documented impacts on populations as well as on natural communities. There are new techniques and research innovations that may help elucidate the processes and causal links for coastal degradation. The applicability of these methods to future monitoring and management paradigms are a critical need for future funding.

2. CHARACTERIZATION OF STRESSORS

The literature is rife with descriptions of coral reef systems that are ecological degraded. It is obvious to define a highly degraded reef with loss of trophic complexity and degraded water quality [1]. The difficulty arises in the characterization of critical levels or thresholds of stressors that initiate or begin this "slippery slope to slime" scenario. Stressors, in the physiological sense, are stimuli that produce damage in some real and measurable manner. Stressors, regardless of source, can produce both short-term responses and long-term adaptations in natural systems (Table 1). To clarify terminology, it is important to understand "stressors" as damaging stimuli to individual species. The net impact of stressors seen across a community includes damage across species from many taxa, across life history stages, and even damage by location or proximity to the stimuli. Fragmentation of back reef habitats is a serious ecological concern for nursery value and fisheries production, but fragmentation is the result of catastrophic stressors (e.g. dredge and fill of mangrove creek systems, marina construction, channels and waterways) that ultimately transform the seascape [7].

Stressors as "damaging stimuli" need to be defined in terms of thresholds linked to specific anthropogenic inputs. Stressors such as toxins are well defined physiologically by characteristics such as "LD- 50", or a dosage or exposure level at which 50% of the population is killed. Stressors in ecological systems can more appropriately be characterized as damage to individuals in the basic processes of growth, reproduction, and recruitment processes. Natural systems are compromised in their resilience and resistance to change when fragmented and reduced in spatial extent by human alterations of coastal areas. The full scope of coastal habitat complexity is necessary to support diversity and production through island connectivity [8]. Herein lays the problems in theoretical and empirical application. A "stressful" level of temperature is that pattern of temperature change (rate and severity) that significantly damages the physiological function of the organism. How is this threshold of damage distinguished from natural variability? The key is defining stressor thresholds, and how that threshold is compromised by overall system-wide alterations by humans.

There are several specific case studies in the coral reef research literature that help elucidate these threshold levels of stressors on back reef systems. For example, the increase or decrease of nutrient input into back reef habitats can affect the quantitative (productivity, abundance) and qualitative (community structure) characteristics of coastal marine communities [9].

Table 1: Description and definition of the abiotic and biotic variables that human actions can alter to stressful levels on back reef systems. Responses of species are particularly critical for benthos (corals, macroalgae, sponges and soft corals) that cannot move to avoid the stressful stimuli. Change in community structure both with the abundance of individuals (population-level changes) and the number of species present (community-level change) will occur with long-term exposure to stressors.

Abiotic and Biotic variables that can become stressors	Short-term response	Long-term adaptation or acclimation
Temperature		
Extreme heat or cold temperatures experienced over a period of time or over a cycle (diurnal, tidal) that is not typical of normal variability.	Increased metabolic costs Reduced deposition of energy reserves Higher food or nutrient demand Bleaching of corals, soft corals, sponges	Selection of more heat tolerant individuals and species and their symbionts Reduction in growth rate and reproductive output Increased susceptibility to disease, lower energy reserves
Oxygen		
Reduce oxygen availability to animals from reduced photosynthesis or restricted amounts of oxygen in the water (e.g. increased BOD)	Death or partial mortality in some individuals and species from hypoxia stress or localized anoxia	Appearance of "dead zones" of acute anoxia Selection towards more hypoxia tolerant species
Energetics–Food and Nutrients		
Stressors that reduce the amount or quality of food energy available to animals or alter the nutrient regime normally experienced by plants	Reduced growth or depletion of food stores Death or partial mortality in some individuals and species from food or nutrient stress Change in diet for some predators or herbivores Increased marine plant growth rate depending on the nutrient regime	Lower coverage of clonal species (but some clonal species might increase) Lower reproductive output Lower recruitment success Dominance of more resistant or "weedy" species
Pathogens		
Increase in the number and severity of disease-causing agents, impacts both plants and animals	Death or partial mortality in some individuals and species from disease	Change in community structure both with types of species present and abundance of species
Toxins and Pollutants		
Toxins that accumulate in sub-lethal levels that can impact selected species and selected life-history stages of species	Change in mortality rates for species or life-history stages of some species Change in abundance of some species Increased susceptibility to diseases or other stress stimuli	Change in community structure both with types of species present and abundance of species

Sewage input in Kaneohe Bay, Oahu, Hawaii, provides a documentation of threshold levels of nutrient runoff on a back reef ecosystem. Sewage discharge directly into the bay over a 25-year period caused a change in the benthic community at the south end of the bay from a system dominated by corals to one dominated by suspension feeders, deposit feeders, and the "green bubble alga" Dictyosphaeria cavernosa [10]. Sewage from two large treatment plants was diverted from the bay to a deep ocean outfall in 1977-1978, and the system began to change back to its original structure. Nutrients, turbidity and chlorophyll a concentrations declined rapidly in the year after sewage diversion, and phytoplankton and D. cavernosa abundance and numbers of filter feeders decreased [11]. However, Kaneohe Bay has not completely returned to "pre-sewage conditions", and initial trends of recovery have slowed or even reversed at some locations in the bay [10], probably due to continued anthropogenic impacts resulting from the growing human population living near the bay. Stresses resulting from long-term impacts of dredging, sedimentation, stream channelization, non-point sources of nutrients, and the introduction of potential toxicants have not been remediated [10].

The complexity of land-based sources of pollutants, especially nitrogen (N), to near shore waters is now approached through a combination of spatial datasets and radioisotopes. Human population growth in coastal watersheds and the resulting run off of nutrients and contaminants to the coastal waters have been documented by spatial models [12]. Rarely have these models been directly related to stressor thresholds for coral reef benthos populations with the end objective to develop ecologically meaningful coastal development limits or carrying capacities.

The Florida Keys presents another case study of the complexity of stressors related to human activities. Changes in land use and landbased sources of pollution (particularly agriculture) have occurred on a dramatic scale throughout the tropical south Florida ecological system for the past 100 years [13]. The unique coastal geomorphology of the Florida Keys combined with the relatively recent sea level rise with the formation of Florida Bay 4000 ybp combined to create a very spatially complex and heterogeneous reef tract. This naturally-variable system contained lagoon, near shore and reef tract environments. Researchers found it difficult to establish a causal link between obvious pollution stressors and degradation of back reef habitats [14,15]. The challenges in identifying stressors come from several factors: 1) Little baseline information existed on processes critical to sustaining back reef and reef communities of the Florida Keys despite over 100 years of land-use change in the Keys and Everglades watershed, 2) The large spatial scale of the back reef systems including Florida Bay was poorly characterized in terms of the natural variability of key abiotic and biotic parameters (e.g. the presence of natural crossshelf and latitudinal gradients along the reef system), and 3) The role of episodic storm events on water quality and sediment deposition was not understood or characterized on meaningful ecological scales. Hurricanes and large tropical storms provide an important disturbance regime for meso-scale sediment or detritus transport on and off of islands in particular (Figure 1). At any given point in time, there is likely a mosaic of varying states of disturbance and recovery across an island archipelago essential for the functioning of the wider ecological system. In the Florida Keys, management and research continue to struggle with the understanding of relatedness in Upper, Middle and Lower Keys systems.

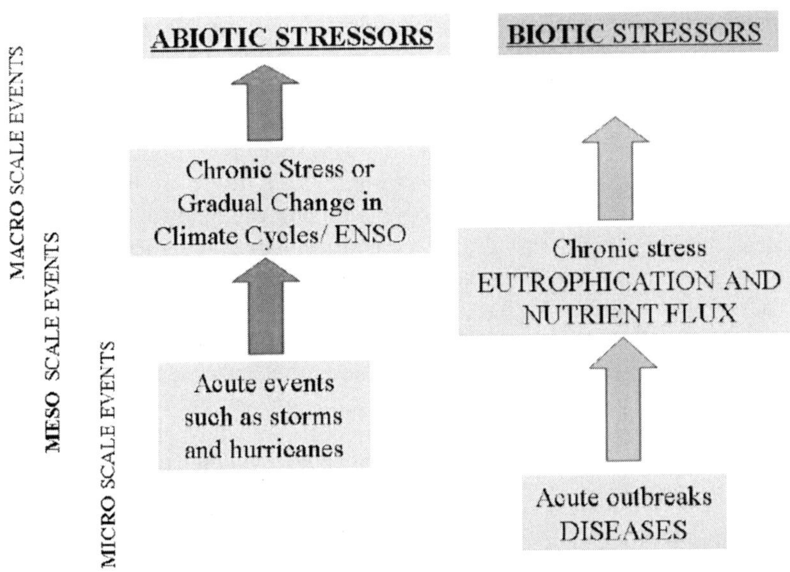

Figure 1: Summary of important stressors to the back reef environment by scale (both temporal and spatial). Some stressors can occur across multiple scales.

The characterization of stressors has hinged on a fundamental characterization of tropical coastal environments [16,17] and will require the combination of system-wide and organism-level approach. The characterization of stressors most likely to impact the long term viability of back reef systems requires then a two-step process of first identification of the coastal system by benthic assemblages and processes, and then, the quantification of anthropogenic impacts (threats) that translate to stressful levels of abiotic and biotic variables (Table 1). The first review paper in this series on back reef environments presented the role of natural variability in back reef environments. Natural variability is a component of initial system characterization, as habitats function ecologically as a mosaic from the coastal zone to the outer shelf.

3. POPULATION-LEVEL CHANGES FROM STRESSORS

Population-level changes in back reef system are defined as changes in the basic population structure and processes by a disturbance to recruitment, growth, fecundity or mortality rates. Population-level studies of lethal and sub-lethal toxicity of heavy metals, petroleum, herbicides, pesticides, synthetic hormones, PCBs, and pharmaceuticals on tropical marine organisms have a long and prolific record in the literature [18-20]. Are these really population-level studies, or just studies of individual species? Also, is the literature really that prolific given there is an awful lot we don't know about these fundamental questions? The earliest studies in aquatic toxicology focused on understanding the effects of chemical pollutants on tropical marine organisms. Chemical contaminants are thought to have chronic, acute, direct and indirect impacts on tropical marine organisms and associated habitats, such as coral reefs, mangroves and seagrasses [21]. However, there remains a significant challenge in understanding the chemical changes and transport mechanisms that get these molecules into the near shore marine environment [12]. Understanding the mode of transport depends on detailed knowledge of the physical environment. For example, the rate of discharge in a coastal region is based on its hydrogeology [22].

A handful of studies have examined the effects of agricultural pesticides on marine organisms. Pesticides, such as organochlorides, organophosphates, and carbamates, can have long-lasting and toxic effects on living organisms due to their slow rate of decomposition and the toxicity of their intermediate breakdown products [22]. Although DDT was banned in the United States in 1973, Singh and Ward [23] found elevated concentrations in the coastal waters of St. Lucia. Since chemical compounds strongly adsorb to soil particles [24], high concentrations have also been detected in sediments. Bioaccumulation of these compounds has been found in animal tissues [19,25,26]. Arukwe [27] found that anthropogenic chemicals disturb reproduction in fish by disturbing the endocrine system.

Pesticides have also been shown to decrease photosynthetic and respiration rates in benthic organisms, such as Halodule sp., a common seagrass in the Caribbean [28,29] and corals [30]. Corals have a thin, lipid-rich layer of tissue that rapidly takes up chemical compounds [21] and a symbiotic relationship with an algal host that is maintained through chemical communication. Therefore, coral colonies may be particularly susceptible to pesticides [22] and heavy metals [31]. Chemical pollutants may have negative impacts at the population level since coral reproduction and recruitment are chemically mediated processes sensitive to coastal pollution and changes in water quality [32].

Methods for evaluating pollutants on tropical marine organisms have slowly been developing. Evaluation of chemical stressors includes exposure tests, bioassays, and ecological response analyses at the individual, population, and community levels [21]. Exposure analyses include determining sources, pathways, and concentrations of toxic chemicals in an ecosystem. For example, the form and bioavailability of toxic chemicals in seawater, and the rate of bioaccumulation in an organism's tissues are unknown for many chemical contaminants. More recently, molecular biomarkers have been developed to examine the direct effects of such stressors on coral reef organisms, however results are not yet conclusive and more work is needed to tease out variability and confounding environmental factors [31,33]. Furthermore, the effect of chemical pollutants on processes such as reproduction and recruitment has yet to be determined. Understanding toxic effects at the species-level as well as at the ecosystem-level is critical. Ultimately, linking biomarkers of exposure within an organism to the population or community level will provide the most useful information to managers [21].

While exposure concentrations and fate and transport processes of chemical pollutants have been determined in temperate regions, parallel research in tropical marine systems is in its infancy. Pollution studies in the future need to incorporate three essential components: 1) Understanding of the fate and transport mechanisms for the chemical in the environment, including bioaccumulation, 2) Elucidating the impact on organisms throughout their life cycle with both acute and chronic exposure levels, and 3) Documenting the long-term scale and scope of these impacts on in situ populations of organisms. Population-level changes in more or less-susceptible species will ultimately change the structure of back reef communities.

4. COMMUNITY-LEVEL CHANGES FROM STRESSORS

The detection of community-level change in highly variable systems is difficult, especially in complex systems [2]. The burden of proof has been placed on the scientific community to not only document the change, but also to establish the chain of events leading to that change (sources-stressors-symptoms as pictured in Figure 2. Established protocols exist for differentiating disturbance-induced changes from natural community changes and variability due to succession, seasonality, and other natural biological events. These protocols are well represented in the environmental impact assessment literature, and only recently are tied to better understanding the overall system ecology and "system outcome" from anthropogenic disturbance. A quick review of established "disturbance detection" protocols for natural communities follows.

A modified BACI (Before/After and Control/Impact) approach has been used to evaluate catastrophic disturbance events or stressors (e.g. grounding events). The BACI approach relies on the assumption that the impact (disturbance event) will cause a pattern of change in the affected area that varies from the natural pattern of change in an unaffected control location. The original BACI design includes two sampling dates (before and after the disturbance event) at two sites (control and impact). There are some inherent flaws in this original design, including the fact that many BACI studies lack replication of sample sites, often due to the impact of natural and even more so anthropogenic impacts being localized and spatially limited. Furthermore, if the control and impact study sites are arbitrarily chosen, it is possible that inherent differences in the sites will make the two sites naturally differ over time, regardless of whether there was a true disturbance effect. Since the BACI approach was first developed, many modifications have been suggested in order to correct for interpretational, logistic, and/ or statistical problems associated with the original approach. A "model study" of the effects of disturbances on near shore marine communities should contain spatial replication, selection of control sites, temporal replication, and simultaneous sampling of control and impacted sites. The rigors of documenting low-level disturbances or chronic pollution impacts are rarely met in tropical reef systems, with impacts most easily understood in hindsight.

New methods in multivariate analysis of a wide suite of abiotic and biotic factors have been most successful is detecting initial community-level changes from stressors. Thus, the inclusion of multivariate techniques can better elucidate patterns of change, since changes in even one target species will likely affect other components. Some multivariate techniques, such as non-dimensional MDS plots can illustrate patterns (directions) of change among chronological samples. Thus, if there were three pre-disturbance sampling times (labeled A, B, C) and three post-disturbance sampling times (labeled D, E, F) at each sample site, comparing the patterns in the paths of the samples over time may highlight similarities and differences following disturbance impacts. Figure 3 illustrates a hypothetical situation in which an MDS plot shows differences between control and impact sites, indicating a disturbance effect.

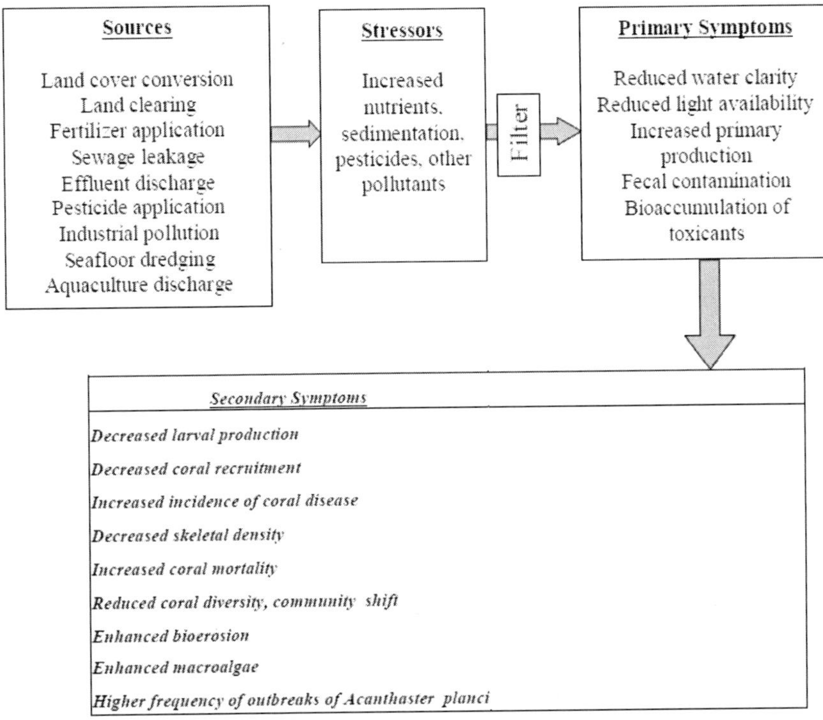

Figure 2: Schema of potential sources, stressors, primary symptoms, and secondary symptoms directly affecting stony corals and coral reef benthos. The "filter" represents local conditions that determine the resistance and resilience of a reef to being affected by terrestrial runoff (from ISRS Briefing Paper #2, 2004).

Stressors produce community changes first recognized by visual "symptoms" including syndromes in benthic invertebrates, most notably corals and sponges. Syndromes can result in dramatic changes in abundances of dominant species. Detection or documentation of community-level change is accompanied by a presumption of change in ecological function or production of that system. Chronic stress to reef (and back reef) environments can lead to a series of system-altering changes that include: 1) Changes in coastal species abundance and diversity (including local extirpation), 2) Changes in natural community structure, 3) Changes in coastal water quality (or the "dynamic" habitat), and 4) Changes associated with exotic species invasion. Coastal development and alteration by humans are likely the key activity giving rise to stressors in back reef environments. Starting from the removal of native coastal vegetation "buffer zones" to the construction of marinas and finger canals has been linked to the demise of corals throughout the tropical oceans [34-36]. Much of the decline in the growth, recruitment and reproduction of corals is attributed to degradation of water quality or fishing impacts. The synergistic effects of pollution and overfishing can dramatically shift the threshold of stressors, and thus amplify the damage to both populations and communities.

Underlying the conspicuous community of benthos in the back reef environment are the complex and poorly characterized microbial communities. Microbial communities play a pivotal role in transformation of nutrients in a largely oligotrophic system. Microbial assemblages can exert a protective influence on benthic organisms, such as corals. However, short term stressors often shift dominance of microbes, and initiate disease, nutrient limitations or abiotic changes in the environment (e.g. dissolved oxygen depletion). The mechanisms by which stressors affect microbial communities are complex; research has elucidated the diversity of responses from stimulation of population growth, initiation of metabolic changes, or induction of prophage (viruses integrated into the host DNA) or changes in behavior of specific microbial populations within the community [37-39].

Specific microbial communities have evolved in association with macro organisms since their appearance on earth. These relationships have become obligate and complex, particularly with aquatic organisms. Studies by Ritchie and Smith [39] have shown that the microbial communities are coral species specific. In addition, they change when exposed to stressors. The net result of this microbial community change is a loss of the specific protective matrix that had evolved with the coral species, thereby, rendering the coral susceptible to disease. Stressors affecting these symbiotic microbial communities change community structure in one of three ways: 1) The differential stimulation of specific populations; 2) The differential inhibition of specific populations and 3) The introduction of new populations (perhaps pathogens) into the community from the water mass. The net effect is destabilization of the overall community as well as the functional relationship with the host.

Recent studies on the function of microbial communities only reinforces the underlying ecological processes, particularly nutrient and energy flux, as critical to community stability, resilience and resistance. The relationship between biological production (ecosystem function) and diversity has become a central focus of environmental science, and must be fundamentally linked to nutrient cycles. Coastal environments represent a model system for examining the impact of anthropogenic disturbance on this balance of species diversity, habitat complexity and ecological function. Central to the ecological function of the back reef is the quality of the medium.

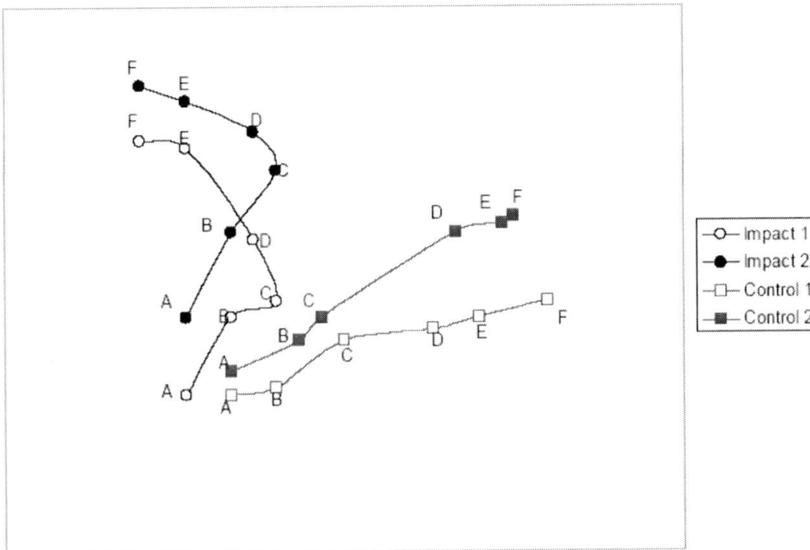

Figure 3: Hypothetical MDS plot showing directionality of chronological temporal replicates (A-F) at two control sites (squares) and two impact sites (circles) before (sample times A-C) and after (sample times D-F) a disturbance event. A line connects the chronological replicates for each unique study site. The MDS shows that prior to disturbance, the four communities had similar patterns and directions of change/ succession over time, while after the disturbance, the impact sites and the control sites varied in their patterns of change.

Water quality is fundamental to both species and natural community distribution in nearshore marine environments, and the impact of water quality degradation is complex. Changes in water quality parameters stem from human alterations of coastal hydrology and runoff patterns. Because water quality parameters such as temperature, salinity and turbidity vary naturally with tidal, diurnal and seasonal cycles, changes in natural variability are difficult to document. Documenting changes in coastal water quality would require some knowledge and monitoring of pre-development conditions, but detection still requires the collection of both abiotic and biotic data to document change [2]. Sewage and wastewater discharge are notorious for long-term changes in tropical marine environments [10,11]. Human activities on land inevitably increase nutrient inputs to coastal waters from deforestation, wastewater, fertilizer, and other sources [40,41].

In fact, water quality studies in Montagu Bay and Nassau Harbour, adjacent to New Providence, do not show elevated nutrient concentrations compared to waters adjacent to undeveloped islands [42]. Despite alteration to 100% of the shoreline, with loss of reef and seagrass habitats, the patterns of water quality variability with tides and seasons are not statistically different when compared to lightly-developed islands in a national park. The ability to detect and document changes in inorganic nutrients in tropical nearshore waters is limited to point sampling on tidal, diurnal or seasonal cycles and only extreme values for parameters such as total nitrogen appear higher in boxplots constructed to capture water quality variability. Most studies have documented the ecological changes to coral reefs rather than establishing pollution sources, quantities and pathways for non-point, land-based pollution [43,44]. The water-quality monitoring program in the Florida Keys National Marine Sanctuary involves a massive sampling effort of over 1400 stations throughout the Florida Keys. This program is only beginning to characterize water quality parameters for Florida Bay, the nearshore Keys waters, and the Keys reef tract, with the only obvious water quality degradation near, and in, dredged residential and commercial canals systems. Detailed water quality monitoring is essential for remediation of problems associated with coastal development, but not necessary for the identification or prediction of ecological impacts. The most significant water quality impacts of coastal development that can be documented are likely to be manifested after severe storms and hurricanes. Anthropogenic alteration of the coastal zone will exacerbate natural disturbances, primarily in the scope and severity of nutrient and sediment transport from land to sea.

5. ADVANCES AND FUTURE CHALLENGES IN STRESSOR CHARACTERIZATION AND DETECTION

New research initiatives are helping to move beyond the basic characterization of back reef environment to understanding the "rates and fates" of key nutrients and microbial complexes. Stable nitrogen isotope analysis is employed to identify this near shore response to anthropogenic nutrient loading. Nitrogen isotope ratios (15N/14N usually expressed as δ15N) can be used to discriminate between marine- and terrestrial-based organic matters and, therefore, to detect terrestrially formed organic matter from sewage effluent in the marine environment.

New methods in detection of pathogens are critical in limiting chronic stressor impacts to wide spread changes in the health of benthos (G. Smith, pers. comm.). These new methodologies that examine the details of nutrient and pathogen flux to back reef environments are still critically dependent on understanding the historical ecology and longterm environmental changes that have already occurred in tropical island systems.

A central research gap is our understanding of variability and fluctuations in benthos and fish under "normal" conditions, particularly the role of the "normal" micro biota in the function of the back reef ecosystem. Threshold levels of certain stressors (e.g., temperature, nutrients, salinity, etc.) that lead to significant change in populations need to be determined. Researchers have begun to look at new ways of modeling the complex interactions of several stressors co-occurring, and how this synergy changes thresholds of mortality and disease in populations. Management demands will require better approaches to understanding restoration needs and the significance of large reductions of species from the back reef environment (e.g., conch, Diadema, any commercially exploited species). Advances in research will utilize a multi-disciplinary approach across the scale of single organism's response to regional systems response.

CONFLICTS AND MANAGEMENT CHALLENGES

For the countries around the world with coral reef resources, sustainable tourism is often part of a national vision for natural resource management. Many of the decisions for locating development are based on a relatively narrow view of the coastal landscape. Although there has been much policy discussion of "sustainable development" of island systems, there remain no real ecological criteria to judge sustainability on islands as isolated and highly variable landscapes. The coastal environment is already fragmented by historical or unplanned alterations. The discussion of the role of marine protected areas has led scientists to look at a network of areas to be set aside as "ecologically intact", meaning no fishing. These "intact" areas can be key control sites for long-term monitoring and disturbance detection. However, this "intactness" and ecological functionality of even protected back reef environments may be, in the long-term, challenged by scales of stressors beyond local sites, regions or even single oceans.

Although we understand the basic ecology of tropical marine systems in a general sense, researchers have rarely looked at integrated processes across the shoreline, particularly in terms of ecosystem function. Protected areas and fisheries reserves adjacent to developed coasts and cities will have an important role in both education and remediation. Restoration of coastal areas that have been altered by development, dredging or loss of wetlands can be expensive, and demonstration projects that are practical outside the funding environment of the US need to be established in conjunction with coral reef research efforts. Restoration and remediation of many coastal environments is inevitable with population growth and the growing need for ecological services. The conservation of coastal fisheries resources as well as the protection of human life and property will motivate the political will to tackle restoration projects. Scientists may need to present information in a more integrated package not dependent on "change-no change" hypothesis to motivate the stakeholders.

REFERENCES

1. Hughes TP (1994) Catastrophes, phase shifts, and large-scale degradation of a Caribbean coral reef. Science 265: 1547-1551.
2. Fabricius KE, De'ath G (2004) Identifying ecological change and its causes: a case study of coral reefs. Ecol Appl 14: 1448-1465.
3. Hatcher BG (1997) Coral Reef Systems: how much greater is the whole than the sum of the parts. Coral Reefs 16: S77-S91.
4. Epstein PR (1999) Climate and Health. Science 285: 347-348.
5. Indian Society of Remote Sensing (ISRS) (2004) The effects of terrestrial run-off of sediments, nutrients and other pollutants on coral reefs. Briefing Paper #3. International Society for Reef Studies p: 18.
6. Connell JH (1978) Diversity in tropical rain forests and coral reefs. Science 199: 1302-1310.
7. Maragos JE (1993) Impact of coastal construction on coral reefs in the U.S.-affiliated Pacific Islands. Coast Manag 21: 235-269.
8. Cowen RK, Lwiza K, Sponaugle S, Paris C, Olson D (2000) Connectivity of marine populations: Open or Closed? Science 287: 857-859.
9. Birkeland C (1988) Second-order ecological effects of nutrient input into coral communities. Galaxea 7: 91-100.
10. Hunter CL, Evans CW (1995) Coral reefs in Kaneohe Bay, Hawaii: Two centuries of western influence and two decades of data. Bull Mar Sci 57: 501-515.
11. Smith SV, Kimmerer WJ, Laws EA, Brock RE, Walsh TW (1981) Kaneohe Bay sewage diversion experiment: perspectives on ecosystem responses to nutritional perturbation. Pac Sci 35: 279-395.
12. United Nations Environment Programme (UNEP) (1992) Environmental problems affecting the marine and coastal environment in the wider Caribbean region. UNEP (OCA)/CAR IG.9/INF.5.
13. Lott, Dye CR, Sullivan KM (1996) Historical overview of the development and natural history of the Florida Keys. Volume 3: Site Characterization for the Florida Keys National Marine Sanctuary and Environs. The Preserver, Farley Court of Publishers. Zenda, WI p: 49.
14. Lapointe BE, Clark MW (1992) Nutrient inputs from the watershed and coastal eutrophication in the Florida Keys. Estuaries 15: 465-476.
15. Lapointe BE, Tomasko DA, Matzie WR (1994) Eutrophication and trophic state classification of seagrass communities in the Florida Keys. Bull Mar Sci 54: 696-717.
16. Sullivan Sealey KM, Bustamante G (1999) Setting geographic priorities for marine conservation in Latin America and the Caribbean. The Nature Conservancy, Arlington, Virginia p: 141.
17. Allee R (2000) Marine and estuarine ecosystem habitat classification. NOAA technical Memorandum NMFS-F/SPO, p: 43.
18. Rees JE, Setiapermana D, Sharp VA, Weeks JM, Williams TM (1999) Evaluation of the impacts of land-based contaminants on the benthic faunas of Jakarta Bay, Indonesia. Oceanol Acta 22: 627-640.
19. Olafson RW (1978) Effect of agricultural activity on levels of organochlorine pesticides in hard corals, fish and mollusks from the Great Barrier Reef. Mar Environ Res 1: 87-107.

20. Randall JE (1972) Chemical pollution in the sea and the crown-of-thorns starfish (Ancanthaster planci). Biotropica 4: 132-144.
21. Peters EC, Gassman NJ, Firman JC, Richmond RH, Power EA (1997) Ecotoxicology of tropical marine ecosystems. Env Toxicol Chem 16: 12-40.
22. Rawlins BG, Ferguson AJ, Chilton PJ, Arthurton RS, Rees JG (1998) Review of agricultural pollution in the Caribbean with particular emphasis on small island developing states. Mar Poll Bull 36: 658-668.
23. Singh NC, Ward RE (1992) Land based pollution sources and marine environmental quality in the Caribbean. Chem Ecol 6: 259-270.
24. Oloffis, PC Albright LJ, Szeto SY, Lau J (1973) Factors affecting the behavior of five chlorinated hydrocarbons in two natural waters and their sediments. J Fish Res Bd Can 30: 1619-1623.
25. Mansingh, A, Wilson A (1995) Insecticide contamination in the Jamaican environment 3: baseline studies on the status of insecticidal pollution of Kingston harbor. Mar Poll Bull 30: 640-645.
26. Glynn PW, Rumbold DG, Snedaker SC (1995) Organochlorine pesticide residues in marine sediment and biota from the northern Florida reef tract. Mar Poll Bull 30: 397-402.
27. Arukwe A (2001) Cellular and molecular responses to endocrine-modulators and the impact on fish reproduction. Mar Poll Bull 42: 643-655.
28. Walsh GE, Hansen DL, Lawrence DA (1982) A flow-through system for exposure of seagrass to pollutants. Mar Environ Res 7: 1-11.
29. Ramachandran S, Rajendran N, Nandakumar R, Venugopalan VK (1984) Effect of pesticides on photosynthetic and respiration of marine macrophytes. Aquat Bot 19: 395-399.
30. McCloskey LR, Chesher RH (1971) Effects of man-made pollution on the dynamics of coral reefs. In: Miller JW, Van Der Walker JG, Waller RA (eds.) Scientists-in-the-sea. Department of Interior, Washington, USA 6: 229-237.
31. Morgan MB, Vogelien DL, Snell TW (2001) Assessing coral stress responses using molecular biomarkers of gene transcription. Env Toxic Chem 20: 537-543.
32. Richmond RH (1993) Coral reefs: present problems and future concerns resulting from anthropogenic disturbance. Am Zool 33: 524-536.
33. Downs CA, Mueller E, Phillips S, Fauth JE, Woodley CM (2000) A Molecular biomarker system for assessing the health of coral (Montastraea faveolata) during heat stress, Mar Biotechnol 2: 533-544.
34. Wittenburg M, Hunte W (1992) Effects of eutrophication and sedimentation on juvenile corals. I. Abundance, mortality and community structure. Mar Biol 112: 131-138.
35. Grigg RW (1995) Coral reefs in an urban embayment in Hawaii: a complex case history controlled by natural and anthropogenic stress. Coral Reefs 14: 253-266.
36. Littler MS, Littler DS, Lapointe BE (1992) Modification of tropical reef community structure due to cultural eutrophication: the southwest coast of Martinique. Proceedings of the 7th International Coral Reef Symposium. Guam 1: 335-343.
37. Proctor LM, Fuhrman JA (1988) Marine bacteriophages and bacterial mortality. EOS 69: 1111-1112.
38. Ritchie KB, Smith GW (1995) Preferential carbon utilization by surface bacterial communities from water mass, normal and white-band diseased Acropora cervicornis. Mol Mar Biol Biotechnol 4: 345-352.
39. Ritchie KB, Smith GW (2004) Microbial communities of coral surface mucopolysacchaide layers. In Coral Health and Disease (Rosenburg and Loya eds.) Springer pp: 259-264.
40. Bell PRF (1992) Eutrophication and coral reefs-some examples in the Great Barrier Reef lagoon. Water Res 26: 553-568.
41. Wolanski E, Richmond R, McCook L, Sweatman H (2003) Mud, marine snow, and coral reefs. Amercian Scientist 91: 44-51.
42. Sullivan Sealey KM (2004) Large-scale ecological impacts of development on tropical islands systems: Comparison of developed and undeveloped islands in the Central Bahamas. Bull Mar Sci.

43. Paul JH, Rose JB, Jiang S, Kellogg C, Shinn EA (1995) Occurrence of fecal indicator bacteria in surface waters and the subsurface aquifer in Key Largo, Florida. Appl Environ Microbiol 61: 2235-2241.
44. McCook LJ, Jompa J, Diaz-Pulida G (2001) Competition between corals and algae on coral reefs: a review of mechanisms and evidence. Coral Reefs 19: 400-417.

CHAPTER 6

Seasonal Dynamics and Community Composition of Epilithic Microalgae in Relation with Environmental Factors at Northwest coast of Jeddah, The Red Sea

Salim Marzoog Al-Harbi*

Department of Marine Biology, College of Marine Sciences, King Abdulaziz University, Jeddah 21589, Saudi Arabia

ABSTRACT

Intertidal epilithic microalgae play very important role as feed of higher organisms. Taxonomical composition of epilithic algae and their seasonal dynamics in relation with environmental factors were studied at middle coast of Saudi Arabia, namely Yanbu, Ar-Rayis, Mastorah, Rabigh and Thuwal, the Red Sea. Temperature, salinity and pH varied from 22.10-32.02°C, 38.72-39.35 psu, and 8.40-8.63, respectively throughout the study. A total of 87 epilithic microalgal species belonging to 4 families was identified: 79 Bacillariophyceae, 5 Cyanophyceae, 2 Dinophyceae, and 1 Cryptophyceae. The epilithic microalgal abundance varied from 2.32×10^5

cells/L to 7.66 × 10^5 cells/L during the study with an average of 4.03 × 10^5 cells/L. The lowest and highest abundance of epilithic microalgae occurred in winter and summer at Mastorah and Rabigh, respectively. Among the microalgae taxa, diatoms were the dominant group which was 91.01%, followed by Cyanophyceae (5.62%), Dinophyceae (2.25%) and Raphidophyceae (1.12%). Among diatom, the centric and pennate diatoms were 74.07% and 25.93%, respectively throughout the study. The occurrence frequency analysis revealed that high frequently observed epilithic microalgae were *Licmophora flabellata, L. paradoxa, L. remulus, Navicula vula, N. incerta* and *N. transitans* throughout the study. Therefore, it could be said that diatoms are the major group of epilithic microalgae which show high abundant in summer at the northwest coast of Jeddah, Saudi Arabia, the Red Sea. Second dominant group was Cyanophyceae, and the dominant species was *Spirulina subsalsa* among epilithic microalgal community

Keywords: Epilithic microalgae; Environmental factors; Seasonal dynamics; Jeddah coast; The Red Sea

1. INTRODUCTION

Microalgae perform a major part of primary production and support food webs from small pond to oceans. Microalgae composition with environmental factors is important to understand the structure and dynamics of any marine ecosystem. Microalgal populations fluctuate from season to season and from region to region due to the interaction among physical, chemical and biological variables [1,2]. Among the microalgae, the benthic microalgae play an important role in intertidal biomass production [3,4]. The intertidal flats are the sites of intense biological activity and often with high productivity [5] though intertidal benthic microalgae have to survive in a harsh environment. To distinguish the groups of benthic organism, different terms such as bioflim, benthos, periphyton, epilithon have been used by several researchers [6]. Aquatic system, these epilithon can comprise the most abundant producers. The epilithon grow on stones and rocks which are generally considered as an important base of the food chain [7]. Natural substrates present advantages to epilithic microalgae because the stones supply nutrients that support community development. Diatoms are the major components of epilithon community because of their continuous presence with different abundances along the aquatic system and their quick response to the environmental changes although the term epilithon encompasses a great diversity of microalgal taxa [8].

A study was conducted on epilithic microalgal composition and dynamics at Jeddah Coast. Jeddah is a Saudi Arabian city located at the coast of the Red Sea (21.50° N 39.1667° E). The Red Sea is a unique area in the tropics for several reasons which having the highest salinity water body in the world's seas with the deep water salinity value of 40.06 psu while the surface water salinity values ranged from 40.10- 40.65 psu [9,10]. The high salinity phenomenon occurs due to its location in an arid and hot climatic zone, isolation from open ocean, without riverine inputs, the high evaporation rate (>210 cm yr-1), and scant rainfall restricted from October to May, resulting in a negative hydrological budget. The hydrological deficit is compensated by inflows from the Indian Ocean through Bab-Al-Mandeb. The Red Sea is seasonally influenced by variable wind-driven monsoon currents such as southwest monsoon in summer (May to September) and north to northwest winds blow on whole axis of the Red Sea in winter. The Red Sea surface waters pile up from December to February along the African coast and some upwelling occurs along the Arabian coast [9], and reversal of summer condition because of the south-southeast winds only affect the southern half of the Red Sea. These two circulation patterns seasonally converge and surface waters sink at Jeddah Coast. In these zones, coastal water downwelling occurs due to the warming of surface layer and nutrient-deficient water layer. Downwelling reduces biological productivity and transports heat, dissolved materials, and dissolved oxygen rich surface waters in to greater depths. The convergence zone is classified as the poorer zone in microalgal species [11].

The Red Sea is considered as an oligotrophic water body due to deficient of several major nutrients such as nitrate, ammonium, phosphate, and silicate [12]. The required nutrients for photosynthetic activity have to come through water intrusion (from the Gulf of Aden), the sub-surface (below the nutricline), or via aerial deposition. Higher nutrient concentrations occur in the southern areas, and gradually decline towards the north. The reservoirs of nutrients in the Red Sea are trapped below the stratified zone due to the persistent pycnocline. The deep water renewal is generally prevented by high stratification levels and the resulting homogeneity of the northern Red Sea deep waters which limits nutrient availability and biological productivity in the euphotic zone. Jeddah is the second biggest city of Saudi Arabia, located at the Coast off the Red Sea and is extending rapidly. It receives huge waste water from fish farms, industries and domestic use which may increase the nutrients concentration. A varying supply of nutrients, among other factors, permits microalgae species richness to increase beyond the number of limiting nutrients, while changes in nutrients are important drivers of change in species composition of primary producers [13,14]. The field studies have shown that episodic nutrient flushing and loading can result in enhancing microalgae species diversity [15-17]. Moreover, temperature values have been increasing abruptly since mid-90's to present, therefore, the oceanic warming may have a direct and/or indirect impact on intertidal microalgae. Studies of macro and microalgae are very few at the Jeddah coast though algae are an important source of food and shelter for aquatic organisms and also indicator of pollution. Presence of new microalgal species was studied at Obhur creek of Jeddah coast [18]. Recently some studies have been performed to monitor the effect of environmental factors on chlorophyll and microalgae abundance near desalination plant, marine fouling and polluted area of Jeddah coast [19,20]. Studies dealing with littoral benthic microalgal ecology are rare compared to those done on water column throughout the world [3,4], and in the Red Sea, the study on benthic microalgae in intertidal zone, especially on epilithic microalgae has not been done. The present study was conducted to monitor and interpret the structure and seasonal dynamics of epilithic microalgae in relation with environmental factors at northwest coast of Jeddah, Red Sea.

2. MATERIALS AND METHODS

2.1 Study area, sampling and sample analysis

The sampling was done seasonally at the middle time of season which was in March (spring), June (summer), September (autumn) and December (winter) in 2015. The sampling was conducted at the northwest coast of Jeddah, Saudi Arabia, the Red Sea. The sampling sites were: (i) Yanbu (23°59′45″N and 38°10′35″ E) that is a major Red Sea port in the Al Madinah province of western Saudi Arabia. It is approximately 300 Km northwest of Jeddah. There are oil refineries and petrochemical industry; (ii) Ar-Ryias (23°33′55″N and 38°36′17″E) is an industrial city of Madinah province; (iii) Mastorah beach city (23°08′80N and 38º 47′14″E) which is famous for fishing; (iv) Rabigh (22°41′35″N and 3900′34″E) that has several industries such as a factory of cement, electricity station, large refinery, Petro Rabigh; and (v) Thuwal (22°16′44″N and 39º 05′21″E) that had long been a fishing center. The town is experiencing an increase in industries due to its proximity to the King Abdullah Economic City (**FIGURE 1**).

15 pieces of gravel stones were collected from each sampling station, then the entire surface area of each stone was cleaned and washed with tooth brush and filtered autoclaved natural seawater, respectively to get the microalgae attached on stones. The whole stones collected from each station were washed three times. Finally, the samples of stones washed and filtered water volume was made 0.5 L and was preserved with Lugol's solution of 2% and was kept for further study. Additionally, the surface area

of the stones was estimated with digital vernier slide caliper after collection of eipilithic microalgae. For taxonomical study, the preserved sample was observed under the phase-contrast microscope (Zeiss Axioplan, Germany) at 400X magnification. Taxonomic identifications were made with reference to Prescott [21-24]. Specfies diversity index (H) was calculated accordingly [25]. For quantitative study, a 1 mL preserved sample was taken after mixing and counted in a Sedgewick-Rafter (S-R) counter chamber with same microscope as mentioned above. Counted results were summarized as cells per liter, and used for graphical presentation. To estimate the concentration level of nutrients (NO_3-N, NO_2-N, NH_3-N, PO4-P and SiO_3-S), the water sample was collected from the respective stations. The concentration of NO_3-N, NO_2-N, NH_3-N, PO4-P and SiO_3 were determined with a spectrophotometer (Shimadzu UV-1201, Japan) according to the methods described by Parsons et al. [26]. The water temperature, salinity and pH were measured on spot with portable meter HACH (Model HQ14d). Correlations between the abundance of epilithic microalgae and environmental factors were determined following MS-Excel statistical program.

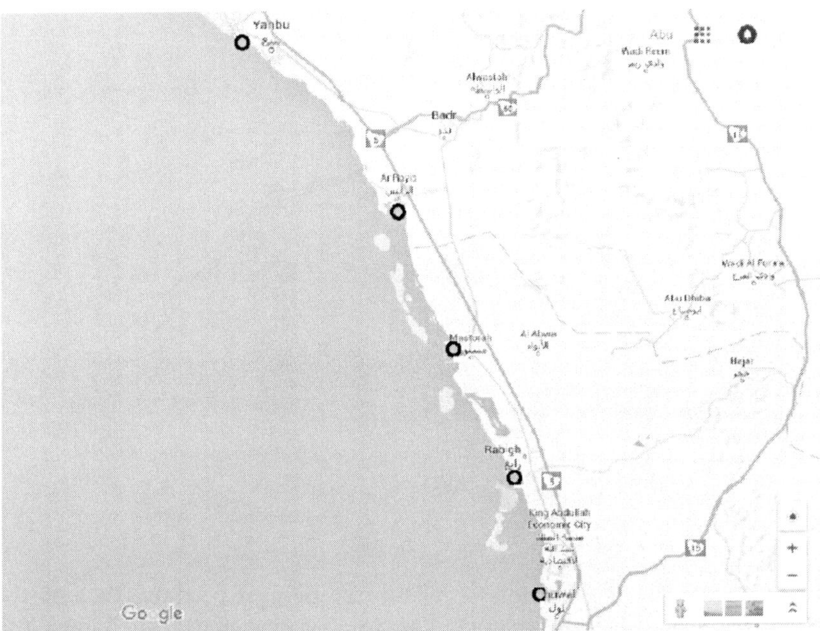

Figure 1: Location of the study sites; Northwest coast of Jeddah, Saudi Arabia, namely Yambu, Ar-Rayis, Mastorah, Rabigh and Thuwal, the Red Sea.

3. RESULTS

3.1 Dynamics of physical factor

The water temperature showed three distinct seasonal cycle characterized by a maximum in summer-autumn, followed by spring and winter. Temperature values varied from 22.10-32.02°C with an average of 28.48°C, showing minimal and maximal in winter and summer at Yanbu and Rabigh, respectively (**FIGURE 2**). Salinity values fluctuated from 38.72-39.35 psu with an average of 38.92 psu. The salinity value was lowest in spring at Ar-Rayis, and highest was in summer at Yanbu (**FIGURE 3**). The pH values varied from 8.40 to 8.63 throughout the study with an average of 8.52. The highest and the lowest pH values were recorded at Rabigh and Mastorah. The highest pH value was counted in spring followed by winter pH values among all the stations (**FIGURE 4**). Among nutrients, the concentrations levels of NO_3-N, NO_2-N, NH_3-N, PO_4-P and SiO_3 were (0.26-2.96, 0.25- 3.24, 0.13-3.18, 0.43-1.47 and 0.25-8.35) µg/L, respectively among the stations of northwest coast of Jeddah, the Red Sea, from winter to autumn 2015 (**TABLE 1**).

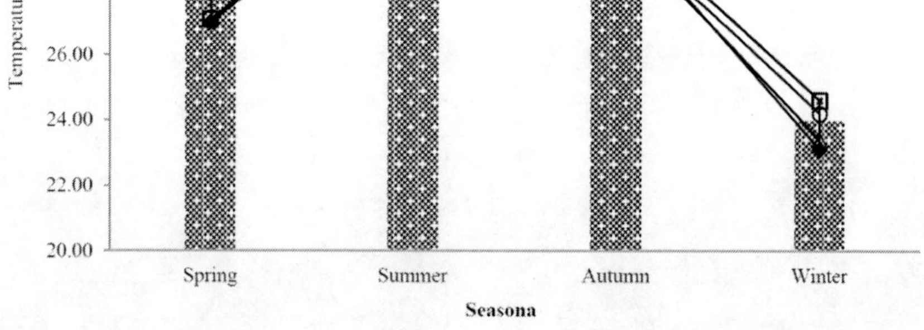

Figure 2: Seasonal variation of water temperature values (°C) at the Northwest coast of Jeddah, namely Yambu, Ar-Rayis, Mastorah, Rabigh and Thuwal, the Red Sea, Saudi Arabia from spring to winter 2015.

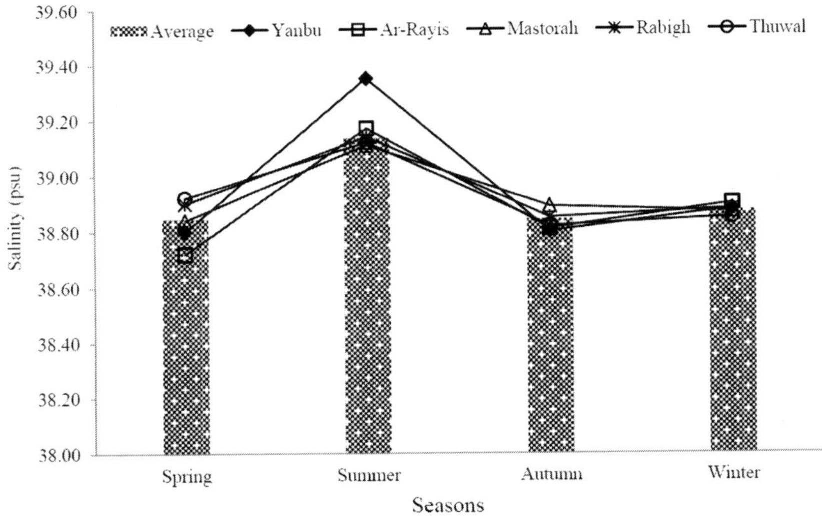

Figure 3: Seasonal variation of salinity values (psu) at the Northwest coast of Jeddah, namely Yambu, Ar-Rayis, Mastorah, Rabigh and Thuwal, the Red Sea, Saudi Arabia from spring to winter 2015.

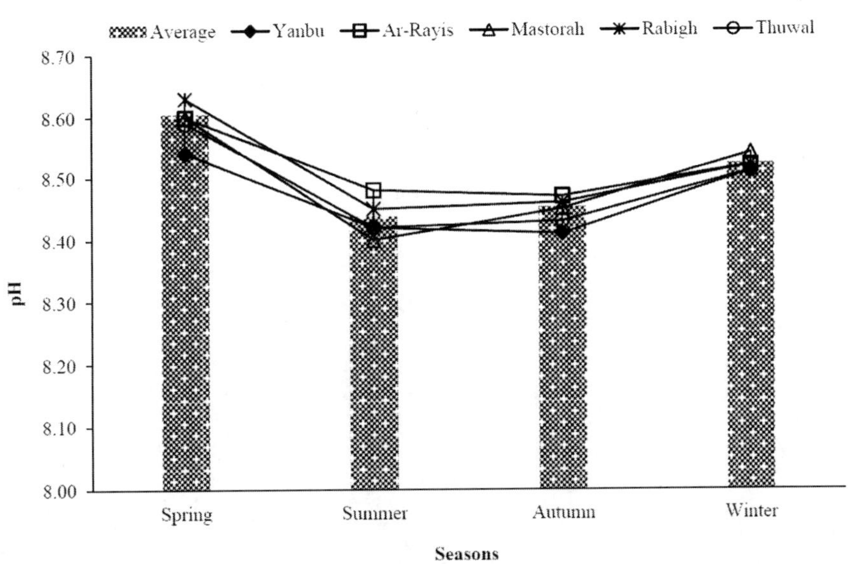

Figure 4: Seasonal variation of pH values at the at the Northwest coast of Jeddah, namely Yambu, Ar-Rayis, Mastorah, Rabigh and Thuwal, the Red Sea, Saudi Arabia from spring to winter 2015.

Table 1: Dynamics of nutrients (μg.L^{-1})at the Northwest coast of Jeddah, Saudi Arabia, namely, Yanbu, Ar-Rayis, Mastorah, Rabigh and Thuwal, the Red Sea, from spring to winter 2015.

Stations	NO_3-N				NO_2-N				NH_3-N				PO_4-P				SiO_3-S			
	Spr.	Sum.	Aut.	Win.	Spr.	Sum.	Aut.	Win.	Spr.	Sum.	Aut.	Win.	Spr.	Sum.	Aut.	Win.	Spr.	Sum.	Aut.	Win.
Yanbu	0.57	0.95	1.43	0.68	0.22	3.02	0.45	0.63	0.78	0.27	0.53	0.38	0.67	1.11	0.49	0.73	1.10	5.79	1.47	0.26
Ar-Rayis	1.02	0.83	1.27	0.72	0.39	1.40	0.52	0.75	0.44	0.20	0.52	0.42	0.83	0.85	0.52	0.42	2.06	4.91	1.81	0.24
Mastorah	0.34	1.91	2.76	1.00	0.28	1.36	0.55	0.71	0.18	0.27	0.41	1.09	0.54	1.43	1.02	0.71	2.11	5.31	2.62	1.22
Rabigh	0.24	0.84	0.54	2.41	0.25	1.77	0.53	0.71	0.44	0.23	0.64	1.11	0.64	0.87	0.57	0.51	2.06	7.05	1.02	1.32
Thuwal	0.61	0.91	1.37	2.30	0.40	1.93	0.65	0.91	0.55	0.37	0.46	2.18	0.48	1.05	0.64	0.49	3.08	6.13	1.80	1.53

The abundance values of epilithic microalgae showed positive correlation with temperature values in winter (r=0.50), spring (r=0.79) and summer(r=0.87). Similarly, the above epilithic microalgae abundance values also showed positive correlation with salinity values in spring (r=0.75) and autumn (r=0.67). The abundance values of epilithic microalgae showed strongly positively correlation with NO_3-N (r=0.97) in winter whereas it was highly negatively correlated in spring NO_3-N (r=-0.79). The concentration values of NH4-N showed positive correlation with abundance values of epilithic microalgae in winter (r=0.67), spring (r=0.65) and in autumn (r=0.54). SiO_2-S and abundance values of microalgae were positively correlated in autumn (r=0.96) and winter (r=0.51).

3.2 Taxonomic composition

Taxonomic community composition of epilithic algal assemblages and their seasonal dynamics were studied for the first time at northwest coast of Jeddah, Saudi Arabia, the Red Sea. A total of 88 epilithic microalgal species were identified, among them 78 Bacillariophyceae, 6 Cyanophyceae, 3 Dinophyceae, 1 Raphidophyceae (**TABLE 2**). Among Bacillariphyceae, the abundance levels of pennate and centric diatoms were 58 and 20 species, respectively throughout the study. The variation in species diversity index value (H) ranged from 1.35 to 3.35 with an average of 2.40 from spring to winter 2015. The lowest and highest H values were at Yanbu in autumn, and in spring at Rabigh (**TABLE 3**).

Table 2: Epilithic microalgal taxa list and occurrence frequency at the Northwest coast of Jeddah, Saudi Arabia, namely Yanbu, Ar-Rayis, Mastorah, Rabigh and Thuwal, the Red Sea, from spring to winter 2015.

Microalgae name	Winter	Spring	Summer	Autumn
Bacillariophyceae				
Pennate diatoms				
Amphiprora Ehrenberg		C		
A. alata Kützing	R			
A. paludosa Smith				R

Amphora sp. Ehrenberg			X	
Amphora arcus Gregory			R	R
A. coffeaeformis Kützing	C			R
A. decussata Grunow	F	H		
A. exigua Gregory	R		R	
Bacillaria paradoxa Gmelin	R	R		
B. paxillifer Müller		R	R	
Closterium closterioides Ralfs				X
Cocconeis placentula Ehrenberg		H		
Cocconeis scutellum Ehrenberg	C		X	
Cylindrotheca closterium Ehrenberg	X	H	R	X
Entomoneis punctulata Grunow		R		
Fragilaria capucina Desmazières		H	C	
Grammatophora marina Lyngbye	X			
Gyrosigma sp.		C		
G. acuminatum Kützing			X	
G. balticum Ehrenberg	C			
G. fasciola Ehrenberg				C
Licmophora sp.		R		
L. abbreviata Agardh	R			X
L. ehrenbergii Kützing	C			H
L. flabellata Agardh	H	H	H	C
L. gracilis Ehrenberg			H	X
L. grandis Kützing		C		

L. hyalina Kützing	H			X
L. paradoxa Lyngbye	R	H	H	H
L. remulus Grunow	H	C	H	H
Navicula delicatula Cleve		H	C	R
N. distans Smith	C		X	R
N. granii Jørgensen	H			
N. incerta Grunow	C	H	H	H
N. peregrine Ehrenberg		H		
N. ramosissima Agardh		H		
N. transitans Grunow	C	H	H	R
N. vula Carter	H	H	H	H
Nitzschia sigma Kützing		R		
N. seriata Cleve			X	
N. hungarica Grunow	H	R	X	C
N. longissima Brébisson	X		X	
Pseudo-nitzschia pungens Hasle	R	H	C	
N. socialis Gregory		H		
Pinnularia gibba Ehrenberg				H
P. viridis Ehrenberg	R		R	
Pleurosigma normanii Ralfs			X	
P. angulatum Smith	X	R	C	X
P. formosum Smith	X			
P. intermedium Smith		H		
Pseudo-nitzschia fraudulenta Cleve		C		
P. lineola Cleve				X
Stauroneis anceps Ehrenberg	C			
Staurosirella pinnata Grunow			C	

Striatella unipunctata Lyngbye	X			
Synedra ulna Nitzsch				X
Tabellaria fenestrate Lyngbye				X
Toxonidea insignis Donkin		C		
CENTRIC DIATOMS				
Biddulphia alternans Bailey			R	
C. costatus Pavillard		C		
C. curvisetus Cleve	X			
C. laciniosus Schütt		C		
C. nodulifer Schmidt		H		
C. radiates Ehrenberg		R		
C. socialis Lauder			X	
C. wailesii Gran & Angst			R	X
Chaetoceros affinis Lauder	X			
Coscinodiscus centralis Ehrenberg				X
Cyclotella sp. Kützing	X			
Guinardia delicatula Cleve	X			
L. adriaticus Schroder		R		
L. danicus Cleve	X	H	X	
Lauderia annulata Cleve	C			
Leptocylindrus minimus Gran	R			
Melosira granulate Ehrenberg				X
R. setigera Brightwell	H	R		X
Rhizosolenia delicatula Cleve	X			
Thalassionema bacillare Heiden	R			X

DINOPHYCEAE				
Bysmatrum arenicola Horiguchi	R			
Cerataulus turgidus Ehrenberg				
Prorocentrum lima Ehrenberg		H	H	
CRYPTOPHYCEAE				
Rhodomonas salina Wislouch		H		
CYANOPHYCEAE				
Merismopedia angularis Thompson		H	R	
Microcystis aeruginosa Kützing				
Oscillatoria formosa Gomont				C
O. limosa Gomont	X		F	H
Spirulina subsalsa Gomont	R		C	X
Trichodesmium sp. Ehrenberg		C	H	R

*Legend of occurrence frequency: Sporadically (X=1-20%), Rarely (R=21-40%), Commonly (C=41-60%), Frequently (F=61-80%) and High Frequently (H=81-100%).

3.3 Abundance dynamics of epilithic microalgae

The epilithic microalgae abundance value varied from 2.32×10^5 cells/L to 7.66×10^5 cells/L during the study with an average of 4.03×10^5 cells/L. The lowest and highest abundance value of epilithic micoralgae occurred in winter and summer at Mastorah and Rabigh, respectively. The second highest abundance value of epilithic micoralgae was 7.20×10^5 cells/L at Mastorah and the third highest was 5.42×10^5 cells/L also at Thuwal in summer, respectively (**FIGURE 5**). Among the microalgae taxa, diatoms were the dominant groups which were 91.01%, followed by Cyanophyceae (5.62%), Dinophyceae (2.25%) and Raphidophyceae (1.12%). Among diatoms, the centric and pennate diatoms were 74.07% and 25.93%, respectively throughout the study. Among diatoms, the frequence analysis of microalgal species occurrence revealed that epilithic microalgal species were *Licmophora flabellata, L. paradoxa, L. remulus, Navicula vula, N. incerta* and *N. transitans* throughout the study. In winter above 15% total epilithic microalgal species contributed the highest abundance value to the epilithic microalgal species communiities. *Licmophora flabellata* contributed the highest abundance value which was 63%, followed by Amphora coffeaeformis (50%) and *Chaetoceros*curvisetus (37%) to the epilithic micoralgal communities. In spring, 8 species were found to be dominated above 15%, and the most dominant species was Toxonidea insignis which contributed 43%, followed by Licmophora remulus (28%) and *Navicula*ramosissima (27%) of epilithic micoralgal communities. Similarly, 6 and 5 species were found to be abundant above 15% in summer and autumn, respectively. *Navicula* delicatula was found to be 79%, followed by, Licmophora remulus (51%) and *Navicula*transitans (26%) among epilithic micoralgae in summer. *Licmophora flabellata* was found to be majority among epilithic micoralgal communities which was 50%, followed by *Microcystis aeruginosa* (39%) and *Trichodesmium* sp. (29%) in autumn (**TABLE 4**).

Table 3: Epilithic microalgal diversity index value (H) at Northwest coast of Jeddah, Saudi Arabia, namely Yanbu, Ar-Rayis, Mastorah, Rabigh and Thuwal, the Red Sea, from spring to winter 2015.

Season	**Yanbu**	**Ar-Rayis**	**Mastorah**	**Rabigh**	**Thuwal**
Spring	2.56	2.39	3.29	3.35	3.07
Summer	1.89	2.31	2.28	2.42	2.08
Autumn	1.80	2.23	2.47	2.39	1.93
Winter	2.38	1.35	2.29	2.92	2.67

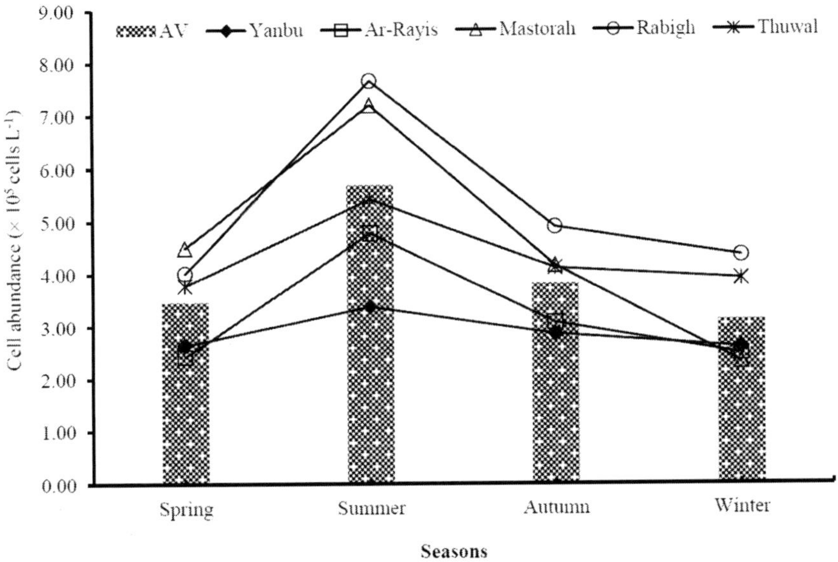

Figure 5: Seasonal variation of epilithic microalgal cell abundance values ($\times 10^5$ cells L^{-1}) at the Northwest coast of Jeddah, Saudi Arabia, namely Yanbu, Ar- Rayis, Mastorah, Rabigh and Thuwal, the Red Sea from spring to winter 2015.

Table 4: List of dominant microalgae species (above 10% abundance) at the Northwest coast of Jeddah namely, Yanbu, Ar-Rayis, Mastorah, Rabigh and Thuwal, the Red Sea, Saudi Arabia from spring to winter 2015.

Species	Yanbu	Ar-Rayis	Mastorah	Rabigh	Thuwal
Spring					
Leptocylindrus donicus Cleve	12	18			
Licmophora grandis Kützing					
L. remulus Grunow	19				
Licmophra sp.					
Merismopedia	12	13			

angularis Thompson					
Navicula delicatula Cleve		10			
N. longissima Brébisson					14
N. transitans Grunow	12				
Prorocentrum lima Ehrenberg		15			
Toxonidea insignis Donkin			16		
SUMMER					
Amphora sp.				10	
Chetoerros socialis Lauder		15			
Cylindrotheca closterium Ehrenberg	15				
Gyrosigma acuminatum Kützing				14	
Licmophora remulus Grunow					45
Merismopedia angularis Thompson		20			
Navicula delicatula Cleve	19		17		
N. transitans Grunow	27		17	15	
N. distans Smith			17		
Nitizschia seriata Cleve	15				
N. sigma Kützing			14	15	

Plearosigma angulatum Smith				13	
Spirulina subsalsa Gomont		18			
AUTUMN					
Coscinodiscus wailesii Gran & Angst				11	
Gyrosigma fasciola Ehrenberg		13			
Licmophora flabellata Agardh	11				44
L. gracilis Ehrenberg				11	
Microcystis aeruginosa Kützing	39		14		
Navicula distans Smith	11				
Nitzschia hungarica Grunow		28		11	
Oscillatoria Formosa Gomont	17				
Pleurosigma normanii Ralfs			10		
Pseudo-nitzschia lineola Hasle		11			
Rhizosolenia setigera Brightwel					
Trichodesmium sp. Ehrenberg			12	25	
WINTER					
Amphiprora alata Kützing			14		
Amphora exigua Gregory	16				14

Chaetoceros curvisetus Cleve			31		
Licmophora abbreviate Agardh	22			14	
L. ehrenbergii Kützing		10			
L. flabellata Agardh		62			
L. paradoxa Gmelin				12	
Navicula granii Jørgensen					10
Nitzschia hungarica Grunow			12	10	
N. longissima Brébisson		12			
Spirulina subsalsa Gomont					17
Stauroneis anceps Ehrenberg	16				

4. DISCUSSION

Epilithic microalgae communities can range from filamentous green algae to silicaceous diatoms to blue-green algae [4,27]. They are very important source of primary producer and contribute energy cycle in inland water and coastal water ecosystem [28]. As epilithic microalgae species of Bacillariophyceae were dominant, followed by Cyanophyceae, Dinophyceae, and Chlorophyceae to the epilithic microalgae at the northwest coast of Jeddah. The average abundance values of those epilithic microalgal species community was high in summer, followed by spring, winter and autumn in the study area. A fluctuation of epilithic species diversity and biomass production can occur due to environmental conditions, geological nature, grazing, resources limitation, habitat disturbance, substrate availability. Moreover, inorganic pollutants can be constrained to epilithic microalgae species diversity, so that the pollutants could consider a limiting factor for growth and dominancy of epilithic microalgal species.

The highest abundance values of the epilithic microalgae were in summer which might be related with the water temperature values, moderate to higher concentration levels of NO_3-N and SiO_3 nutrients in correlation with high temperature values might have created a suitable substrate for proliferation and growth of epilithic microalgal species communities, since epilithic microalgae abundance values were high in summer. The abundance of diatoms was the dominant group among epilithic microalgae throughout the study. Similarly, diatoms are the dominant group of microalgae in Arabian Gulf and Shatt Al-Arab estuary [11,29,30]. Agusti and Sanchez [31] reported that temperature values and nutrients (N, P and Si compounds) concentrations level, diatoms at different regimes is related with their adaptive ability. Abundance values are controlled by the environmental factors and their adaptive ability, although diatom grazing is another important controlling factor [32].

Among diatoms, the different taxa were found to be dominant in different seasons and different stations. The small size diatoms were found to be occurred and dominated commonly in different sampling stations throughout the year. Among diatoms, two species of pennate diatoms and three species of centric diatoms were found to be dominant during the study. Among pinnate diatoms, Licmophora spp., and *Navicula* spp., were found to be dominant. *Licmophora flabellata* was 61.78% of total epilithicmicroalgal community at Ar-Rayis in spring (**TABLE 4**). Abundance of *Navicula* spp., was high in summer and the highest cell abundance value was for *Navicula* transitans which was 27% of epilithic microalgae species community at Yanbu (**TABLE 4**).

Temperature values were identified as the most effective parameter on the epilithic microalgae abundance values and species composition variation in the Homa Lagoon Izmir, Turkey where *Navicula* species were dominant [33,34]. Three taxa of centric diatoms were found to be available throughout the study, which were *Chaetoceros* spp., *Coscinodiscus* spp., and *Leptocylindrus*spp.

Among centric diatoms, only *Chaetoceros* spp., contributed 30.02% to the epilithic microalgae as a dominant species in winter. The species composition and abundance values of all microalgae follow seasonal temperature cycle patterns in the Red Sea [35] that Dowidar et al. [36] studied on water column microalgal species communities and found that diatom was dominant in summer. Similarly, among epilithic microalgal communities, diatoms are the dominant group and they reach at an abundance peak in summer. There were Dinophyceae and Cyanophyceae and they did not show any dominancy though two species (Spirulina subsalsa and *Trichodesmium* spp.) of Cyanophyceae contributed high cell abundance among epilithic microalgal community. The epilithic microalgal abundance followed the seasonal pattern of temperature and salinity as their abundance was high in summer when the salinity and temperature were also high. The diversity of epilithic microalgae was high in spring in all stations.

CONCLUSION

This study contributed to the knowledge improvement of the epilithic microalgal species communities at Jeddah, the Red Sea, in Saudi Arabia, by providing the information of seasonal dynamics of taxonomical composition and abundance in relation with environmental factors from the study. It could be concluded that diatoms were the most dominating group of epilithic microalgae and their abundance values became high in summer.

ACKNOWLEDGEMENT

This research was supported by the 2015 deanship research project No.1434/150/355 scientific promotion program funded by King Abdulaziz University, Jeddah

REFERENCES

1. Litchman A, Klausmeier CA (2001) Competition of microalgae under fluctuating light. Am Nat 157: 170-187.
2. Reynolds CS, Dokulil M, Padisak J (2000) Understanding the assembly of microalgae in relation to the trophic spectrum: where are we now? In: Reynolds CS, Dokulil M, Padisak J (eds.) The Trophic Spectrum Revised: the Influence of Trophic State on the Assembly of Microalgae Communities. Development in Hydrobiology 150. Kluwer Academic Publishers, London, pp: 147-152.
3. Cadee GC (1980) Reappraisal of the production and import of organic carbon in the western Wadden Sea. Neth J Sea Res 11: 305-322.
4. Moncreiff CA, Sullivan MJ, Daehnick AE (1992) Primary production dynamics in seagrass beds of Mississippi Sound: the contributions of seagrass, epiphytic algae, sand microflora, and microalgae. Mar Ecol Prog Ser 87: 161-171.
5. Herman PMJ, Middelburg JJ, Vande-Koppel J, Heip CHR (1999) Ecology of estuarine macrobenthos. Adv Ecol Res 29: 195-240.
6. Mac-Intyre HL, Geider RJ, Miller CD (1996) Microphytobenthos: the ecological role of the "secret garden" of unvegetated, shallow-water marine habitats. I. Distribution, abundance and primary production. Estuar Coast 19: 186-201.
7. Kelly MG, Whitton BA (1998) Biological monitoring of eutrophication in rivers. Hydrobiol 384: 55-67.
8. Salomoni SE (2004) Diatomáceasepilíticasindicadoras da qualidade de águanabacia do Rio Gravataí, Rio Grande do Sul, Brasil. PhD thesis. Ppge Ufscar, São Carlos, São Paulo.
9. Marcos SA (1970) Physical and chemical oceanography of the Red Sea. Oceanogr. Mar Bio Ann Rev 8: 7-202.

10. Manasrah R, Badran M, Lass HU, Fennel WG (2004) Circulation and winter deep-water formation in the northern. Red Sea Oceanol 46: 5-23.
11. Halim Y (1984) Plankton of the Red Sea and the Arabian Gulf. Deep-Sea Res II 31: 969-982.
12. Acker J, Leptoukh G, Shen S, Zhu T, Kempler S (2008) Remotely-sensed chlorophyll a observations of the northern Red Sea indicate seasonal variability and influence of coastal reefs. J Mar Sys 69: 191-204.
13. Grover JP (1989) Phosphorus-dependent growth kinetics of 11 species of freshwater algae. Limnol Oceanogr 34: 341-348.
14. Reolke DL, Eldridge PM, Cifuentes LA (1999) A model of microalgae competition for limiting and non-limiting nutrients: implication for development of estuarine and nearshore management schemes. Estuar 22: 92-104.
15. Padisak J (1993) The influence of different disturbance frequencies on the species richness, diversity and equitability of microalgae in shallow lakes. Hydrobiol 249: 135-156.
16. Hambright KD, Zohary T (2000) Microalgae species diversity controlled through competitive exclusion and physical disturbances. Limnol Oceanogr 45: 110-122.
17. Lovejoy C, Legendre L, Price NM (2002) Prolonged diatom blooms and
18. microbial food web dynamics: experimental results from an Arctic polynya. Aquat Micro Ecol 29: 267-278.
19. Aleem AA (1978) Contributions to the study of the marine algae in the Red Sea. III. Marine algae from Obhur in the Vicinity of Jeddah, Saudi Arabia. Bull Facul Sci KAU Jeddah 2: 99-118.
20. Khomayis HS, Harbi SMA (2003) Pcriphyton flora in the coastal water of Jeddah, Saudi Arabia. J KAU Mar Sci 14: 3-18.
21. Khomayis HS (2002) The annual cycle of nutrient salts and chlorophyll a in the coastal waters of Jeddah, Red Sea. J KAU Mar Sci 13: 131-145.
22. Prescott GW (1973) Algae of the western Great Lakes Area. Dubuque, W.C Brown Company, p: 977.
23. Tomas CR (1993) Marine Microalgae, a guide to naked flagellates and coccolithophorids. Academic Press, Harcourt Brace, San Diego, California, USA.
24. Tomas CR (1997) Identifying marine microalgae. Academic Press, Harcourt Brace, San Diego, California, USA.
25. Shim JH (1994) Illustrated encyclopedia of fauna and flora of Korea, Vol. 34. Marine microalgae, Ministry of Education, Seoul, Korea.
26. Washington HG (1984) Diversity, biotic and similarity indices: A review with special relevance to aquatic ecosystems. Water Res 18: 653-694.
27. Parsons TR, Maita L, Lalli CMA (1984) Manual of Chemical and Biological Methods for Seawater Analysis. Pergamon Press, New York, USA.
28. Chung MH, Lee KS (2008) Species composition of the epiphytic diatoms of the leaf tissues of three *Zostera*species distributed on the southern coast of Korea. Algae 23: 75-81.
29. Kassim TA, Saadi HAA (1995) Seasonal variation of epiphytic algae in a marsh area (southern Iraq). Acta Hydrobiol 37: 153-161.
30. Al-Saadi HA, Hadi RAM (1987) Ecological and taxonomical studies on microalgae in Arab gulf. J Biol Sci 18: 7-31.
31. Subba-Rao DV, Yamani FA (1998) Microalgae ecology in waters between Shatt Al-Arab and straits of Hormuz, Arabian Gulf: Review. Plankton Biol Ecol 45: 101-116.
32. Agusti S, Sanchez MC (2002) Cell viability in natural microalgae communities quantified by a membrane permeability probe. Limnol Oceanogr 47: 818-828.
33. Aberle N, Wiltshire KH (2006) Seasonality and diversity patterns of microphytobenthos in a mesotrophic lake. Arch Hydrobiol 167: 447-465.
34. Busse S, Snoeijs P (2003) Gradient responses of diatom communities in the Bothnian Sea (northern Baltic Sea), with emphasis on responses to water movement. Phycologia 42: 451-464.

35. Fatima ÇS, Sapanc M, Koray T, Büyükışık B (2011) Relationship of epilithic diatom communities to environmental variables in Homa lagoon (Izmir, Turkey). Aquat Biol 13: 233-241.
36. Lindell D, Post AF (1995) Ultra microalgae succession is triggered by deep winter mixing in the Gulf of Aqaba (Eilat), Red Sea. Limnol Oceanogr 40: 1130-1141.
37. Dowidar NM, Aleem AA, Raheem SAED (1978) Microalgae population in the Obhur (Jeddah, Saudi Arabia). Bull Facul Sci KAU 2: 271-292.

CHAPTER 7

Embryonic development of anemone fishes in captivity

Muthuramalingam Uthaya Siva, Mohideen Abdul Badhul Haq

Centre of Advanced Study in Marine Biology, Faculty of Marine Sciences, Annamalai University, Parangipettai-608502. Tamilnadu. India.

ABSTRACT

Broodstock development, spawning, fertilization, embryonic development, fecundity, "hatching rate and deformation rates" of five species (*Amphiprion ocellaris, A. percula, A. melanopus, A. clarkii and A. nigripes*) of anemone fish were studied in captive condition. The egg diameter, parental care, fecundity rate, fertilization rate, hatching rate, deformation rates and size of eggs were quantified. In addition, the courtship behaviors and parental care of each broodstock were noted. The day wise embryonic developments from 1st day to 8th day were observed under microscope and major morphological and functional features of embryos were documented. The egg size ranged from 2.2 to 3.1 mm in length and width of 1.1 to 1.3 mm. The numbers of spawned eggs were varied between 405 to 810 numbers. Spawning was achieved in every 15 to 20 days of interval in summer and in monsoon it was increased ranging from 20 to 40 days. The clutch diameter ranged from 5.2 to 14. 3 cm in length and width of 4.8 to 11.5 cm.

Keywords: Anemone fish, embryonic development, fecundity, fertilization, hatching, deformation.

1. INTRODUCTION

Tropical and subtropical countries are exports marine ornamental species for the aquarium trade (Olivotto et al., 2003). Today, owing to the increasing petition, many studies was carried on ornamental fishes, particularly, larval rearing and nutrition have been performed (Avella et al., 2007) and due to their notable colour and some remarkable behavioral features provides the important place in aquarium trade such as symbiotic with sea anemones (Fautin and Allen, 1997), formation of a group consisting monogamous pairs and protandrous hermaphrodites (Ross, 1978). Because of their main characteristic features, they won hearts of millions of people such as adaptability, easiness to be fed with artificial diets and their fascinating display behavior (Ignatius et al., 2001). However, among the coral reef fishes, anemone fishes belonging to the family, Pomacentridae and subfamily Amphiprioninae are abundant and about 30 species have been recognized under two genera, *Amphiprion* and *Premnas* (Allen et al., 2010).

The development of hatchery production techniques, it is necessary to understand the entire ontogeny of a particular fish in the aquarium. Ontogeny of fish can be divided into several periods of time such as embryonic (laying to Hatching), larval, juvenile, sub-adults and adult. For researchers, embryonic development of a particular fish is very important in terms of reproduction biology (Balon, 1981: Urho, 2002). Dhaneesh et al. (2009) documented that breeding patterns and behavioural aspects of many tropical and subtropical Pomacentrid fishes have been recognized well and the description of eggs, embryological development, larval rearing and juvenile production are available for relatively few Pomacentrid fishes, and Ajith Kumar et al. (2012) were resulted his breeding experiments in Lakshadweep waters. Falk-Petersen (2005) resulted that, development of early life stage, from fertilization to embryo formation among Teleost fish, generally follows the same pattern.

Spawning and behavioural aspects of many Pomacentrid species have been well documented by previous researchers (Alva and Gomes, 1989; Wilkerson, 2001; Overton et al., 2008; Dhaneesh et al., 2009; Ajith Kumar and Balasubramanian, 2009; Swagat et al., 2012; Uthaya Siva et al., 2014). However, description of ova, embryonic development, larval rearing and production of juveniles are available only for a few Pomacentrids and are scarce in the case of species dwelling in Indian waters. Hatching and chronological flow of embryonic developmental stages were differing depending on the species and specific environmental conditions in which they take place (Dhaneesh et al., 2009). Chen stated that the free embryo phase begins when the embryo is free of egg membranes and the beginning of the larval period is indicated when the larvae are able to capture feed objects (Chen, 2005).

Culture practice is the only eco-friendly way to conserve ornamentals. It will hopeful to understand about egg development, importance of parental and their high trade value. Since the study notably on anemone fish embryonic development are not much popular. In the present study provide basic information on day wise embryonic development to enhance captive production of anemone fishes.

2. MATERIALS AND METHODS

2.1 Collection and pair formation of anemone fish

Anemone fish species (10 individuals for each species) of *Amphiprion ocellaris, A. percula, A. melanopus, A. clarkii and A. nigripes* were (5.2 to 7.2 cm TL) procured from the ornamental fish traders. Immediately after fish arrival in the hatchery, they were acclimatized in 5000 L rectangular cement tank filled with 1000 L water and sea anemones, *Stichodactyla haddoni* and *Heteratics magnifica* were also introduced in the same tank. Pair of *A. ocellaris, A. percula, A. clarkii* and *A. nigripes* with *Stichodactyla haddoni* and *Amphiprion melanopus* with *Heteratics magnifica* was obtained, after three to eight months of rearing and they were transferred into the Fiberglass Reinforced Plastics (FRP) tanks with 750 L (Swagat et al., 2012).

2.2 Water quality and feeding habits

The estuarine water was filtered through sand filter and followed with UV treatment and quality water was purified by flowing. The water quality parameters were maintained as salinity 24-30 ± 1 ppt, pH 8.1 ± 0.2, temperature 26.5 ± 0.5°C, dissolved oxygen <5.0 ± 0.4 ml/L, and ammonia 0.01 ppm. The fishes were fed three times in a day at 07:00, 11:00, 16:00 at 5% of their body weight and they were fed with mixed diet (boiled oyster, shrimps meat). After feeding, excess feeds and waste materials were confiscated from the tank through siphoned. The water exchange was done once in a week (30-40%) for elimination of fouling organisms like barnacles, algal communities and other dust materials settled in the tank. Sea anemones *S. haddoni* and *Heteratics magnifica* were fed with post larval form of shrimps in a day interval.

2.3 "Embryonic development"

The eggs were sampled randomly (5 eggs/day) to document the embryonic development. They were placed on sterile glass slide with UV filtered brackish water for record morphological development based of method of Swagat et al. (Ignatius et al., 2001; Swagat et al., 2012). The photographs were taken with a digital camera (Sony cyber-shot G, China) under the light microscope (Novex, Holland) from day 1 to 8 for documenting major morphological and functional features of embryos. Milky white eggs found in the sample were considered as dead or unfertilized and were removed.

2.4 "Estimation of fecundity, fertilization, hatching and deformation rate"

Developed broodstock were tested for their efficiency of fecundity, fertilization, hatching and deformation rate using the formula (Gunasekara et al., 1996):

Fecundity rate =π (R_1+R_2) ×Number of eggs/1^{cm}

Where R_1= clutch width and R_2= clutch length.

Fertilization rate = No of eggs fertilized / No of total eggs laid × 100

Hatching rate = No of eggs hatched / No of total eggs fertilized ×100

Deformation rate= No of deformed (dead or unhealthy) larvae / No of larvae hatched ×100

3. RESULTS

3.1 "Broodstock development"

After obtaining successful pair formation of five species (Figure 1) in the period of 6 to 8 months and brooders were spawned every 15 days intervals. It was observed that bigger fish (female) in the tank always dominant and occupied the earthen pots, which they defended aggressively and also fed first during the feeding. Females of *A. nigripes* and *A. percula* matured shortly and it stimulated the male by bruising (cleaning) many locations near by hiding place such as sea anemone to showed the willingness. Each spawning tank was provided with substratum like tiles, dead corals and pipes for egg deposition which copies the natural environment. Before spawning, the pairs show their courtship behaviours which are precise in the male showed morphological and behavioral changes. Courtship began 2 days before spawning especially *A. Sebae* and *A. clarkii*, with the female initiating by displaying a range of behaviours that included vertical dives, ferocious swimming signals and clinging to the bottom of the rearing tank and no colour changes were observed during spawning.

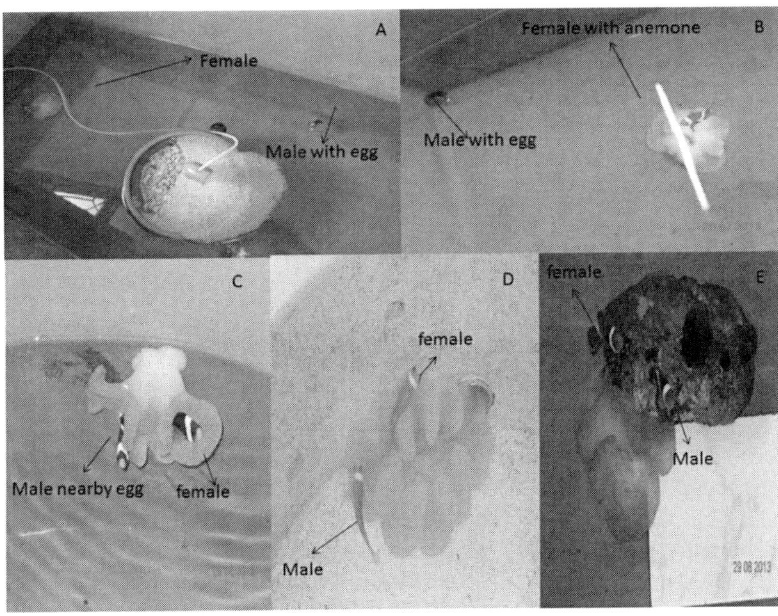

Figure 1. Matured five species of egg laying brooders (A- *Amphiprion melanopus*, B- *A. percula*, C- *A. clarkii*, D- *A. nigripes*, E- *A. ocellaris*).

3.2 "Spawning and parental care"

The brooders were brushed the algae and debris, when thought to lay their eggs, which is more closely to the sea anemone. There was a result all the anemone fishes preferred nearby sea anemone and at time of before spawning, the abdomen of female became larger and its took large quantity of feed. Most of the spawning occurred in the morning (8:00-9:00 h) for *A. melanopus* and *A. nigripes* and in some fishes were in the afternoon (15:00-16:00 h) in case of *A. clarkii, A. percula* and *A. ocellaris*. While female deposit their eggs and followed by the male releasing spermatozoa for fertilization. The spawning process was extended 50 min to 1 h for completing the entire clutch fertilization. The colour of newly laid eggs was varied from bright orange to yellowish in most of the anemone fishes but in case of *A. melanopus* dark orange to reddish in colour was observed on first day.

The development of eggs were noted morphologically on color changes, were yellow/orange on 1st and 2nd days, black/brown from 3rd day and silvery during pre-hatching. The eggs were adhesive, covering with transparent chorion with narrow perivitalline space. Egg diameters were measured and surface of the egg capsule was smooth. Thereafter major role in parental care (fanning or mouthing) was mainly demonstrated by males in all the species and partially done by female, which created a cooling effect as well as suitable environment to the clutch that reduced the damage of egg mass. Continuous fanning was observed by male but in case of female only 2 attempts / hour (during first 3 days) was observed. The longest and shortest duration of single time fanning by the males (includes mouthing) were 1.25 and 0.21 min. In the case of females, maximum and minimum fanning duration were 0.57 and 0.40 min respectively. During the fanning and mouthing the unfertilized, dead eggs and dust particles were also removed by the process of mouthing.

3.3 "Egg size, shape and clutch diameter"

All the five species laid their eggs in spherical shape of clutches. The egg size ranged from 2.2 to 3.1 mm in length with a width of 1.1 to 1.3 mm (Table 1). The number of eggs produced varied between 405 and 810 numbers. Per spawning for per pair and spawning was archived at every 15 to 20 days in summer and in monsoon it was varied between 20 and 40 days. The clutch diameter ranged from 5.2 to 14. 3 cm in length and width of 4.8 to 11.5 cm.

Table 1. Fecundity, egg size and newly hatched larvae size. Species

Species name	Number of spawning/year	Egg diameter (mm)	Newly hatched larva size (mm)
Amphiprion ocellaris	18	2.4	3.8
A. nigripes	16	2.2	3.7
A. melanopus	14	3.1	4.7
A. clarkii	17	3.0	4.4
A. percula	19	2.6	4.1

3.4 "Fecundity and fertilization rate"

The estimated fecundity and fertilization rate was presented in the Figures 7 and 9. Maximum fecundity rate was observed in *A. clarkii* and minimum was observed in *A. melanopus* after spawning. Fertilization rate was more or less similar in all the five species of anemone fishes and it was maximum in *A. clarkii* and minimum in *A. percula.*

3.5 "Embryonic development" of anemone fishes

The general pattern of day wise embryonic development of five anemone fishes (Figure 2 to 6), from fertilization to hatching was classified into 7 to 8 days. During the development of the embryo is entirely depends on the nutrition provided by the mother, mainly by the form of yolk.

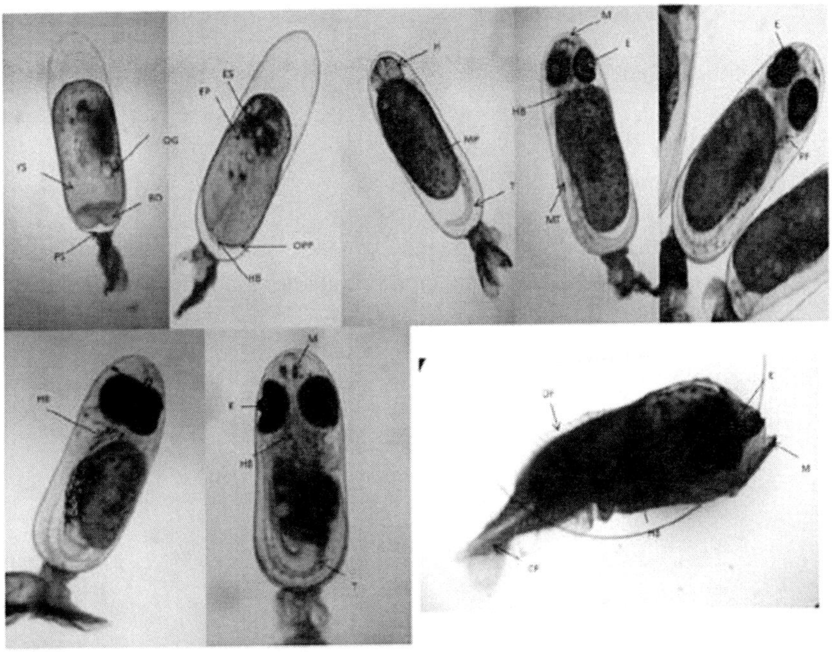

Figure 2. Day wise embryonic development of anemone fish *A. clarkia.*

3.5.1 First day

The cytoplasm of the fertilized egg was clear in day 1 embryo, yolk and different sizes of maximum number of oil globules were dispersed in it. One end of the egg was identified as animal pole attached to the substrata (earthen pot and tiles) with a tuft of filamentous stalk or cilia with glutinous substance and cell cleavage observed at animal pole (Dhaneesh et al., 2009).

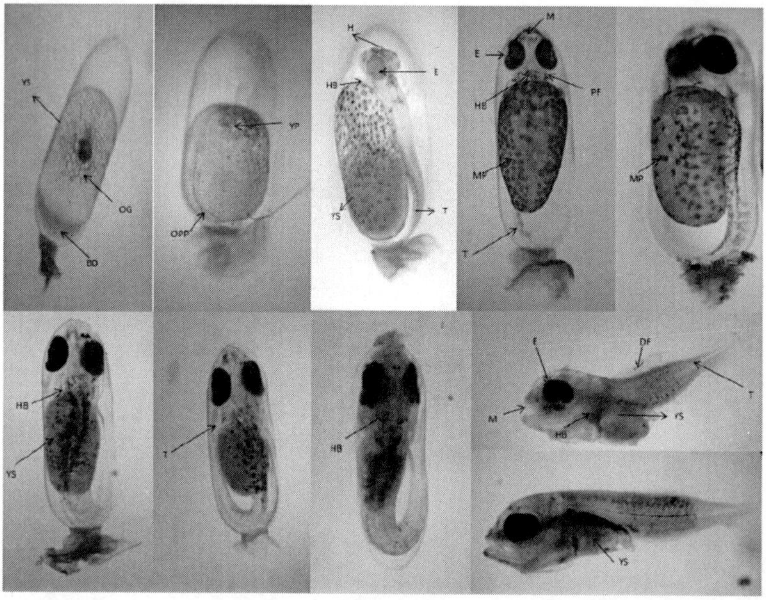

Figure 3. Day wise embryonic development of anemone fish *A. melanopus.*

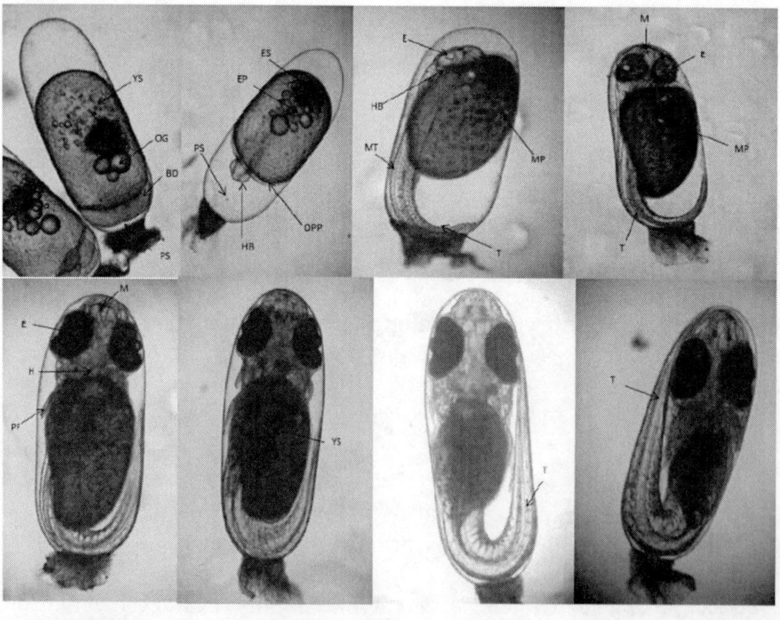

Figure 4. Day wise embryonic development of anemone fish *A. nigripes.*

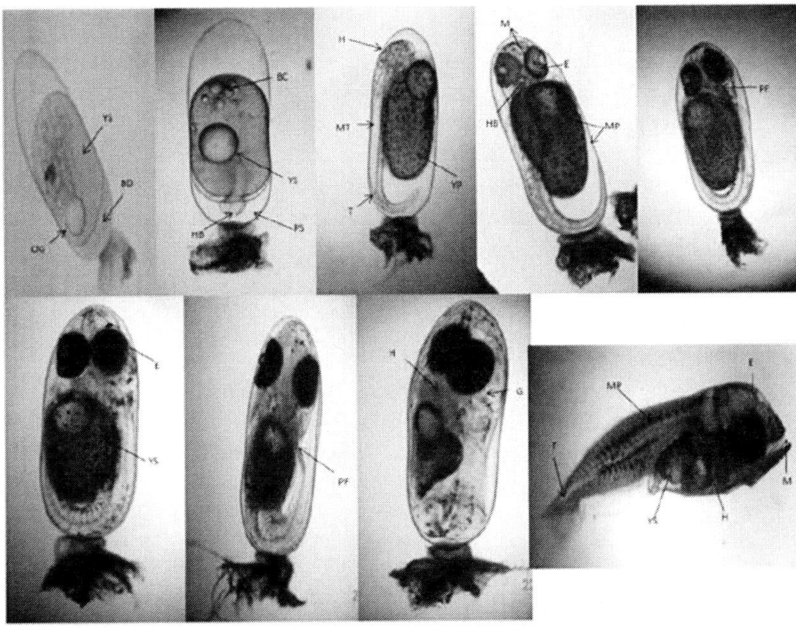

Figure 5. Day wise embryonic development of anemone fish *A. ocellaris.*

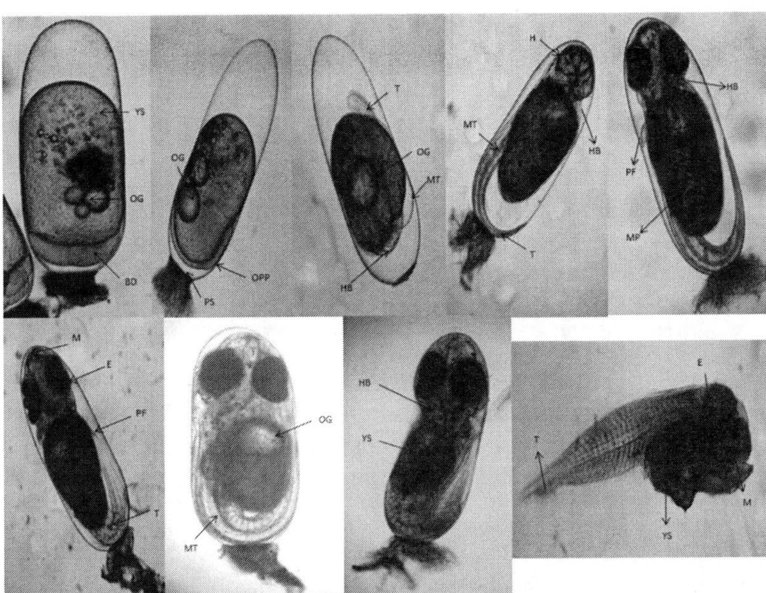

Figure 6. Day wise embryonic development of anemone fish *A. percula.*

3.5.2 Second day

After 24 h, the cleavage was completely stopped and oil globules were moved at opposite end from the animal pole. On the bottom of yolk, head bud, below head bud and perivitalline space was clearly observed in *A. ocellaris, A. nigripes,* and *A. clarkii* but in case of *A. melanopus* and *A. percula*, head bud was not observed. It is the gastrula stage in which the blastomeres were extended towards the vegetal pole and the numbers of fat globules decreased.

3.5.3 Third day

The embryo showed the clear visible head, mytosomes and tail. The body was transparent having no muscular structure and heart beet was clearly observed in all the four species except *A. percula* in which head bud and perivitalline space only developed.

3.5.4 Fourth day

The embryo exhibited eye, melanopores development and the tail became separated from the yolk and moved freely but the body was still attached to the yolk. On the microscope, utilization of yolk was clearly observed while, clear blood circulation.

3.5.5 Fifth day

The mouth and eye were evident for blood circulation through the vessels could be observed. It indicated the functioning of circulating system of the embryo.

3.5.6 Sixth day

In embryo,head and tail were clearly separated from the yolk. Thereafter, embryo was further enlarged and occupied most of the space within the capsule; pectoral fin development was also observed.

3.5.7 Seventh day

On the day, yolk sac became quite small and covered by the abdomen of the embryo. The melanophores are distributed throughout the body. Fins and eyes are well developed in *A. clarkii* and egg clutch showed silvery in colour. On the day night 19:00 to 20:00 h and the larva are free from the capsule.

3.5.8 Eighth day

Fins and eyes are well developed; in this stage the embryo began to hatch by moving itself vigorously to break the capsule.

3.5.9 Ninth day

Embryo was free from the capsule and became a larva on day 9. The dorsal fin, caudal fin and anal fin were continuous in a longitudinal line.

3.6 "Hatching"

Hatching was took place on 8 to 9th days of post- fecundation during the first 2 h of darkness. The dorsal, caudal and anal fins were continuous in a longitudinal line. The distal end of the chorion becomes soft and pliable and the movements of the embryo became more frequent and violent until the weakness chorion gives way. The newly hatched larva had a transparent body, large eyes, open mouth and a small yolk sac. Immediately after hatching, the larvae were found floating on the surface vertically with up head position and larval size ranging from 3.7 to 4.4 mm in length. After 3 to 5 h, the larvae were transferred to 50 L FRP larval rearing tanks with stocking density of 5 individuals l^{-1} and it could consume rotifer (*B. plicatilis*-130-340 μm) and zooplankton as its primary food.

3.7 "Incubation time, hatching rate and deformation rate"

The estimated incubation time, hatching rate and deformation rate were presented in the Figure 8, 10 and 11. The incubation time was varied between the species, *A. clarkii* showed minimum incubation time than others and maximum incubation time was noted in *A. melanopus*. Hatching rate was maximum in *A. percula* and minimum in *A. nigripes*. Deformation rate was maximum in *A. nigripes* and minimum in *A. melanopus*.

4. DISCUSSION

Breeding of anemone fishes in hatchery holds minimum challenges as compared to the other marine ornamental fishes and many breeding experiments conducted successfully in hatcheries on different species of anemone fishes (Alva and Gomes, 1989; Wilkerson, 2001; Overton et al., 2008; Dhaneesh et al., 2009; Ajith Kumar and Balasubramanian, 2009; Swagat et al., 2012; Uthaya Siva et al., 2014). However, there would be certain limitations in culture practices of anemone fish in private aquaria. Therefore, the present study was analyzed about embryonic development and spawning efficiency of five different anemone fish as fecundity rate in captivity. Boby Ignatius et al. (2001) observed that the brooders fed with clam meat and marine polycheate attained the sexual maturity within 3 months of maintenance and spawned repeatedly when compared with those pairs fed with minced fish meat alone (Dehasque et al., 1995). In the present study, boiled clam, shrimp meet and commercial pellet feeds were provided for maturation of brooders to balance the broodstock nutrition.

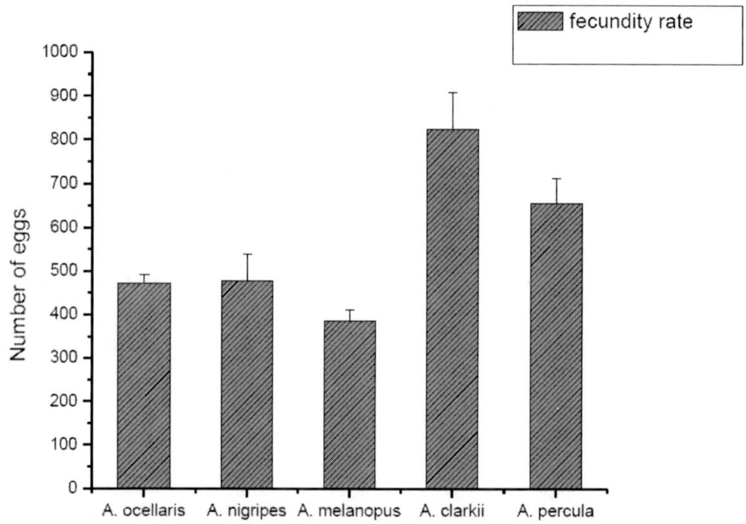

Figure 7. Percentage of fecundity or spawning frequency among the species.

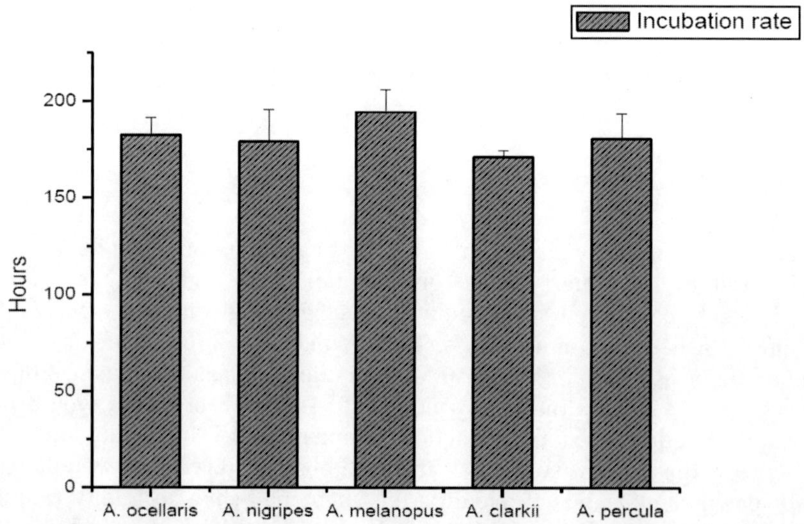

Figure 8. Percentage of incubation tame among the species.

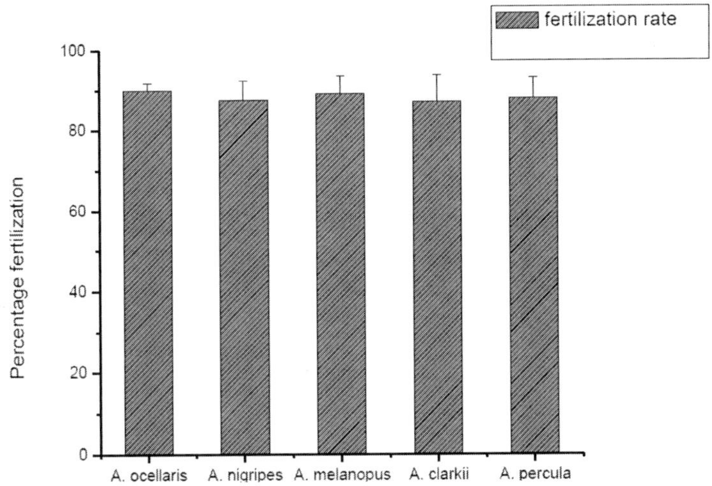

Figure 9. Percentage of fertilization rate among the species.

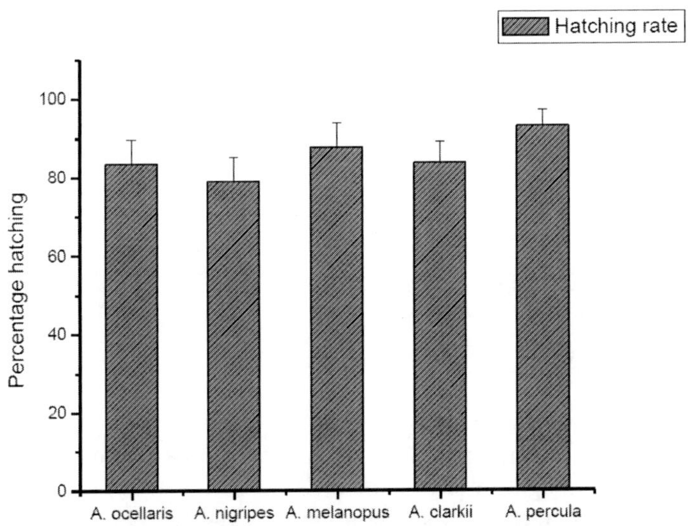

Figure 10. Percentage of hatching rate among the species.

Dehasque et al. (1995) stated from his result that feed consumption showed the capacity of egg laying power of brooders. On the other hand, co-feeding of live feeds supplemented with lipid and vitamin formulation helped to increase the nutritional reserves of broodstock and hence to improve the overall egg and larval quality (Allen, 1974) because in aquaculture spawning and proper embryo development was related to broodstock nutrition and high quality breeders were essential for successful larval rearing (Fautin and Allen, 1997; Jobling, 2002). The present study resulted that fertilization, hatching and deformation rate of larvae were depends on several factors (feed quality, brooder health and environmental parameters) not only because of brooders. Although, the brooders were maintained in the backwater (with salinity variation) showed similar fecundity and spawning frequency to those kept in running seawater (Alva and Gomes, 1989; Wilkerson, 2001; Overton et al., 2008; Dhaneesh et al., 2009; Ajith Kumar and Balasubramanian, 2009; Swagat et al., 2012; Uthaya Siva et al., 2014).

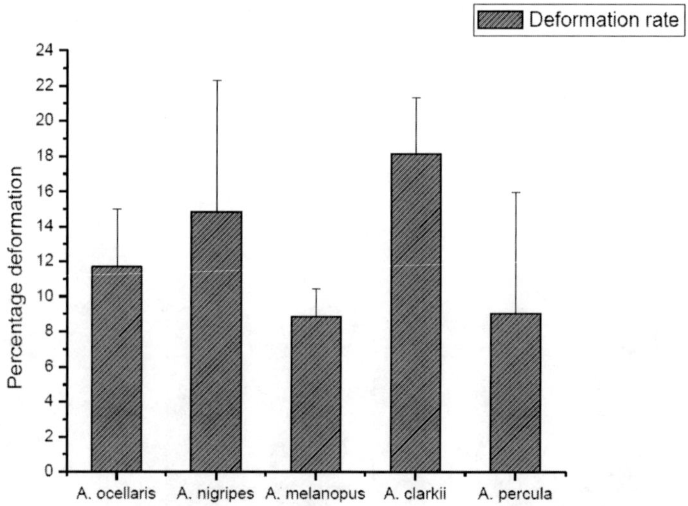

Figure 11. Percentage of deformation rate among the species.

In addition, several factors such as temperature, salinity, ammonia level, dissolved oxygen, pH, light intensity and photoperiod were known to affect the growth and development of anemone fish (Alva and Gomes, 1989). In the extant study also said that the water quality parameters were influenced on growth as well as breeding of anemone fish thus, present study maintained in optimum level as temperature 27±2°C, salinity 29±3 ppt, ammonia level 0.001 ppm, dissolved oxygen 4±2 ppm, pH 7.7±0.2, light intensity 18 μ $mol^{-1}sec^{-1}$and photoperiod 13 L: 11 D. Unless other parameters, temperature was mainly greatly influenced on embryo development. During summer, incubation was observed at 7 days but in monsoon it was extended to 8 days because development of embryo taken additional time than summer, it indicated temperature (summer-27°C, monsoon-24°C) was the important environmental parameter involved in embryonic development. Spawning ratio was maximum in summer because normal temperature was observed and minimum in monsoon.

According to Jobling (2002) and Moyle and Cech (2004) the embryonic developmental period was the phase between fertilization and commencement of organogenesis. The contour of eggs in different species of Pomacentridae fish varies from oval to capsule shape (Moyer and Nakazono, 1978; Pathiyasevee, 1994; Balamurugan et al., 2013; Hoff, 1996; Uthayasiva et al., 2014). Similarly, the clutches of anemonefishes were in oval shape. The clutch diameter demonstrated the fecundity rate of a particular species was found as per formula. Balamurugan et al. (2013) documented the development of eggs were noted morphologically on color changes, were yellow/orange on 1st and 2nd days, black/brown from 3rd day and silvery during pre-hatching. No color change in the eggs on 3rd and 4th days was considered immature/decayed (Hoff, 1996; Melville and Griffiths, 1997). Similarly, the egg clutches of anemone fish have been found as light/dark orange in color on1st and 2nd days. Light/ dark brown was observed from 3rd to 4th/6th days and silvery in pre-hatching (Figure 12).

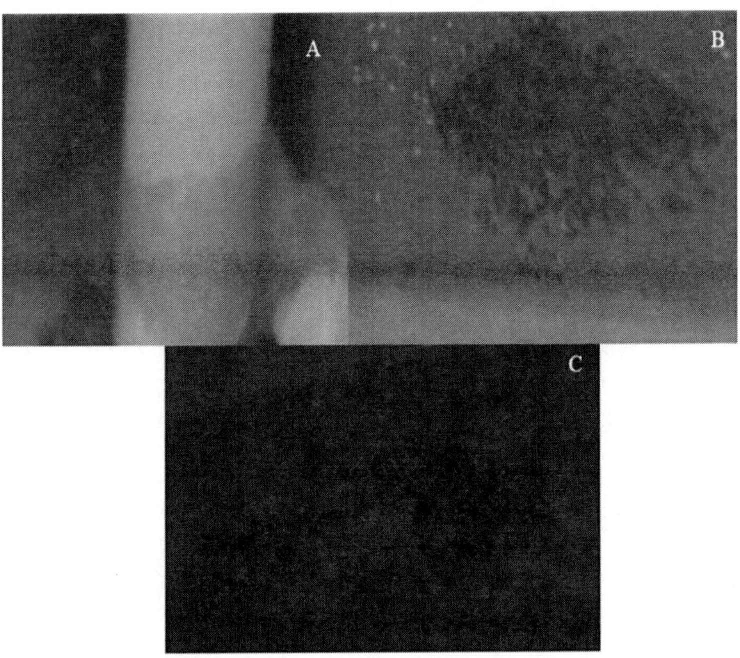

Figure 12. The development of eggs morphology was as yellow/orange on 1st and 2nd days (A), black/brown from 3rd day (B) and silvery during pre-hatching (C).

Hoff (1996) reported that the length of anemone fish eggs were ranged from 2.0 to 2.4 mm but in the case of present study, it varied from the earlier reports (Table 1). While, Swagat et al. (2012) documented that other species of cultured anemone fish in aquaria make different size of clutches, such as *A. sebae* (8.5-9 cm), *A. percula* (6.5-7.0 cm), *A. clarkii* (8.5-9.5 cm), *A. ocellaris* (4.5- 5.5 cm) and *P. biaculeatus* (7.5 cm) but in this study have been observed clutch diameter was frequently high then previous reports (Table 1). The above two statement was occurred because of the size of females and age. Earlier Dhaneesh et al. (2009) researches demonstrated that the fecundity rate was increased with size of the female. Same observation has recorded in current study also in *A. percula* and *A. clarkia* female was found bigger size than *A. melanopus, A. ocellaris* and *A. nigripes.* However; fecundity rate of *A. melanopus, A. ocellaris* and *A. nigripes* were lesser than *A. percula* and *A. clarkii.* Hence, the fecundity rate of the anemone fishes only depends on the size of female and also its age.

After fertilization, especially the male taken care of eggs by fanning with their pectoral fins and cleaned the clutch gently through fanning and mouthing until hatching (Swagat et al., 2012). Similar observations were reported in other anemone species also (Overton et al., 2008; Melville and Griffiths, 1997; Madhu et al., 2006a) however, there were some variations in fanning and embryonic development periods related to anemone fish species. Mariscal (1970) reported the embryonic development rate of fertilized eggs were varied depending on ambient temperature and dissolved oxygen content of water in the tank. Similarly, the present result demonstrated that hatching occurred in different types of incubation hours due to the temporal variations. From this observation it could be suggested that, embryonic development of anemone fishes were depending on temperature and parental care.

Wilkerson (2001) reported that the *Amphiprion* embryo formed a heart, brain and spinal cord on the 3 day after deposition, and blood circulation was clearly visible on day 4 in *A. chrysopterus* (Jobling, 2002). Whereas in the study, it was stated that blood circulation was observed in 3 days after the fertilization in all the species. In the study of Ignatius et al. (2001), Pomacentrids fish male parent also exhibited culling by unfertilized or dead embryos and was made to produce other embryos but in case of present study, in anemone fishes male alone cared for the eggs during incubation. In some species, female (*A. clarkii* and *A. melanopus*) cared in early stages of egg development and other does not. Silvery coloration with distinct visible eyes is usually a good indication for the hatch out within 12 h, because the matured eyes showed this color. These observations are also similar to those reported for other anemone species (Pathiyasevee, 1994; Madhu et al., 2006a). In this study, on 7th day *A. clarkii* eggs have got silvery and 8th day completely silvery of all other four species (*A. percula, A. clarkii, A.*

melanopus, A. nigripes and *A. ocellaris*) because *A. clarkii* eggs got matured within 7 days than other species.

Appearance was noted and it is due to variations in the fertilization and incubation time from the beginning of spawning and also it indicates the parental care of the anemone fishes during incubation. Incubation time varied because of seasonal variations. So, temperature is important factor for development of embryo (Uthayasiva et al., 2014). The embryonic development of five different species provides a better idea on their development stages and this information helped the mass production of this tropical coral reef ornamental fish in captivity. Briefly, the increasing popularity and demand of marine ornamental fish, especially anemone fish in market sustainably increased worldwide.

The present study conclude that there is less influence of size variations of breeders on the fecundity, parental care, incubation periods, hatching, and deformation rate of anemone fishes would help the stockholders to culture the particular species to earn high price. Estuarine water is adequate for sustainable production, this will help the coastal fisher folk to enhance their livelihood through setting up of backyard hatcheries and significantly, it will help conserve the precious marine ornamental fish biodiversity.

ACKNOWLEDGEMENTS

Authors are thankful to authorities of CAS in Marine Biology, Annamalai University for providing necessary facilities.

REFERENCES

1. Ajith Kumar TT, Balasubramanian T (2009). Broodstock development, spawning and larval rearing of the false clown fish, *Amphiprion ocellaris* in captivity using estuarine water. Curr. Sci. 97(10):1483-1486.
2. Ajith kumar TT, Gopi M, Dhaneesh KV, Vinoth R, Ghosh S, Balasubramanian T, Sanmugaraj T (2012). Hatchery production of the clownfish Amphiprion nigripes at Agatti Island, Lakshadweep, India. J. Environ. Biol. 33:623-628.
3. Allen GR, Drew J, Fenner D (2010). *Amphiprion pacificus*, a new species of anemonefish (Pomacentridae) from Fiji, Tonga, Samoa, and Wallis Island. Aquaculture 16:3-15.
4. Allen GR (1974). The Anemonefishes: The classification and biology, 2nd edn. TFH Publication, Inc. USA. 352.
5. Alva VR, Gomes LAO. (1989). Breeding marine aquarium animals: the anemone fish. The Nega, ICLARMQ. pp. 12-13.
6. Avella MA, Olivotto I, Gioacchini G, Maradonna F, Carnevali O (2007). The role of fatty acids enrichments in the larviculture of false percula clownfish. *Amphiprion ocellaris* Aquaculture 273:87-95.

7. Balamurugan RJ, Ajith Kumar TT, Balasubramanian T (2013). Influence of host anemone (*Stichodactyla haddoni*, Saville-Kent, 1893) locomotion on its resident anemone fish reproduction. Anim. Reprod. Sci. 140:103-107.
8. Balon EK (1981). Additions and amendments to the classification of reproductive style in fishes. Environ. Biol. Fishes 6:377-389.
9. Boby ignatius G, Jagadis I, Kandasami D, Victor ACC. (2001). Spawning and larval rearing technique for tropical clown fish *Amphiprion sebae* under captive condition. J. Aquac. Trop. 16(3):241-249.
10. Chen YF (2005). Induced ovulation and embryonic development of ocellated puffer, *Takigugu ocellatus.* J. Appl. Ichthyol. 1(2):136-140.
11. Dehasque M, Candreva P, Carrascosa M, Lavens P (1995). Comparison of an artificial diet and live food as broodstock maturation diets for sea bream (Sparusaurata): effects on spawning and egg quality In: Lavens, P., Jaspers, E. and Roelants, I. (Eds.) Larvi' '95 Gent, Belgium. Eur. Aquac. Soc. 24:16-21.
12. Dhaneesh KV, Ajith Kumar TT, Shunmugaraj T (2009). Embryonic
13. Development of Percula Clownfish *Amphiprion percula* (Lacepede, 1802). Middle-East J. Sci. Res. 4(2):84-89.
14. Falk-Petersen IB (2005). Comparative organ differentiation during early life stages of marine fish. Fish Shellfish Immunol. 19:397-412.
15. Fautin DG, Allen GR (1997). Anemone fishes and their host sea anemones. Western Australian Museum-160.
16. Gunasekara RM, shim KF, lam TJ (1996). Effect Protein Level on Spawning Performance and Acid Amino Composition of Eggs of *Nile tilapia, Oreochromis niloticus.* Aquaculture 146:121-134.
17. Hoff FH (1996). Conditioning, spawning and rearing of fish with emphasis on marine clown fish. Dade city: Aquaculture Consultants 212.
18. Ignatius B, Rathore G, Jagdis I, Kandasami D, Victor ACC (2001). Spawning and larval rearing technique for tropical clownfish *Amphiprion sebae* under captive conditions. J. Aquac. Trop. 16(3):241-249.
19. Jobling M (2002). Environmental factors and rates of development and growth. In: Hart PJB and Reynolds JD. (Eds.). In: Hand book of Fish Biology and Fisheries, vol.1. Fish Biology Blackwell Science Ltd. Blackwell Publishing Company.
20. Madhu K, Madhu R, Krishnan L, Sasidharan CS, Venugopalan KM (2006a). Spawning and larval rearing of *Amphiprion ocellaris* under captive conditions. Mar. Fish. Infor. Serv. T&E Ser. 188.
21. Madhu K, Rema M, Retheesh T (2012). Broodstock development, breeding, embryonic development and larviculture of spine-cheek anemonefish, *Premnas biaculeatus* (Bloch, 1790). Indian J. Fish. 59(1):65-75.
22. Mariscal RN (1970). The nature of symbiosis between Indo-Pacific anemonefishes and sea anemones. Mar. Biol. 6:58-65.
23. Melville K, Griffiths S (1997). Recent developments in disease management. Bull. Aquac. Assoc. Canada 2:1-19.
24. Moyer JT, Nakazono A (1978). Protandrous hermaphroditism in six species of the anemonefish genus *Amphiprion* in Japan. Jap. J. Ichthyol. 25:101-106.
25. Moyle PB, Cech JJJR (2004). Fishes: An introduction to ichthyology, 5th ed. Pearson, Prentice-Hall, Inch. Upper Saddle River, N J. 726.
26. Olivotto I, Cardinali M, Barbaresi L, Maradonna F, Carnevali O (2003). Coral reef fish reading: The secrets of each species. Aquaculture 224:69-78.
27. Overton JL, Bayley M, Paulsen H, Wang T (2008). Salinity tolerance of cultured Eurasian perch, Percaflu viatilis L: Effects on growth and on survival as a function of temperature. Aquaculture 277:282-286.
28. Pathiyasevee U (1994). Egg-laying behaviour and growth of false clown anemonefish *Amphiprion ocellaris* (Cuvier, 1830). Annual seminar on 1994. National Institute of Coastal Aquaculture Department of Fisheries, Bangkok, Thailand pp. 393-412.
29. Ross RM (1978). Territorial behaviour and ecology of the anemone fish *Amphiprion melanopus* on Guam. Z. Tierpsychology 46:71-83.

30. Swagat G, Ajith Kumar TT, Nanthinidevi K, Balasubramanian T (2012). Reef fish breeding and hatchery production using brackishwater, a sustainable technology with special reference to clarkii clownfish, *Amphiprion clarkii* (Bennett, 1830). Int. J. Environ. Sci. Dev. 3:1.
31. Urho L (2002). Characters of larvae-what are they? Folia Zool. 51:161-186.
32. Uthaya Siva M, Balamurugan J, Ajith Kumar TT, Badhul Haq MA (2014). Domestication of Anemone Fishes with High Fecundity and Spawning Efficiency under Captive Condition for Indian Scenario. Theriogenology Insight. Int. J. Reprod. Anim. 4(1):9-16.
33. Uthayasiva M, Balamurugan J, Kumar TTA, Haq MAB (2014). Seasonal Disparity in Fecundity and Embryonic Development of Cinnamon Anemone Fish, *Amphiprion melanopus* (Bleeker, 1852) In Captivity. J. Mar. Biol. Oceanogr. 3:3.
34. Wilkerson DJ (2001). Clownfishes: A guide to their captive care breeding and natural history, 1stedition., Microsom Ltd. USA. pp. 240-260

CHAPTER 8

Recruitment pattern and catch per unit effort of the Black Tiger Shrimp (Penaeus monodon) in Andoni River System, Niger Delta Region of Nigeria

Komi, W. Gentle[1]* and Francis, Amiye[2]

[1]*Department of Extension and Innovation Studies, College of Agricultural and Environmental Sciences, Makerere University, Uganda.*

[2]*Department of Fisheries, Faculty of Agriculture, University of Port Harcourt, Nigeria.*

ABSTRACT

The investigation of recruitment pattern and catch per unit effort of the black tiger shrimp (Penaeus monodon) was conducted from March 2012 to February 2013 in Andoni River System, Niger Delta region of Nigeria. Total weight and length of sampled P. monodon, number of fishers, fishing duration and temperature were sampled monthly. Total weight of P. monodon landed, number of fishers and fishing duration was used to determine catch per unit effort in kg/man/h. The recruitment pattern was analyzed using the length frequency data of the stock. The result shows that recruitment pattern was continuous and higher during the rainy season, that is from May to October. Catch per unit effort was lowest

(46.09Kg/man/h) in January and highest (78.11 kg/man/h) in September. Mean temperature ranged from 27±0.32°C to 31±0.89°C which was suitable for the stock. A positive moderate correlation existed between recruitment and catch per unit effort (r = 0.50056) while a positive weak correlation occurred between recruitment and temperature (r = 0.21102). It is concluded from the study that a rise in the recruitment of P.monodon translated into a rise in catch per unit effort and that temperature was not the only factor necessary for recruitment.

Keywords: Shrimp, recruitment, catch per unit effort (CPUE), Andoni River, Penaeus monodon, Nigeria.

1. INTRODUCTION

It is popular that changes in shrimp populations are primarily driven by variability in recruitment-a concept that is defined as the number of shrimps surviving to enter into the fishery or that has grown to maturity which eventually can be harvested at the fishing ground. Recruitment is influenced by the number of offspring produced as well as the effect of density independent (temperature and food condition) and density dependent (competition for food or refuge) factors which occur at different pre-recruit stages (Pablo and Bodmer, 2004).

Recruitment, being a determining factor on the abundance of a species, influences the efforts exerted by shrimpers in catching the target species of shrimp. Hence, catch per unit effort (CPUE) in fisheries and conservation biology is an indirect measure of the abundance of a target species; and changes in the catch per unit effort are inferred to signify changes to the target species' true abundance (Pablo and Bodmer, 2004). There is paucity of knowledge on how recruitment of the black tiger shrimp relates with CPUE and temperature largely for the fact that the species is none native to Andoni River systems and such study had never been conducted. This correlation study intends to unravel the underlying relationships. Therefore the study provides a baseline data for the exotic Penaeus monodon in Andoni River Systems, Niger Delta region of Nigeria.

The black tiger shrimp (P. monodon) is a marine crustacean that is highly reared for food (FAO, 2011). It is an invasive species to Nigerian coastal water, first reported in 1999 by a trawl fishing company (FAO, 2000). The natural distribution is Indo-West-Pacific, ranging from the eastern coast of Africa, the Arabian Peninsula, South-east Asia and the sea of Jordan as well as eastern Australia and the Mediterranean Sea. Further invasive populations have become established in Hawaii and the Atlantic coast of the

USA --Florida, Georgia and South Carolina (Suthep, 2015; Fuller et al., 2014).

The limitations in this study lie in the assumptions that all shrimp/fish in a given set of data grows as described by a single set of growth parameters; that recruitment occurs in eleven months only, in a-twelve-month data sampling (Pauly, 1986; Gayanilo et al., 1989). Furthermore, sample collection was not replicated to necessitate mean and standard error, and that all artisanal shrimpers operate four active boat days (being the average number of fishing days in a week). When recruitment is juxtaposed with fishing effort otherwise referred to as catch per unit effort, (described in terms of quantity of fish landed by a fisherman over a specified time) an idea of the abundance of the stock, and productivity of a river can be revealed. However, productivity of a river is not in isolation of physical properties such as temperature of the river.

Water temperature is critical in aquatic ecosystems. It limits the migration, spawning, egg incubation, growth and metabolism of aquatic organisms (Kelly and Linda, 1997). The Andoni River, being a system of flowing river tributaries, maintain a suitable surface water temperature for P. monodon at the fishing grounds, though it is not clear what relationship exists between recruitment and temperature of the River. Most aquatic organisms are exothermic and are not able to control their body temperature except by behaviour such as temperature selection. Metabolic rate increases two or four fold for every increase of 10°C (Waterwatch/Namoi CMA, 2013). Increased metabolic rate leads to higher oxygen consumption and waste production of CO2 and ammonia. Temperature is therefore a significant physical property of water because it affects the amount of dissolved oxygen in the water. Water at 0°C will hold up to 14.6 mg/L while at 30°C it will hold only up to 7.6 mg/L (Waterwatch/ Namoi CMA, 2013).

Temperature also affects the rate of photosynthesis of plants, the metabolic rate of aquatic animals, rates of development, timing and success of reproduction, mobility, migration patterns and the sensitivity of organisms to toxins, parasites and disease. Life cycles of aquatic organisms are often related to changes in temperature (Water watch/Namoi CMA, 2013). Temperature which naturally occurs as a result of solar energy transfer may be affected by man-made structures including dams and weirs, and releases of water from them. Many factors influence the temperature of pond and lake including seasonal air temperature, water depth (Abowei, 2010), ground water inflow, stream flow, mixing due to wind and water currents, and the amount of sunlight and shade (Kelly and Linda, 1997).

As with dissolved oxygen, each aquatic organism has an optimal temperature range for its metabolism. Warmer water promotes higher metabolism and respiration rates. Water temperature indirectly affects lake by influencing the dissolved oxygen concentration. Warm water has less dissolved oxygen than cold water (Water watch/Namoi CMA, 2013; Kelly and Linda, 1997).

This paper therefore aims at filling the knowledge gap on how recruitment of the black tiger shrimp relates with CPUE and temperature in Andoni River Systems, providing a baseline data for the exotic P. monodon in Andoni River Systems of Niger Delta region of Nigeria.

2. MATERIALS AND METHODS

2.1 Study area

The Andoni River system (Figure 1) lies between latitudes 4°28' to 4°45' N and longitudes 7°45' E. It is a major fish nursery in the Niger Delta region of Nigeria, West coast of Africa (Francis, 2003) most probably due to the abundance of mangroves within this brackish water ecosystem.

The climate of the Andoni area consists of a long wet season which stretches from April to October and a short dry season from November to March dotted with rains. Ebbing and flood tide durations within aquatic system are six hours each. The fishers use tide movement to aid their transport during fishing since they primarily use dugout canoe (unmotorized) in fishing. The Andoni River is brackish with annual salinity range of 10 to 28% (Yoloye, 1976), 5 to 22% (Francis, 2003).

Nypa fruticans (an exotic palm), Rhizophora mangle (Red mangrove) and Avicenia marina (white mangrove) and flood plains border the river and its numerous creeks and tributaries and these are well exposed at low tides. The Andoni River system has two main landing sites viz; Kaa water front and Oyorokoto fishing settlement. Kaa water front is located between 4°33.915'N to 4°34.248'N and 7°21.746'E to 7°22.637'E (Figure 1). Kaa (Ika) fish market is the major point of sales of live, fresh and smoked fish products.

2.2 Research design

Landing survey, otherwise known as catch assessment survey described by Stamatopoulos (2002) was conducted at Kaa landing site once in every first week of the month for a period of twelve months (from March, 2012 to February, 2013). The data collected included quantity of catch, species composition, associated effort, and other secondary data such as prices, weight and length of the black tiger shrimp landed (Sobo, 2004).

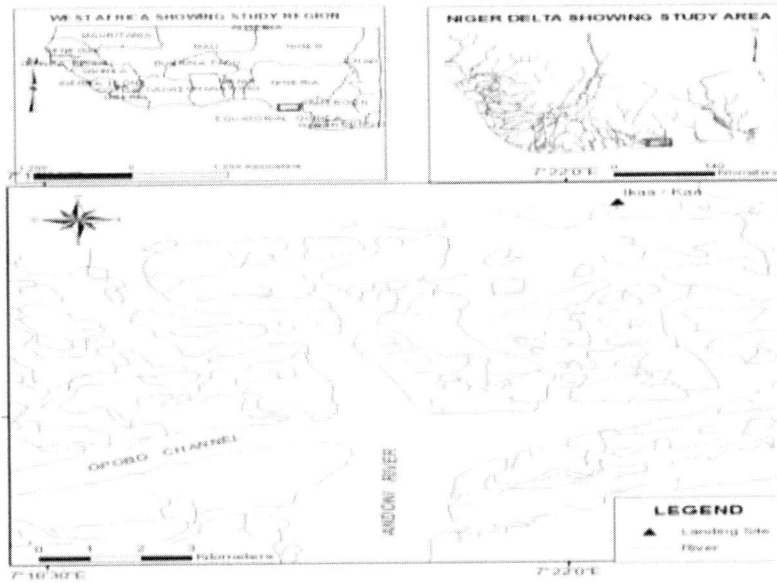

Figure 1. Map of Andoni River system (Francis et al., 2007).

2.3 Recruitment pattern

In determining the recruitment pattern of the black tiger shrimp, length-frequency data was produced from the recorded individual weight and total length of the species and used. The length –frequency data collected from sub-sample of P. monodon landed by artisanal fishers of the Andoni River system at Kaa landing site and representative of whole catch were subjected to recruitment procedure. FAO-ICLARM Stock Assessment Tool (FISAT II), a program package developed mainly for analysis of fishery data, was used. The recruitment routine in FISAT was employed for the estimation of the recruitment pattern using the number of recruitment pulses per year and evaluating the importance of this pulses when compared to each other.

The recruitment patterns were analyzed using NORMSEP (SEParation of the NORMally distributed components of length-frequency samples) to fit into Gaussian distribution on length-frequency data for the year, pooled together as described by Pauly (1983), Moreau and Cuende (1991) and Gayanilo and Pauly (1997).

2.4 Catch Per Unit Effort (CPUE)

In determining the catch per unit effort of the black tiger shrimp landed in Kaa water front (landing site) of Andoni River system, the daily catch landed were weighed using a weighing balance and measured in kilogram. The numbers; of fishing communities, boats, fishers and fishing duration were noted.

Twenty-five fishing communities land their catches at Kaa water side each day of the survey. An average of seven boats go fishing from each of the fishing communities. Number of boats that land shrimp per day equal 25x7=175 boats. Each boat was operated by two shrimpers; therefore the total numbers of shrimpers per day equals 175x2=350, meanwhile shrimpers spend an average of eight hours for actual fishing of P. monodon, this happened by the early hours of the morning and late evening depending on tidal movement (Awajikarem, Andoni fisherman, personal Communication). Catch per unit effort was determined by dividing the total monthly catch by the effort (number of fishers per boat) and by the number of hours of fishing.

$$\text{CPUE} = \text{Total catch/number of fishers/fishing hours} \qquad (1)$$

CPUE=kg/man/h (King, 1995; Francis, 2003). This formula was multiplied by four days being the average number of fishing days in a week and further multiplied by four weeks being the average number of weeks in a month as estimated by artisanal fisher folks.

Therefore,

Monthly Catch per unit effort (MCPUE) = (kg/man/h) x 16 (2)

Where M = monthly active boat days = 16 (number of active boat days).

kg = Weight of black tiger shrimp landed

Man = No. of fishers involved

h = fishing hour.

2.5 Temperature

A total of five sampling stations were established and their water temperature determined using Mercury-in-glass Celsius Thermometer with the probe inserted into the river; and the value at the pointer of the scale was read off and recorded. Units were in degree Celsius (°C).

2.6 Correlation analyses

Correlation between recruitment and catch per unit effort as well as between recruitment and temperature was determined using 2007 micro soft excel package.

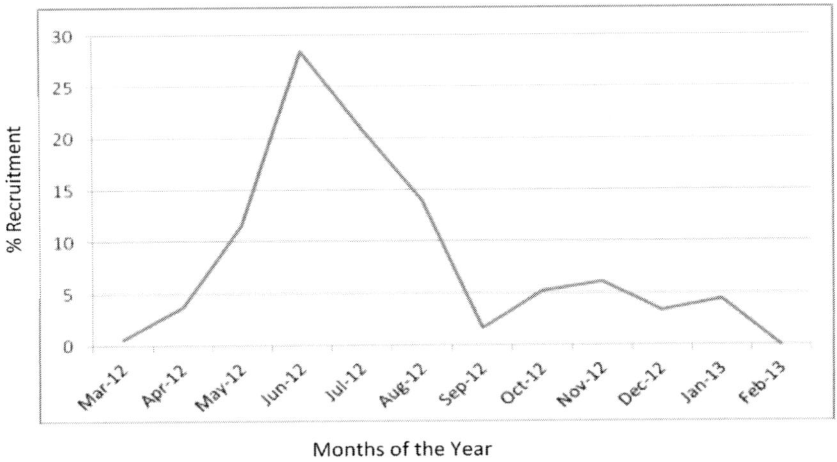

Figure 2. *P. monodon* recruitment pattern in Andoni River system.

3. RESULTS

3.1 Recruitment pattern

There was all-year-round recruitment (Figure 2) of the black tiger shrimp with peak during the rainy season from May to September, though the rate of recruitment over the period of study does not automatically determine the catch per unit effort of the fishers but (recruitment and CPUE) were both higher during the rainy season and lower in dry season. The highest catch per unit effort was recorded during the rainy season from May to October with a value of 78.11 kg/man/h in September while 46.09 kg/man/h was the least recorded and it was in the month of January. The percentage recruitment of P. monodon into the fishing ground was lowest (0.61%) in March 2012 and highest (28.41%) in June 2012. The recruitment pattern (Figure 2) can be described to be continuous with a single peak in June. The absence of replicated monthly data of tiger prawn landed limits the calculation of monthly mean and standard error, however, Figure 2 shows the trend of recruitment over a period of twelve months.

3.2 Catch per unit effort

There were an average of 350 artisanal shrimp fishers, fishing for an average of 8 hrs on each active boat day (4 active boat days per week) in the Andoni estuaries. The highest catch per unit effort was recorded in September with a value of 78.11 kg/man/h while 46.09 kg/man/h (Table 1) was the least recorded for the month of January.

3.3 Temperature

The highest temperature 31± 0.89°C was recorded in September and the least 27± 0.32°C in August (Figure 3). Fluctuation was higher in the rainy season and relatively stable during dry season. The temperature range was within acceptable WHO (2006) and FEPA (1991) limits.

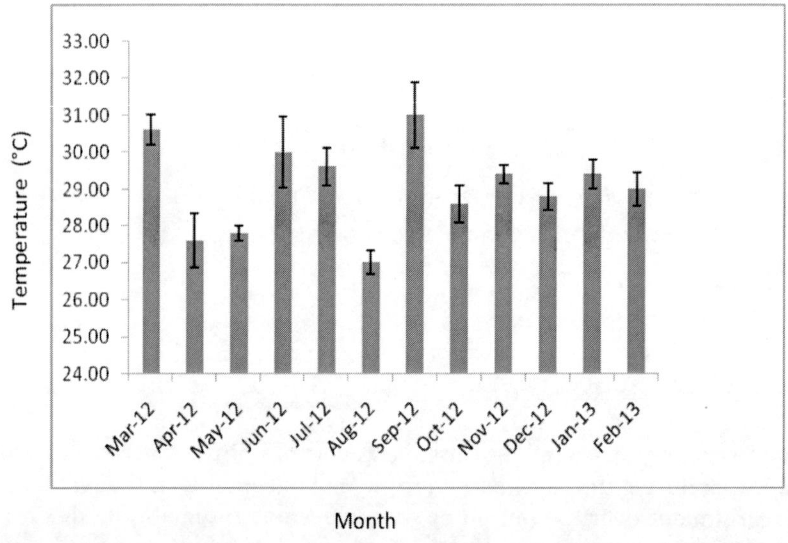

Figure 3. Mean temperature values of the Andoni River system from March 2012 to Feb 2013 (Source: Komi and Sikoki, 2013).

3.4 Correlation analyses

The correlation between percentage recruitment and monthly Catch per unit effort (MCPUE) of P. monodon in of Andoni River system. Correlation coefficient (r) = 0.21102.

Table 1. Catch data of *P. monodon* of the Andoni River system (March 2012 to February 2013).

Month	Catch per Month (kg/month)	MCPUE (kg/man/h)	Catch per day (kg/day)	DCPUE (kg/man/h)	Catch in ton/month
2012 March	2625.41	60.01	164.09	3.75	2.63
2012 April	3148.48	71.97	196.78	4.49	3.15
2012 May	3201.44	73.18	200.09	4.57	3.2
2012 June	3330.37	76.12	208.15	4.76	3.33
2012 July	3252.91	74.35	203.31	4.65	3.25
2012 August	3112.32	71.14	194.52	4.45	3.11
2012 September	3417.28	78.11	213.58	4.88	3.42
2012 October	3229.28	73.81	201.83	4.61	3.23
2012 November	2276.96	52.04	142.31	3.25	2.28
2012 December	2421.44	55.35	152.34	3.46	2.42
2013 January	2016.67	46.09	126.04	2.88	2.02
2013 February	2305.41	52.69	144.09	3.29	2.31
TOTAL	34337.97				34.35

MCPUE = Monthly catch per unit effort, DCPUE = Daily catch per unit effort. The mean and standard error of data were not calculated since data was collected from one landing site (Kaa water front) and once a month but for temperature, mean and standard error were calculated because water quality was analyzed at five sampling stations.

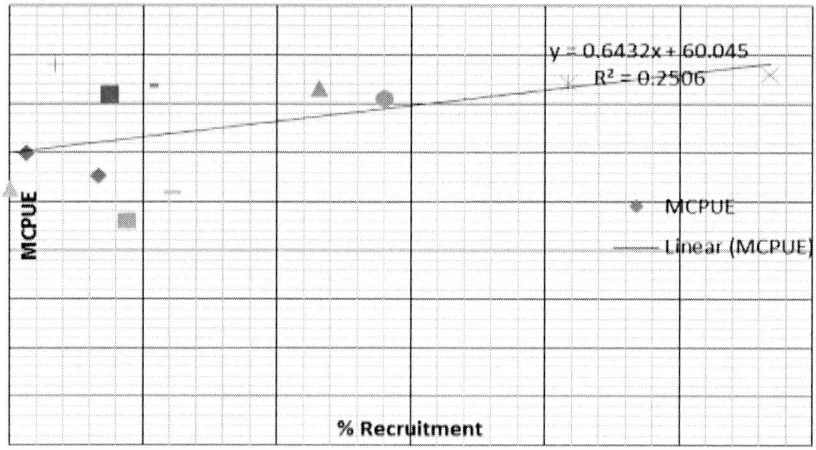

Figure 4. Correlation of % recruitment and catch per unit effort of *P. monodon* in Andoni River system. MCPUE= monthly catch per unit effort; Correlation coefficient (r) = 0.50056.

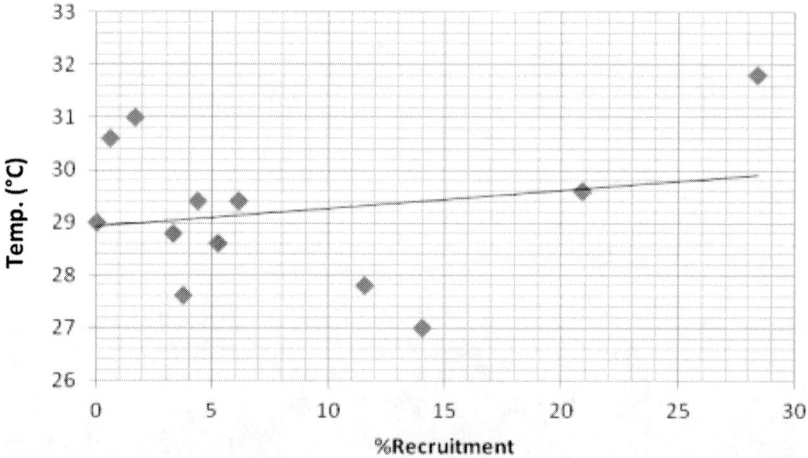

Figure 5. Correlation between % recruitment of *P. monodon* and surface water temperature

Andoni River system (Figure 4) showed that there was a linear relationship between both parameters. The coefficient of correlation (r) being 0. 5, indicated a moderate positive correlation but a weak positive correlation (r = 0.2) occurred between temperature and recruitment.

In a similar vein, the correlation between temperature and percentage recruitment (Figure 5.) was positive, suggesting that as temperature rose within acceptable range for shrimp growth, percentage recruitment equally rose. The correlation coefficient (r) being 0.2 showed a weak positive correlation betwee both parameters indicating that temperature was not a major factor that influences recruitment.

4. DISCUSSION

4.1 Recruitment pattern

The recruitment of P. monodon can be described to be continuous with only one peak in June. This corresponded to the findings of Amin et al. (2009) who reported the highest recruitment of Acetes japonicus in the coastal waters of Malacca between April and June. Dinh et al. (2010) recorded recruitment of eight shrimps species, having single peak; in May for Haliporodes sibogae and Parapenaeus gracillima; in June for H. harpax, Metapenaeus affinis and M. tenuipes while M. breviconis had its peak in July. Parapenaeus cultrirostris and P. maxillipedo were at their peaks in August; all having their peak during the rainy season which is in consonance with the result of the current study.

The trend of recruitment of *P. monodon* in the Andoni river system also shows seasonality. Recruitment increased to its highest value of 28.41% in the first three months of the rainy season and declined steadily in the dry season to its least value of 0.61%. It is phenomenal that one month out of the twelve months be assumed to be zero (0) recruitment (Gayanilo et al., 1989). Amin et al. (2009) also recorded least recruitment of 0.37% in the dry season. Low recruitment of the invasive black tiger shrimp during the dry season can be attributed to plankton availability (Oketoki, 2015). Shrimp fishers therefore, are likely to have better livelihood due to bumper harvest during the rainy season.

In literature, the recruitment of *A. japonicus* in the coastal waters of Malacca was described to be continuous throughout the year indicating one major peak. The highest recruitment peak occurred between April and June (Amin et al., 2009). The least percentage recruitment was 0.37% in November and highest was 27.47% in May (Amin et al., 2009). The study in the Mekong River Southeast Asia further revealed that during rainy season from May to August, estuarine ecosystem provided a habitat for nursery of shrimps and other aquatic organisms (Dinh et al., 2007). Francis and Sikoki (2005) reported recruitment pattern in Andoni River system to be higher from May to October (which coincided with the rainy season) and low in dry season.

This agrees with the findings in the current study. The trajectory of recruitment pattern suggests that more P. monodon migrate to fishing grounds during the raining seasons. This suggests that the rains wash off food through water runoff or erosion and increase plankton growth for the shrimps during the rainy seasons.

4.2 Catch per unit effort

Catch per unit effort was higher during the rainy season (March to October) and declined in dry season (November to February/March). It is however not clear why the month of peak recruitment was different from CPUE peak. This calls for further research. Daily CPUE of P. monodon in Andoni estuary ranged from 2.88 to 4.88 kg/man/h. Khan et al. (2003) however, reported CPUE ranging from 0.22 to 16.23 kg/h^{-1} though only the weight (Kg) and fishing duration (hr) were considered. The relatively higher CPUE of P. monodon is indicative of its abundance in the Andoni River system.

Catch per unit effort (CPUE) on the other hand is important in determining maximum sustainable yield, potential yield and abundance of a species (Francis, 2003). Many inshore waters of West African coast are rich in fisheries resources on a commercial scale (Tobor, 1991) including shrimps.

Khan et al. (2003) reported the highest mean catch per unit effort as 16.23 kgh-^{1} and the least 0.22 kg/h^{-1} of shrimps between 1985 and 1987. P. monodon was 0.40 kg/h^{-1}; P. merguiensis was 0.03 kg/h^{-1} while P. semisulcatus recorded 0.09 kg/h^{-1}. Catch per unit effort varied among species from 1.94, 0.01, 0.34, 0.04, 0.13 and 1.64 kg/h^{-1} for Metapenaeus monoceros, M. spinulatus, Parapenaeopsis stylifera, P. indicus, M. brevicornis and other shrimp respectively.

4.3 Temperature

Andoni River system maintained a temperature range of 27±0.3to 31.8±0.95°C, typical of tropical estuarine water. Earlier researchers had reported 26.2 to 32.4°C (Ansa et al., 2007) and 26.05 to 32.1°C on the same Andoni Estuary. Temperature influences migration, spawning, egg incubation, growth and metabolism of aquatic organisms (Kelly and Linda, 1997). The temperature of Andoni River system is favourable to the growth and metabolism of P.monodon, hence their adaptation and seeming prolific nature in their new environment. This finding is in consonance with research work in Niger Delta by earlier authors, including, Zabbey (2002), Ansa (2004), Sikoki and Zabbey (2006), Jamabo (2008). Ansa et al. (2007) reported temperatures of 26.2 to 32.4°C in Andoni flats of Niger Delta.

Francis et al. (2007) who also worked on Andoni River system reported a temperature range of 26.05 to 32.1°C.

CONCLUSION

Recruitment being continuous with a single peak is a good index for fisheries management as this will enable law makers to make informed decisions on when to impose a closed season, that is when shrimp fishers will not be allowed to fish. Recruitment had a single peak in June, implying that more matured adults entered fishing grounds in June, though higher throughout the rainy season and low in dry season. Catch per unit effort being higher during the rainy season could be attributed to abundance of food for the shrimp during the season and is reflective of the season for bumper harvest.

The correlation between recruitment and CPUE being moderate and positive suggested that catches of the stock were relatively dependent on the recruitment pattern. If recruitment is interfered by factors such as pollution, habitat destruction and adverse weather conditions, catch per unit effort will be affected negatively. The moderate positive correlation implied that as recruitment of P. monodon increases, catch per unit effort tends to rise. The correlation coefficient (r) of 0.2 between temperature and % recruitment showed weak positive correlation between both parameters indicating that temperature was not a major factor that influenced recruitment. There are therefore, other factors rather than only temperature that affect recruitment of the black tiger shrimp in Andoni River.

Further research is needed to determine those factors that influence recruitment of the stock significantly in Andoni River.

ACKNOWLEDGEMENTS

The authors are grateful to Professor Francis David Sikoki and Professor I.P. Aleleye-Wokoma for supervising the research; and to Miss Dorcas Agi who typed the manuscript.

REFERENCES

1. Abowei JFN (2010). Salinity, dissolved oxygen, pH and surface water temperature conditions in Nkoro River, Niger Delta, Nigeria. Adv. J. Food Sci. Technol. 2(1):36-40.

2. Amin SM, Arshad S, Siraj S, Japar Sidik B (2009). Population structure, growth, mortality and yield per recruit of segestid shrimp, Acetes japonicus (Decapoda: Sergestidae) from the coastal waters of Malacca peninsular, Malaysia. Indian J. Mar. Sci. 38(1):57-68.
3. Ansa EJ (2004). Studies on the benthic macrofauna of the Andoni Flats in the Niger Delta Area of Nigeria. A PhD Thesis Submitted to the Department of Animal and Environmental Biology. University of Port Harcourt. 242:20-22.
4. Ansa EJ, Sikoki FD, Francis A, Allison ME (2007). Seasonal variation in interstitial fluid quality of the Andoni flats, Niger Delta Nigeria. J. Appl. Sci. Environ. Manage. 11(2):123-127.
5. Dinh TD, Moreau J, Van MV, Phuong NT, Toan VT (2010). Population Dynamics of shrimps in Littoral Marine waters of the Mekong Delta, South of Viet Nam. Pakistan J. Biol. Sci. 13:683-690.
6. Dinh TD, Ambak MA, Anuar H, Phuong NT (2007). Population biology of the goby Pseudapocrypte elongates (Cuvier, 1816) in the coastal mud flat areas of the Mekong Delta, Vietnam. J. Asian Fish. Sci. 20:165-179.
7. FAO (2011). Penaeus monodon (Fabricius, 1798). Cultured Aquatic Species Information Programme. Food and Agriculture Organisation.
8. FAO (2000a). Report of the four GEF/UNEP/FAO Regional Workshops on Reducing the Impact of Tropical Shrimp Trawl Fisheries. Lagos, Nigeria, 15–17 December 1999; Puntare-as, Costa Rica, 15–17 January 2000; Teheran, Islamic Republic of Iran, 28 February–1 March 2000; Denpasar, Bali, Indonesia, 6–8 March 2000.FAO Fisheries Report 627 Rome.
9. Federal Environmental Protection Agency (1991). Guidelines and Standards for environmental pollution control in Nigeria. FG Press 238 p.
10. Francis A, Sikoki FD (2005). Recruitment pattern and fish production within the Andoni River system in the Niger Delta of Nigeria.
11. Francis AF, Sikoki D, Ansa EJ (2007). Physico-chemical parameter of the Andoni River system- Niger Delta, Nigeria. J. Fish. Int. 2(1):27-31.
12. Francis A (2003). Studies on the Ichthyofauna of the Andoni River System in the Niger Delta of Nigeria. A PhD Thesis submitted to the Department of Animal and Environmental Biology. University of Port Harcourt pp. 13-39.
13. Fuller PL, Knott DM, Kingsley-Smith PR, Morris JA, Buckel CA, Hunter ME, Hartman LD (2014). Invasion of Asian shrimp, Penaeus monodon Fabricius, 1798, in the western north Atlantic and Gulf of Mexico. Aquat. Invas. 9(1):59-70.
14. Gayanilo FC, Pauly D (eds.) (1997). 'FAO-ICLARM Stock Assessment Tools (FiSAT)' Reference Manual. FAO Computerised Information Series (fisheries), Rome 8:262.
15. Gayanilo FCJr, Soriano M, Pauly D (1989). 'A draft guide to the Compleat ELEFAN' ICLARM software 2, P 70.
16. Jamabo NA (2008). Ecology of Tympanotonus fuscatus (Linnacus, 1958) in the mangrove swamps of the upper Bonny River, Niger Delta, Nigeria. Ph.D Thesis Rivers State Univ. Sci. Tech. 340 p.
17. Kelly A, Linda G (1997). Dissolved oxygen and temperature. Natural Resources Facts, University of Rhode Island, Fact sheet No. 96-3.
18. Khan MA, Sada NU, Chowdhury ZA (2003). Status of the demersal fishery resources of Bangladish, p.63-82 In G. Silvestre, L. Garces, I. Stobutzki, M.Ahmed, R.A. Valmonte-Santos, C.Luna, L. Lachica-Alino, P. Munro, V. Christensen and D. Pauly (eds.) Assessment, Management and Future Directions of Coastal Fisheries in Asian Countries. World Fish Center Conf. Proc. 67:1120.
19. King M (1995). Fisheries Biology, Assessment and Management. Fishing News Books, Surrey, 336p.
20. Komi GW, Sikoki FD (2013). Physico-chemical characteristics of the Andoni river and its potentials for production of the giant tiger prawn (penaeus monodon) in nigeria. J. Nat. Sci. Res. 3(12):83-89.
21. Moreau J, Cuende FX (1991). On improving the resolution of the recruitment pattern of fishes. ICLARM. Fishbyte 9:45-46.

22. Oketoki TO (2015) Survey on Phytoplankton Biomass and Water Parameters in the Habitats of Invasive Tigers Shrimps (Penaeus Monodon) in Nigeria. Fish. Aquac. J. 6:145.
23. Pablo EP, Bodmer RE (2004). Hunting effort as a tool for community-based wildlife management in Amazonia. In Kirsten M. Silvius, Richard E. Bodmer and José M. V. Fragoso. People in Nature: Wildlife Conservation in South and Central America. Columbia University Press. pp. 123–136.
24. Pauly D (1986). On improving the operation and use of the ELEFAN programs. Part III: Correcting length -frequency data for effects of gear selection and /or incomplete re-enactment. ICLARM Fishbyte, 4(2):11-13.
25. Pauly D (1983). Some simple methods for the assessment of tropical stocks. FAO Fish. Tech. Pap. 234, 52 p.
26. Sikoki FD, Zabbey N (2006). Environmental gradients and benthic community of the middle reaches of Imo River, South-East Nigerian. Environ. Ecol. 24(1):320-36.
27. Sobo FAS (2004). Strengthening of Artisanal Fisheries data collection and management in Tanzania, Fisheries Training Programme. The United Nations University, Iceland. 8-16.
28. Stamatopoulos C (2002). Sample-based Fishery Surveys: A technical handbook. FAO Fisheries Technical Paper.No.425, Rome 87-102.
29. Suthep P (2015). Black Tiger Shrimp (Penaeus monodon): A Practical Hatchery and Farm Management Information Handbook". National Programme for Food Security, Lagos, Nigeria.
30. Tobor JG (1991). Marine Fish Resources of West Africa: Potential Management, development and constraints to their utilization to satisfy increasing demand. NIOMR Tech. Paper 70:1-25.
31. Water Watch/Namoi Catchment Management Authority (2013).
32. World Health Organization (2006). Guidelines for drinking water quality (1) Geneva, Switzerland.
33. Yoloye V (1976). The ecology of the bloody cockle Anadara (Senilia Senilis (L) Bulletin d' I.I.FAN T. 38. 1:25-56.
34. Zabbey N (2002). An ecological survey of benthic Micro invertebrates of Woji creek, off the Bonny River system Rivers State. M.Sc Thesis, University of Port Harcourt.102 p.

CHAPTER 9

Estimating the Augmented Reflectance Ratio of the Ocean Surface When Whitecaps Appear

Zhantang Xu [1], Wen Zhou [1], Zhaohua Sun [1,*], Yuezhong Yang [1], Junfang Lin [1], Guifen Wang [1], Wenxi Cao [1] and Qian Yang [2]

[1] *State Key Laboratory of Tropical Oceanography, South China Sea Institute of Oceanology, Chinese Academy of Sciences, Guangzhou 510301, Chin*

[2] *Institute of Oceanographic Instrumentation, Shandong Academy of Sciences, Qingdao 266001, China*

ABSTRACT

The presence of foam influences the accuracy of satellite-derived water-leaving radiance. A model has been developed to estimate the augmented reflectance ratio (A(λ,U)) due to differences in the fraction of whitecap coverage (w) on the ocean surface. A(λ,U) can be calculated from the product of w and ρ(λ,U), where ρ(λ,U) is the augmented ratio of the reflectance of background water $R_b(\lambda)$) caused by the presence of whitecaps. Our results showed that the average A(400~700, U) in the visible region was approximately 1.3% at U = 9 $m{\cdot}s^{-1}$, 2.2% at U = 10 $m{\cdot}s^{-1}$, 4.4% at U = 12 $m{\cdot}s^{-1}$, 7.4% at U = 14 $m{\cdot}s^{-1}$, 19% at U = 19 $m{\cdot}s^{-1}$ and 37.9% at U = 24 $m{\cdot}s^{-1}$, making it is necessary to consider the augmented reflectance ratio for remote sensing applications. By estimating remote sensing augmented reflectance using A(λ, U), it was found that the result was in good agreement with previous studies conducted in other areas with

U from 9 to 12 m·s^{-1}. Since $R_b(\lambda)$ is temporally and spatially variable, our model considered the variation of $R_b(\lambda)$, whereas existing models have assumed that $R_b(\lambda)$ is constant. Therefore, the proposed model is more suitable for estimating the augmented reflectance ratio due to whitecaps.

Keywords: whitecaps; wind speed; remote sensing reflectance; augmented ratio

1. INTRODUCTION

The presence of foam at the ocean surface should be taken into consideration not only in radiation budget studies (less solar energy absorbed by the ocean, more solar energy reflected back into space), but also in aerosol and ocean color remote sensing from space [1,2,3,4,5,6]. Since foam on the ocean surface contains many air bubbles and increases upwelling irradiance, the ocean surface no longer provides a negligible background for atmospheric correction [7]. Actually, the additional or augmented spectral upwelling irradiance from whitecaps must be quantified not only at the wavelength used for atmospheric correction, but also at other wavelengths, particularly in the visible region, because spectral information provides a means of determining water content [8,9]. Therefore, in both the visible- and long-wavelength regions, the influence of foam, whitecaps, and streaks on the optical properties of the ocean should be studied.

When whitecaps were first studied, it was thought that the reflectance of the foam would not change with increasing electromagnetic wavelength. They used a wavelength-independent foam reflectance of 50%–100% in conjunction with estimates of fractional coverage to determine the augmentation effect [10,11]. In 1977, Gordon and Jacobs used a radiation-transfer model to demonstrate that for totally reflecting whitecaps, a wind speed increase from 6 to 14 m·s^{-1} would double the local planetary albedo [12]. However, as experimental evidence began to accumulate, researchers found that the assumption that foam reflectance was independent of wavelength was not appropriate. For example, laboratory measurements carried out by Whitlock *et al.* [13] yielded a value of 55% in the visible part of the spectrum for thick foam in clear water and showed that this value diminished with increasing wavelength beyond 800 nm, with decreases of 5% at 850 nm, 10% at 1020 nm, and 50% at 1650 nm. Observations on foams were also conducted *in situ*. In 1996, field reflectance measurements of foam generated by breaking waves in the surf zone indicated that the reflectance decreased by 40% at 870 nm, 50% at 1020 nm, and 95% at

1650 nm relative to the reflectance at 440 nm [2]. Furthermore, to understand the influence of oceanic whitecaps on the global radiation budget, Frouin *et al.* [4] obtained radiative forcing values reaching 0.7 $W \cdot m^{-2}$ in the Indian Ocean during summer. As a result, we found that the above studies were all performed at wavelengths > 700 nm, because their main purpose was to determine the atmospheric correction needed when whitecaps appeared. More recently, the water-leaving radiance in the visible wavelength has been used by researchers to determine seawater contents such as colored dissolved organic matter (CDOM), suspended particulate matter (SPM), chlorophyll a, and particulate organic carbon (POC) [14,15,16]. Researchers have even used remote sensing data in the visible wavelengths to assess diffuse attenuation coefficients and euphotic depths [17,18]. Therefore, it is important to assess the influence of whitecaps on the radiance received by satellite sensors in the visible wavelengths. Moore *et al.* [8,19] have designed a measurement system to determine the spectral reflectance of whitecaps in the visible wavelengths. During their experiments, however, the remote sensing augmented reflectance results were derived assuming a single constant reflectance of background seawater. Because the reflectance of seawater changes in different ocean regions, the results obtained by Moore *et al.* did not satisfy the needs of global research. Furthermore, since there were only five wave bands in the visible wavelength during Moore *et al.*'s experiment, a hyperspectral investigation also needs to be conducted for the future research.

In this research, a hyperspectral measurement system has been designed specifically to measure the reflectance of the sea surface, including whitecaps and background water. During the expedition, whitecaps were frequently present at the measurement sites, enabling reflectance values for natural foam to be collected at a wavelength resolution of 1.6 nm. Based on the measured reflectances, a new model suitable for different ocean waters was developed to estimate the augmented reflectance ratio when the surface was covered with whitecaps.

2. MATERIALS AND METHODS

2.1. Description of the Model

The reflectance of the sea surface can be given by:

$$R(\lambda) = \frac{L_u(\lambda, \theta, \varphi)}{E_d(\lambda)} \quad (1)$$

where $L_u(\lambda, \theta, \varphi)$ is the upwelling radiance from the sea surface and $E_d(\lambda)$ is the downwelling irradiance above the seawater. When whitecaps appear, the reflection characteristics of the sea surface can be described as:

$$R_t(\lambda,U) = (1-w(U))R_b(\lambda)+w(U)R_w(\lambda) \tag{2}$$

where $R_t(\lambda,U)$ is the total reflectance of the sea surface including whitecaps and the whitecap-free surface, w(U) denotes the fractional coverage of whitecaps, and (1−w(U)) denotes the fraction of the surface without whitecaps. $R_b(\lambda)$ is the reflectance of the whitecap-free surface, which can be easily obtained because it simply involves measuring the background seawater, and $R_w(\lambda,U)$ is the weighted average whitecap reflectance. $R_w(\lambda,U)$ is difficult to obtain because of the varying coverage of whitecaps on the sea surface. As wind increases, the sea surface is roughened by intermediate short gravity waves, which break after reaching a critical steepness. This breaking produces foam patches and bubbly streaks, and finally the sea becomes almost completely covered by a layer of foam [20,21]. When whitecaps occur, it is generally assumed that the sea surface consists of spume, foam streaks, surface foam, bubble clouds, and background sea surface. Air bubbles in seawater substantially scatter incident solar light. Different surface characteristics have different bubble distribution, which leads to different upwelling radiance.

Therefore, the weighted average whitecap reflectance $R_w(\lambda,U)$ was introduced to represent the reflectance for a particular whitecap area (because the reflectance of whitecaps changes at different whitecap positions). $R_w(\lambda,U)$ is described as

$$R_w(\lambda,U) = [1+\rho(\lambda,U)]R_b(\lambda) \tag{3}$$

where $\rho(\lambda,U)$ is the augmented ratio of the reflectance of background water (*i.e.*, the augmented reflectance ratio for a particular whitecap area. This parameter is used to describe the discrepancy between the reflectance of a surface with whitecaps and a whitecap-free surface). Then the augmented reflectance ratio for a whitecap-free surface A(λ,U)) (*i.e.*, the augmented reflectance ratio of the whole ocean surface. This parameter is used to describe the discrepancy between the whole surface with no whitecaps and when whitecaps appear on the surface; it depends on the fractional coverage of whitecaps, as Equation (2) shows) is given by

$$A(\lambda,U)=R_t(\lambda,U)-R_b(\lambda)R_b(\lambda) \tag{4}$$

Whitecaps are typically related to environmental parameters such as wind speed, wind friction velocity, and the temperature difference between the atmosphere and the sea surface. For the sake of simplicity, it is assumed that whitecaps are related only to wind speed because they are primarily determined by wind speed [8]. Therefore, by combining Equations (2) and (3), Equation (4) is simplified to

$$A(\lambda, U)=w(U)\rho(\lambda,U) \tag{5}$$

where U is the wind speed at 10 m elevation. All the notations used in this study are given in Table 1. A flowchart of the model procedure is shown in Figure 1. As the flowchart shows, w(U) and ρ(λ,U) are the primary parameters that must be obtained. The model is based on the data of $R_b(\lambda)$ and $R_x(\lambda)$. Therefore, the first step is to obtain the reflectance of the sea surface, including whitecaps and unbroken seawater *in situ*.

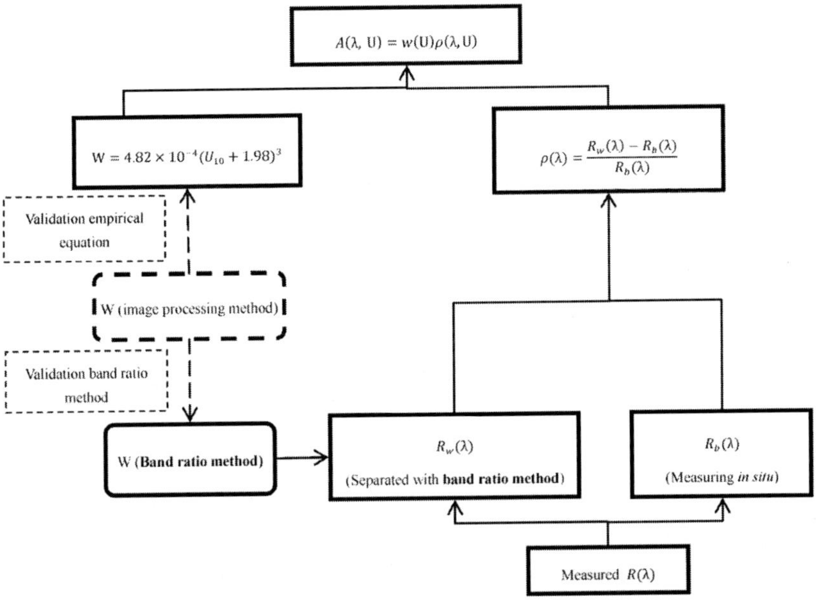

Figure 1. Flowchart of the model process. The dashed lines indicate a standard w to validate the w obtained by the band ratio and empirical methods.

Table 1. Notations used in this paper.

Notation	Description	Unit
$A(\lambda, U)$	augmented ratio of reflectance for the surface	None
$Ed(\lambda)$	downwelling irradiance above the seawater	$\mu W \cdot cm^{-2} \cdot nm^{-1}$
$Lu(\lambda, \theta, \varphi)$	upwelling radiance from sea surface	$\mu W \cdot cm^{-2} sr^{-1} \cdot nm^{-1}$
$L_w(\lambda, \theta, \varphi; \theta_0, \varphi_0)$	water-leaving radiance	$\mu W \cdot cm^{-2} \cdot sr^{-1} \cdot nm^{-1}$
$L_{sky}(\lambda, \theta, \varphi; \theta_0, \varphi_0)$	reflection radiance of incident sky radiance	$\mu W \cdot cm^{-2} sr^{-1} \cdot nm^{-1}$
$R(\lambda)$	reflectance of sea surface including $R_b(\lambda)$ and $R_x(\lambda)$	sr^{-1}
$R_w(\lambda, U)$	the weighted average whitecap reflectance	sr^{-1}
$R_x(\lambda)$	reflectance of whitecaps	sr^{-1}
$R_b(\lambda)$	reflectance of unbroken background seawater	sr^{-1}
U_{10}	wind speed at 10 m elevation	$m \cdot s^{-1}$
$w(U)$	fractional coverage of whitecaps and foam	None
$y(\lambda)$	equal to $\rho(\lambda, U)/\rho(400, U)$	None
$\rho(\lambda, U)$	the augmented ratio of the reflectance of background water	None
ΔT	difference between the temperature of the atmosphere and the sea surface water	°C
λ	wavelength	nm

2.2. A Description for the Observation Instrument

To obtain reflectance values, a whitecap measurement system was designed specifically for this study. A schematic overview of this system is shown in Figure 2a. The system has two hyperspectral spectrometers (Avaspec-2048FT-3-DT, Avantes, The Netherlands) which are connected separately to two kinds of fiber optic detectors. These optical detectors can monitor different light fields simultaneously. Furthermore, a high-definition camera is mounted beside the optical detectors to record simultaneously with the radiometer measurements to provide a visual reference. A lithium battery is used to provide power; it can keep the system working for about 24 hours. The effective integration time ranges from 3 to 3000 ms. In this research, to minimize measurement time, the integration time was set to a constant value, and all the measurements were saved automatically. The system has a wavelength resolution of 1.6 nm. The signal-to-noise ratio for the instrument is 250:1. Optical error is within 0.5% over the 400–700 nm spectral range. All the spectrum graphics given in this paper are for the wavelength range from 400 to 700 nm.

During an observation, the upwelling radiance includes mainly water-leaving radiance L_w (λ, θ, φ; θ_0, φ_0) and reflection of incident sky radiance on the seawater surface:

$$L_u(\lambda,\theta,\varphi;\, \theta 0,\varphi 0) = \mathrm{L_w}(\lambda,\theta,\varphi;\, \theta 0,\varphi 0) + \sigma\cdot \mathrm{L_{sky}}(\lambda,\theta,\varphi;\, \theta 0,\varphi 0) + \Delta \qquad (6)$$

where σ is the effective reflection coefficient depending on measurement geometry and wind speed [22], and Δ is the glitter (specular reflection of solar light) and incremental radiance introduced by whitecaps. The error caused by reflection of whitecaps can be ignored when the sea is calm.

As the wind increases, the sea surface becomes complex, and glitter becomes an optical contaminant which must be considered during the observation. Two methods were used to avoid glitter contamination. First, glitter depends on the zenith and azimuth angle of the optical detector and the solar incident radiance. By setting the radiance detector to a suitable zenith and azimuth angle, the effect of glitter can be minimized. The suitable nadir angle for the radiance sensor facing the sea surface has been suggested as 40°, and the azimuth angle relative to the Sun's azimuth angle has been suggested as 135° [22]. During the observations in this research, the Sun zenith angle varied from 51° to 67°. The detector was set at a zenith angle of 10° in consideration of the large Sun zenith angle and the complex seawater surface. The azimuth angle was set to values up to 135°. A description of the positions of the observation detectors is shown in Figure 2b. As wind speed increases, the surface geometry becomes complex. This complex surface reflects light in different directions, which somewhat contaminates the measurements regardless of the observation angle of the detector. Fortunately, the effects on observations of whitecap-free and whitecap-covered surfaces were similar and therefore were not of any significant consequence in determining the spectral variations in whitecap reflectance [2]; second, glitter contamination is also dependent on sky condition. Cloud around the Sun will result in less solar glitter to contaminate measurements. Therefore, in this research, the optical contamination caused by glitter was successfully minimized by setting a suitable observation angle under cloudy sky.

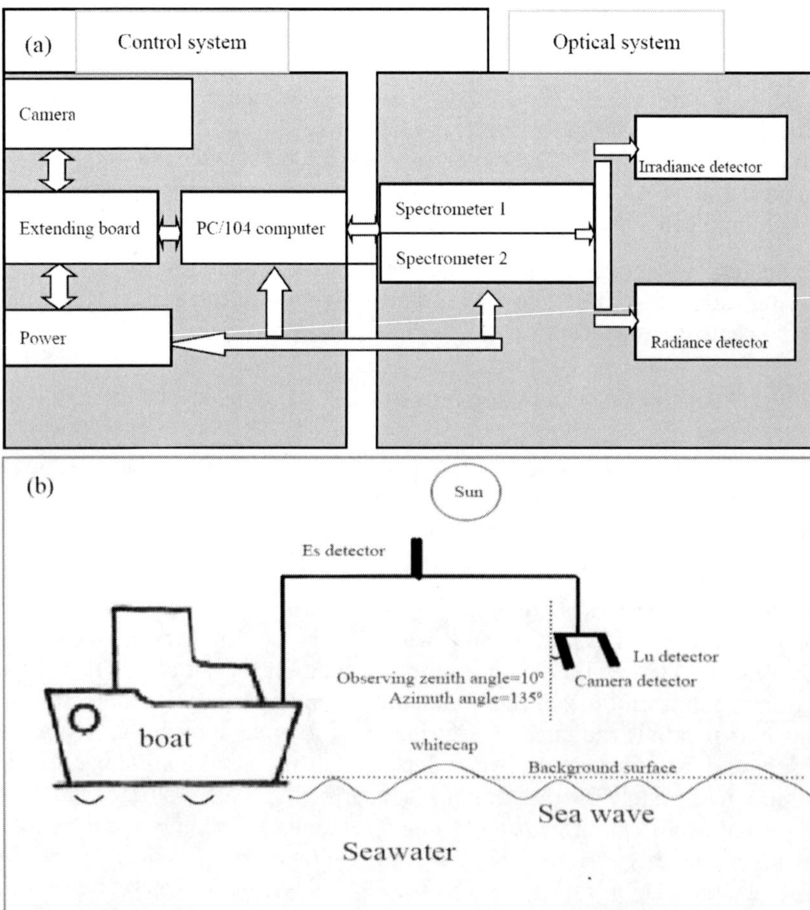

Figure 2. **(a)** Schematic overview of the whitecap measurement system. **(b)** Positions of observation detectors.

Table 2. Description of cloud condition, observation time and wind speed.

Sites	Regions	Time	Sky	Wind Speed ($m \cdot s^{-1}$)
1	South China Sea	30 September 2012	Some clouds	10.9
2	South China Sea	8 October 2012	Overcast	12.8
3	South China Sea	9 October 2012	Overcast	11.7
4	South China Sea	10 October 2012	Overcast	12.2
5	East China Sea	27 November 2014	Some clouds	15.5
6	East China Sea	27 November 2014	Some clouds	10.4
7	East China Sea	28 November 2014	Some clouds	8.6
8	East China Sea	29 November 2014	Some clouds	12.5
9	East China Sea	29 November 2014	Some clouds	17.6

2.3. Observation of Sea Surface Reflectance in situ

Measurement was conducted on the *Shiyan-3* research vessel during a 25-day voyage in the South China Sea in 2012. The reflectance of seawater, sea surface images and videos were all recorded on this cruise. Unfortunately, only four ideal sites were available for the recording to be conducted. In order to analyze the augmented reflectance ratio of the ocean surface when whitecaps appear, more recording of background seawater and whitecap reflectance was required. This was done at another five sites during a 36-day voyage on the *Xinshijian* research vessel in the East China Sea in 2014. During this experiment, all the reflectances were obtained, but no associated image or video data was recorded due to instrument failure. The information for the nine sites is given in Table 2. Clear sky conditions were avoided to eliminate glitter contamination. For safety reasons, the highest wind speed at 10 m elevation was 17.6 $m \cdot s^{-1}$.

There are large discrepancies in the reflectance spectrums of coastal and ocean seawater [23]. It is clearly more important to study ocean seawater than coastal waters, because it is the primary component of global seawater, occupying about 60% of the global ocean surface [24]. Thus, the purpose of this experiment was to generate a model for estimating the augmented reflectance ratio of ocean seawater. Figure 3 shows the reflectance obtained at nine sites in ocean seawater, in which all the spectral trends are similar in the visible wavelength. At the four sites in the South China Sea research area, the reflectance at 400 nm show little difference: 0.0084 at Site 1, 0.0103 at Site 2, 0.01 at Site 3, and 0.0096 at Site 4. In the East China Sea research region, the reflectance discrepancy was large, with values from 0.0056 to 0.0105, and at Site 7 where the reflectance was largest, with a value of 0.019 at 400 nm. Generally, sites in ocean seawater have similar reflectivity spectrum trends and a large reflectance range, consistent with the present study. Furthermore, the similar sky conditions at all nine sites also met the experimental requirements of this study.

During observations, a camera about 12 m above the sea surface was used to take sea photographs of the sea surface for estimating w using image processing methods. The surface images and the values of $R(\lambda)$, including $R_b(\lambda)$ and $R_x(\lambda)$in the South China Sea study area are shown in Figure 4.

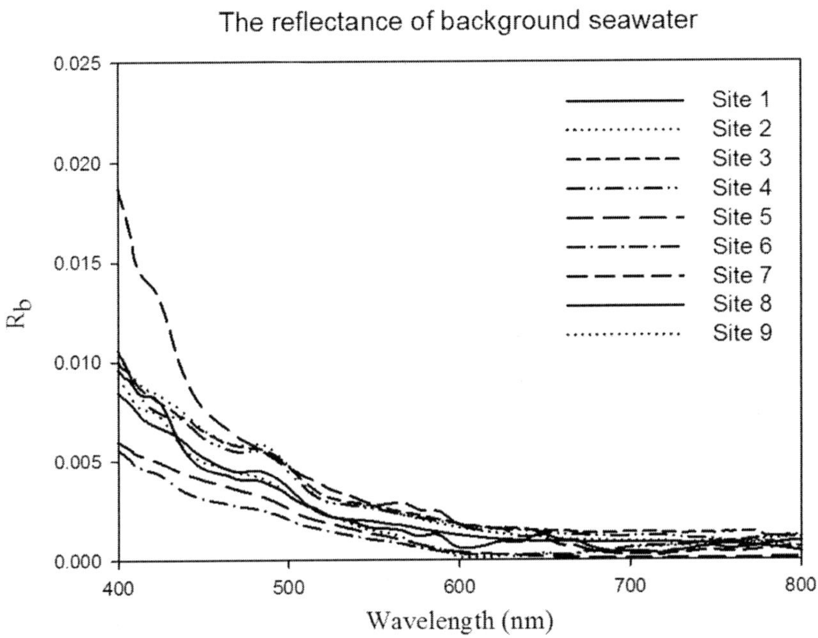

Figure 3. Reflectance of whitecap-free surface at nine sites obtained by measuring the background seawater with the whitecap reflectance measurement instrument.

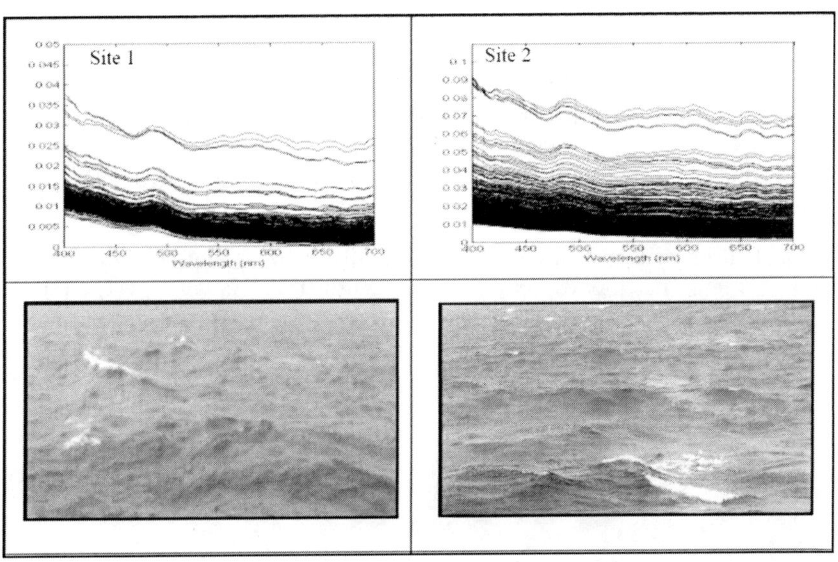

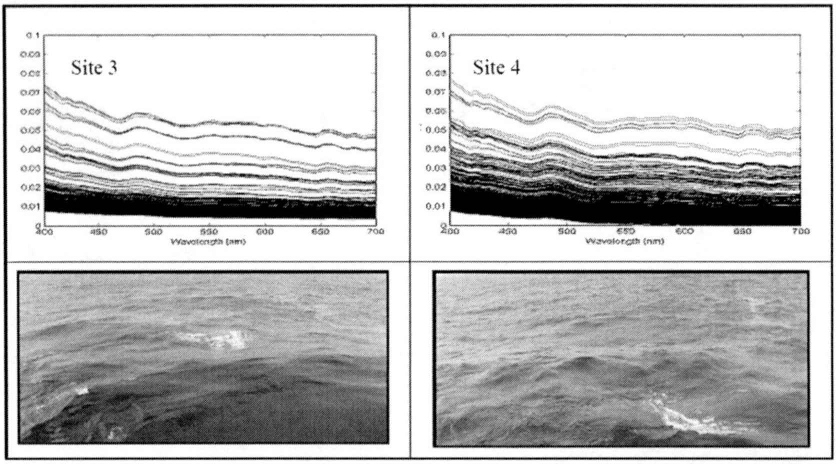

Figure 4. Reflectance spectra at wavelengths from 400 to 700 nm in South China Sea study area. The photographs of the sea surface at each site were used to estimate the fractional coverage of whitecaps using an image processing method.

3. RESULTS

3.1. Calculating the Fractional Coverage of Whitecaps (w(U))

3.1.1. Distinguishing $R_b(\lambda)$ and $R_x(\lambda)$

As shown in the flowchart (Figure 1), $R_b(\lambda)$ and $R_x(\lambda)$ are the required input parameters. $R_b(\lambda)$ has already been obtained, as shown in Figure 3. $R_x(\lambda)$ should be separated from $R(\lambda)$ at each site, and a band ratio method is introduced here.

In most cases, seawater flow velocity and direction ($v(t)\cdot\varphi(t)$, where t is the time and φ is the flow direction) change with time. However, $v(t)\cdot\varphi(t)$ can be assumed constant when the time is not too long. In this research, $v(t)\cdot\varphi(t)$ was fixed because each observation lasted <15 minutes. Then w can be given as

$$w = \frac{v(t)\cdot Ø(t)(t_w(1)+t_w(2)+t_w(3)+\cdots+t_w(n))}{v(t)\cdot Ø(t)\left(\left(t_w(1)+t_w(2)+t_w(3)+\cdots+t_w(n)\right)+\left(t_B(1)+t_B(2)+t_B(3)+\cdots+t_B(n)\right)\right)}$$

$$= \frac{(t_w(1)+t_w(2)+t_w(3)+\cdots+t_w(n))}{\left(\left(t_w(1)+t_w(2)+t_w(3)+\cdots+t_w(n)\right)+\left(t_B(1)+t_B(2)+t_B(3)+\cdots+t_B(n)\right)\right)} \quad (7)$$

$$= \frac{(t_w(1)+t_w(2)+t_w(3)+\cdots+t_w(n))}{t_{total}}$$

where t_W is the appearance time of whitecaps, tB is the appearance time of unbroken surface, and t_{total} is the total time, which is equal to the sum of tW and tB. A schematic diagram for obtaining w was shown in Figure 5. The data sampling rate was 2.9 times per second. Because t(n) is a constant, the recorded data can be used to describe the sampling time. Therefore, Equation 7 can be simplified as

$$w = \frac{numbers(whitecaps)}{numbers(total)} \quad (8)$$

where *numbers(whitecaps)* is the total recorded whitecap times and *numbers(total)* is the total recorded times. To obtain w, then, it is necessary to determine whether the recorded spectrum represents $R_b(\lambda)$ or $R_x(\lambda)$. Here, band ratios were used to separate $R_x(\lambda)$ from $R(\lambda)$:

$$B(\lambda_C, \lambda_D) = \frac{R(\lambda_C)}{R(\lambda_D)} \quad (9)$$

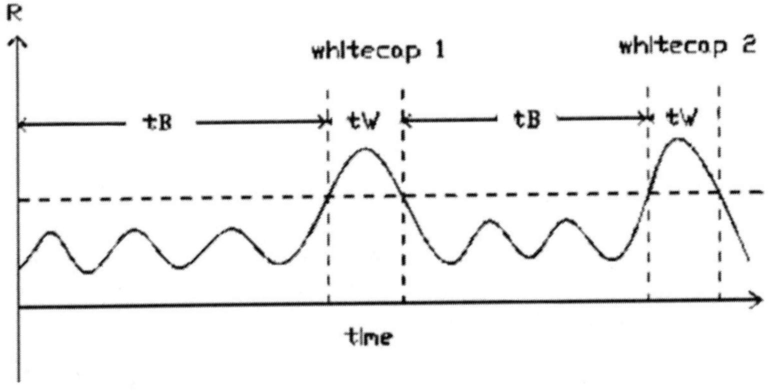

Figure 5. Schematic diagram for estimating the fractional coverage of whitecaps, where t_W is the time during which whitecaps appear and tB is the time of whitecap-free sea surface.

By comparing $B(\lambda_C,\lambda_D)$ values at every site, the difference in $B(\lambda_C,\lambda_D)$ between whitecaps and background seawater may be found. For this purpose, λ_C and λ_D should be determined. If suitable values of λ_C and λ_D are chosen, a larger discrepancy in $B(\lambda_C,\lambda_D)$ between whitecaps and unbroken background seawater can be obtained.

The important wavelengths for ocean color remote sensing are 412, 443, 490, 520, 555, 620, and 665 nm. Considering that this field work might be extended to the study of ocean color remote sensing, the choice of λ_C and λ_D from these bands is good. By setting λ_C and λ_D as two of these wavelengths, $B(\lambda_C,\lambda_D)$ was obtained. At every site, all the $B(\lambda_C,\lambda_D)$values were sorted in descending order. Generally, larger discontinuities appearing in the ordered lines indicate that more suitable values of λ_C and λ_D have been selected. As a result, it was found that the best value of λ_C was 620 nm and that of λ_D was 412 nm. Figure 6 shows the sorted line of $B(\lambda_{620},\lambda_{412})$. Discontinuities occurred at approximately 0.59 at Site 1, 0.71 at Site 2, 0.70 at Site 3, and 0.70 at Site 4. These points can be assumed to be thresholds. Most of the thresholds were close to 0.7 except for Site 1, which had a threshold of approximately 0.59. There was no obvious discrepancy in $R_b(\lambda)$between Site 1 and the other sites, as shown in Figure 3. The spectral trend of $R_b(\lambda)$ was similar at these sites. The reflectance at Site 1 was only slightly smaller than that at the other sites. Because the observation conditions at these sites were not completely the same, the introduced glitter at Site 1 was somewhat different from that at other sites, which caused a different threshold to be determined at Site 1. However, in this study it was considered that a threshold value of 0.7 $B_0(\lambda_{620},\lambda_{412}) = 0.7$) was reasonable for ocean seawater. All $R_x(\lambda)$ could be separated when $B(\lambda_{620},\lambda_{412})$ exceeded the threshold. The instrument incorporated a camera and an optical radiance detector to observe a single seawater surface location. Comparison between the spectrum and image data showed that the band ratio method effectively separated $R_x(\lambda)$ from $R(\lambda)$.

3.1.2. Validation of the Band Ratio Method and Empirical Equation

By separating $R_x(\lambda)$ from $R(\lambda)$, w(U) was obtained using Equation (8), as shown in Table 3. Image processing is widely used to estimate w(U) [25,26,27,28,29]. During this observation, another camera located at about 12 m above the seawater surface was used to take sea photos. RGB information was then extracted using MATLAB software. R_0, G_0, and B_0were defined as thresholds. A pixel in the photo was assumed to be a whitecap pixel when its R, G, and B values were larger than the thresholds (R_0, G_0, and B_0). Then w can be calculated a w = Pixel(*whitecaps*)/*Pixel*(total) where *Pixel*(*whitecaps*) is the total number of whitecap pixels and *Pixel*(*total*) is the total number of pixels in a photo. Table 3 shows the values of w obtained from photograph processing used to validate the values from band ratio and empirical methods.

A comparison of the values of w obtained using the band ratio and image processing methods shows that the smallest absolute percentage difference is 4.7% at Site 2, and the largest value is as high as 56.1% at Site 4. The mean absolute percentage difference at the four sites is 29.9%. Despite attempts to locate the observation position as far from the vessel as possible, it nevertheless was considered that the hull of the ship may have affected the wave distribution over the sea surface to some extent; if so, this would have affected the accuracy of the w value derived by the band ratio method. However, this result shows that the band ratio method gave a reasonable estimation of w, which is suggested could be used in future remote sensing.

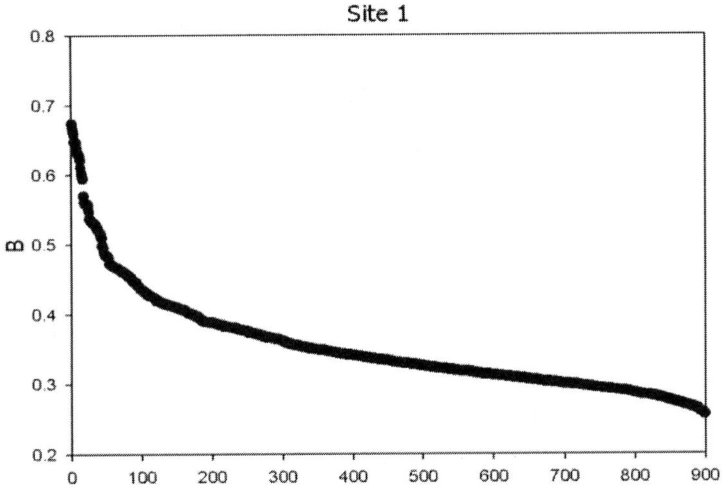

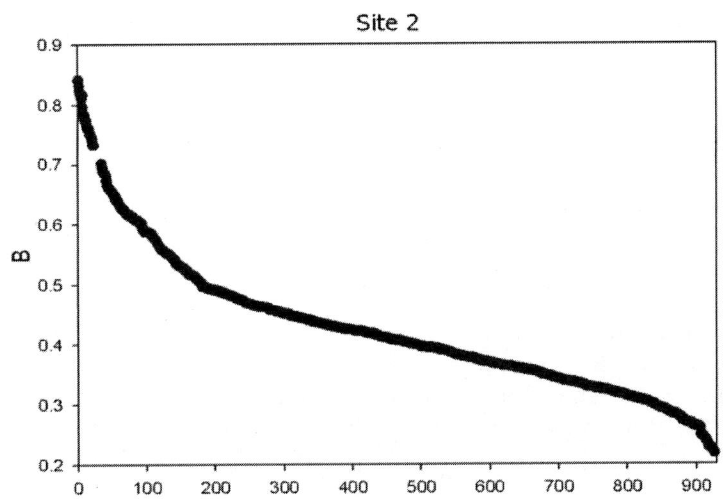

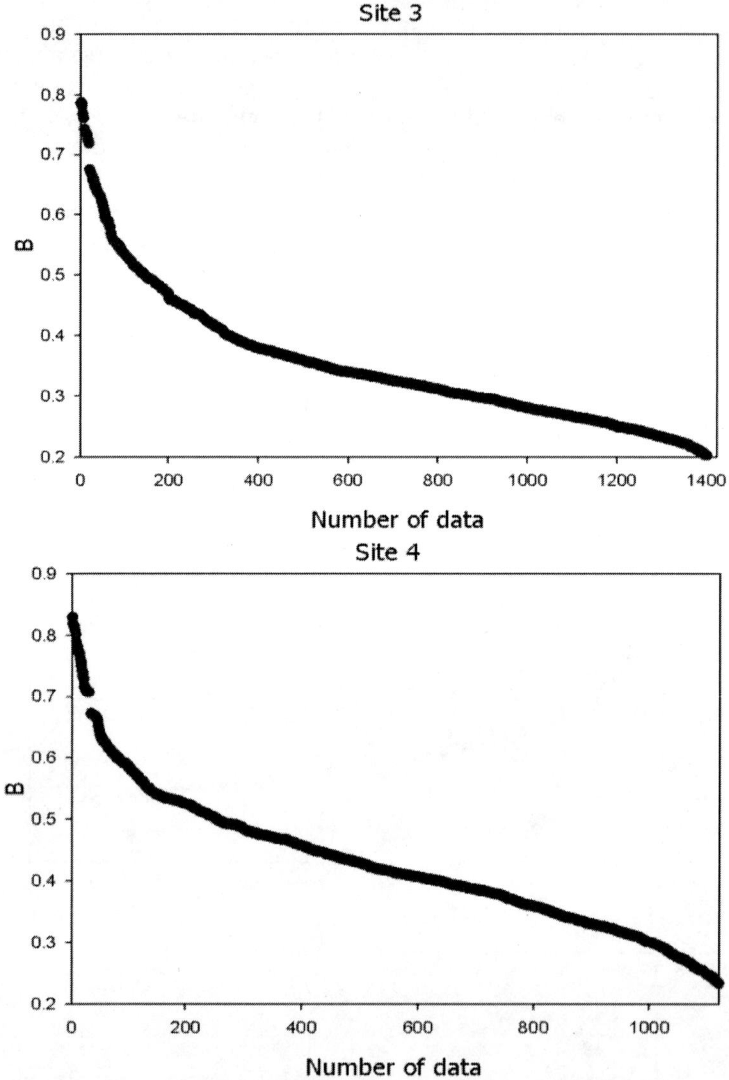

Figure 6. Sorted plots of B(λ_{620},λ_{412}) at Sites 1, 2, 3 and 4. In most cases, discontinuities in the sorted lines occur close to 0.7.

To estimate A(λ, U), as the flow chart shows, a formula is needed to estimate *w*(U). Because the band ratio method has not been validated by remote sensing, an empirical equation should be introduced to estimate w. The most common w parameterizations take the form of the power law relation between w and U_{10} in the form $W = aU_{10}^{b}$, in which the values of a and b may be derived from a best-fit procedure for a set of observations. The values for a and b vary widely with different locations and conditions. In a review of tabulated examples in the whitecap literature, it was found that b ranges from 2.0 to 5.16, with an average of roughly 3 [30,31,32]. The term w has also been reported to be related to the cube of the wind speed, given as $W = a(U_{10}+b)^3$ [28,32,33]. Two possible equations are then obtained either for $U_{10}<11.25 m\cdot s^{-1}$, or for $U_{10}>9.25\ m\cdot s^{-1}$. Because the wind speed is greater than $9.25\ m\cdot s^{-1}$ at most sites $m\cdot s^{-1}$, the empirical equation that is applicable when $U10>9.25\ m\cdot s^{-1}$ was adopted to estimate w in the present work [28]:

$$w=4.82\times10^{-6}(U_{10}+1.98)^3 \tag{10}$$

Table 3. Fractional coverage of whitecaps from band ratio, image processing, and empirical equations.

Site	Wind Speed ($m\cdot s^{-1}$)	$B_0(\lambda_{620}, \lambda_{412})$	W (Band Ratio Method)	W (Image Processing)	W (Empirical Formula)
1	10.9	0.59	0.01967	0.0164	0.0103
2	12.8	0.71	0.02688	0.0282	0.0156
3	11.7	0.7	0.01405	0.0101	0.0123
4	12.2	0.69	0.02482	0.0159	0.0137

In assessing the suitability of Equation (10) for the study area, it was found that the calculated value of w agreed well with the result estimated by the photograph processing method (see Table 3). Almost all of the values of w were consistent with wind speed. The mean absolute percentage difference was 29.4%. The smallest absolute percentage difference was 13.8% at Site 4, and the largest was 44.7% at Site 2. In most cases, the value of w derived from processed images was larger than that estimated by the empirical equation. Since no prior wind-speed data was available for the study areas, a certain response delay was assumed between increase in wind speed and the formation of whitecaps. Furthermore, the photographs did not cover large area of the sea surface; in addition, some w errors were also expected in the image processing method. As a result, Equation (10) was used to estimate w in the proposed model.

3.2. Extraction of $R_w(\lambda)$

As discussed above, whitecap reflectance was extracted from the overall recorded reflectance data by the band ratio method. Finally, the weighted average whitecap reflectance ($R_w(\lambda,U)$) was obtained. Figure 7 shows $R_w(\lambda,U)$ and $R_b(\lambda)$ at sites 1 to 4, in which it is seen that the reflectance decreases with increasing wavelength, decreasing sharply at short wavelengths from 400 to 600 nm, in a similar manner to the reflectance of the background seawater. The decreasing trend in the reflectance of whitecaps differs somewhat from the background seawater at longer wavelengths. To describe the decreasing trend, values of $R_w(\lambda)/R_w(443)$ are listed in Table 4. For example, at Site 1, $R_w(490)/R_w(443)$ is 0.90, $R_w(520)/R_w(443)$ is 0.59, $R_w(620)/R_w(443)$ is 0.38, and $R_w(700)/R_w(443)$ is 0.34. These ratios are similar to those at Sites 2, 3, and 4 at the visible wavelength. Generally, the average R_w at 443 nm was about 11% less than at 490 nm, 40% less than at 520 nm, 44% less than at 555 nm, 62% less than at 620 nm, and 69% less than at 700 nm. A similar trend for the changes in $R_w(\lambda)$ with increasing wavelength also occurred at 400 to 700 nm in the East China Sea study area. As a result, it was concluded that the reflectance decreased slowly with increasing wavelength when whitecaps appeared. These results indicate that the reflectance of whitecaps in the open ocean was dependent on wavelengths between 400 and 700 nm.

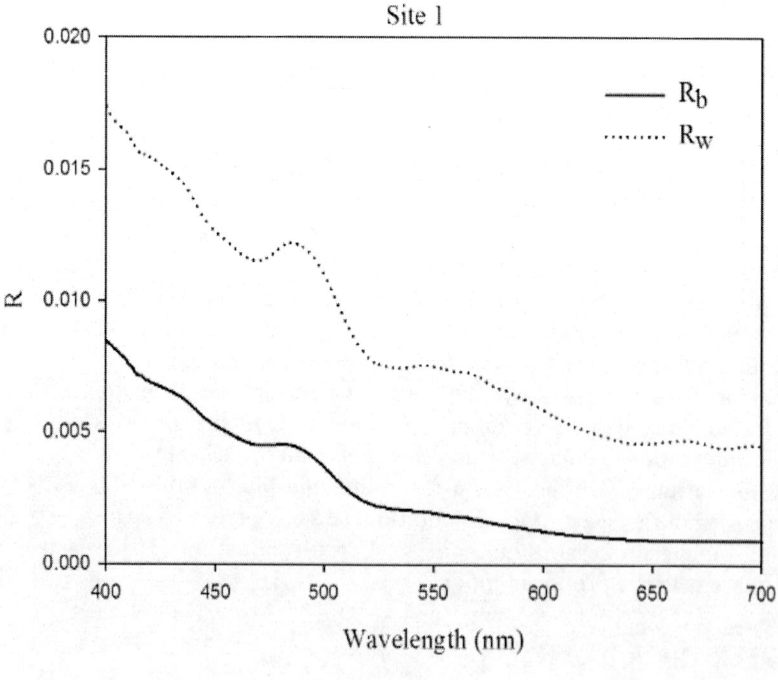

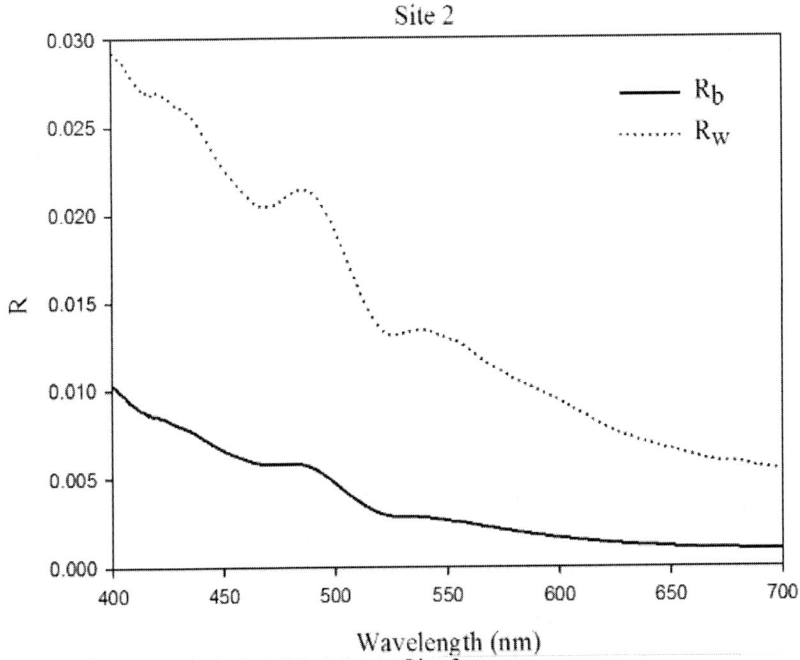
Site 2
R
0.030
0.025
0.020
0.015
0.010
0.005
0.000
400
450
500
550
600
650
700
Wavelength (nm)
R_b
R_w

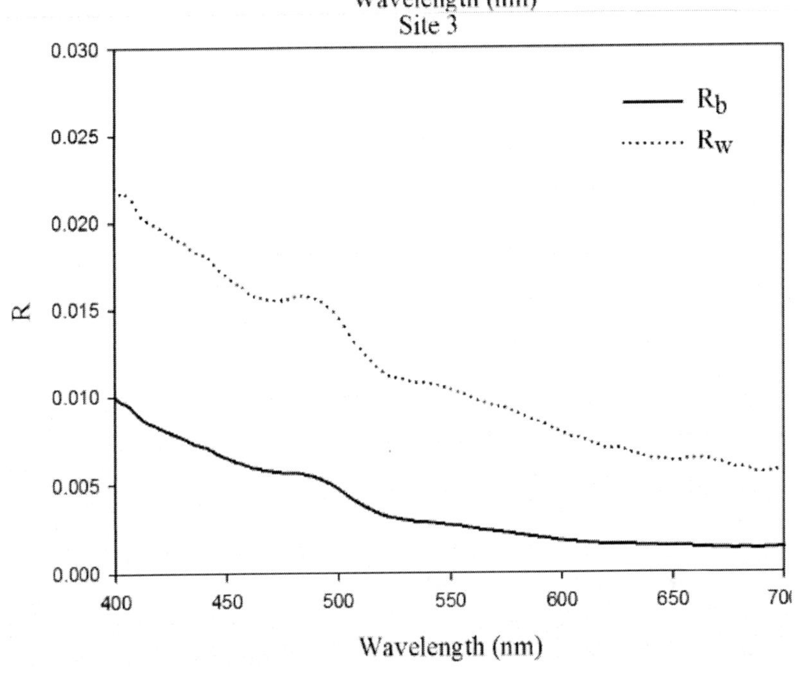
Site 3
R
0.030
0.025
0.020
0.015
0.010
0.005
0.000
400
450
500
550
600
650
Wavelength (nm)
R_b
R_w

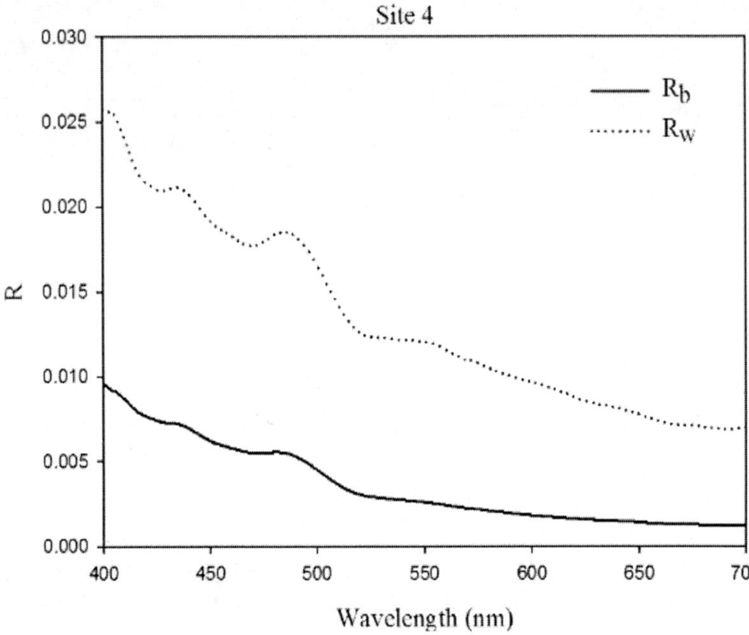

Figure 7. $R_b(\lambda)$ and $R_w(\lambda,U)$ at sites (1–4). (Solid line = $R_b(\lambda)$; broken line = $R_w(\lambda,U)$).

Table 4. Reflectance at 490, 520, 555, 620, and 700 nm divided by that at 443 nm. Generally, Rw at 443 nm is reduced by 88% at 490 nm, 67% at 520 nm, 63% at 555 nm, 51% at 620 nm and 46% at 700 nm.

Site	R_w	490/443	520/443	555/443	620/443	700/443
1	R_w	0.90	0.59	0.55	0.38	0.34
2	R_w	0.88	0.56	0.53	0.33	0.23
3	R_w	0.88	0.64	0.57	0.39	0.32
4	R_w	0.90	0.62	0.59	0.43	0.34

3.3. Variation of ρ(λ,U) at Different Wind Speeds

The $\rho(\lambda,U)$ can be obtained by substituting the weighted average whitecap reflectance $R_w(\lambda,U)$ and the reflectance of background water $R_b(\lambda)$ into Equation (3). Figure 8 is a plot of $\rho(\lambda,U)$, shown to increase with increasing wavelength. The spectra are similar except at Site 8 (Figure 9). It was supposed that this large discrepancy may have been caused by the hull of the ship. During the observation period, the ship tended to yaw considerably. The sky condition at Site 8 was recorded as "some clouds", which suggests that the ship's shadow may have influenced the observed light field. Figure 8a shows that $\rho(\lambda,U)$ increased sharply when the wavelength increased from 400 to 620 nm, then increased slowly and began to decrease at 620–800 nm.

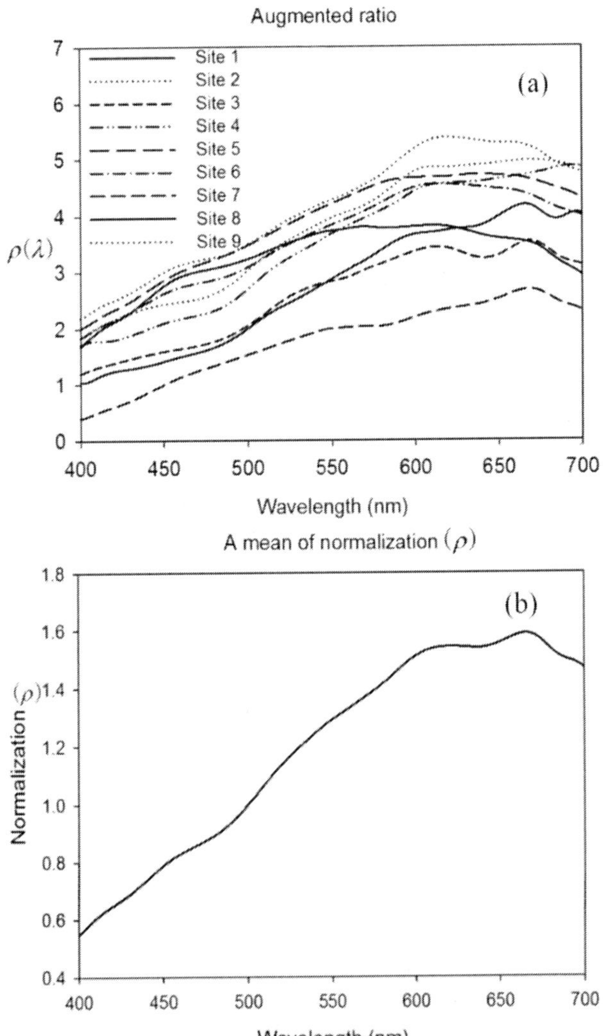

Figure 8. (**a**) $\rho(\lambda,U)$ *vs.* increasing wavelength from 400 to 700 nm. The spectra are similar except Site 8. When wind speed increased, generally ρ also increased. (**b**) Average $\rho(\lambda)/\rho(500)$ for the nine sites.

To predict $A(\lambda,U)$, it is necessary to know the relationship between $\rho(\lambda)$ and U_{10}. In a study of whitecap coverage in extreme hurricanes by Holthuijsen *et al.* [3], it was reported that the coverage approximated to a power law when $U_{10} \leq 24 m\cdot s^{-1}$, and was constant when $U_{10} > 24\ m\cdot s^{-1}$. Since Equation (10) was used to describe w for values of $U_{10} > 9.25\ m\cdot s^{-1}$, wind speeds of 9.25 and 24 $m\cdot s^{-1}$ were set as the bounds of the augmented ratio. In the present study, data from the nine sites was used to study the relationship between measured $\rho(\lambda)$ and U_{10} (these include the study areas in the South China Sea and the East China Sea). Two processes were conducted on $\rho(\lambda)$. First, all the $\rho(\lambda)$ were normalized with $\rho(500)$. Second, by averaging the normalization $\rho(\lambda)$ at the nine sites, a mean $\rho(\lambda)$ was obtained as shown in Figure 8. The median wavelength of 500 nm was chosen and a fit equation was obtained to describe the relationship between $\rho(\lambda)$ and U10:

$$\rho_{simulated}(500) = -283.3U_{10}^{-2.2} + 4 \tag{11}$$

Figure 9c shows the measured and simulated $\rho(500)$. To obtain values of $\rho(\lambda)$ in the visible spectrum, an average $\rho(\lambda)/\rho(500)$ was adopted, giving the simulated relationship

$$\rho_{simulated}(\lambda) = \frac{\rho(\lambda)}{\rho(500)}\rho_{simulated}(500) = \frac{\rho(\lambda)}{\rho(500)}(-283.3U_{10}^{-2.2} + 4) \tag{12}$$

where $\rho(\lambda)/\rho(500)$ is obtained from Figure 8b. For example, λ = 400 nm gives the ratio $\rho(400)/\rho(500) = 0.55$; therefore $\rho_{simulated}(\lambda)$ is $\rho_{simulated}(400) = 0.55(-283.3U_{10}^{-2.2} + 4)$. The results for 400, 500, 600, and 700 nm are shown in Figure 9a, b, c and d, respectively. The values of R^2 and RMSE indicate that Equation (12) gives a reasonable estimate of $\rho(\lambda)$. Generally, $\rho(\lambda)$ increases sharply with increasing wind speed from 9.25 to 14 $m\cdot s^{-1}$ then begins to increase more slowly, even approaching a constant value with increasing wind speed.

4. DISCUSSION

Once $w(U)$ and $\rho(\lambda,U)$ had been obtained as described above, then $A(\lambda, U)$ could be calculated using Equation (5). The result can be stated as

$$A(\lambda, U) = w(U)\rho(\lambda, U) = 4.82 \times 10^{-6}\frac{\rho(\lambda)}{\rho(500)}(U_{10} + 1.98)^3(-283.3U_{10}^{-2.2} + 4) \tag{13}$$

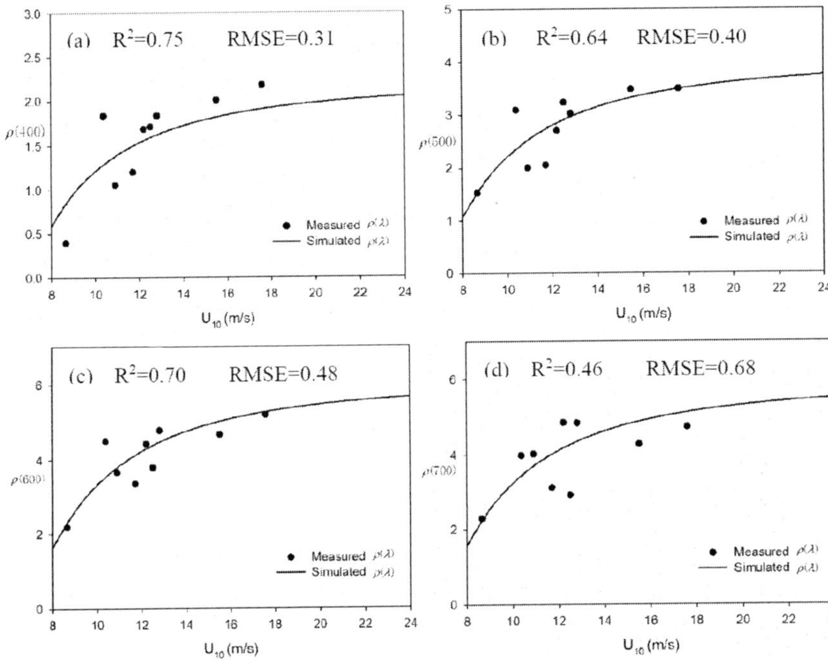

Figure 9. Measured and simulated ρ(λ) at 400, 500, 600, and 700 nm were given in (**a**–**d**), respectively.

At wavelengths from 400 to 700 nm, A(λ, U) is obtained by substituting U and the average value of ρ(λ)/ρ(500) from Figure 8b into Equation (13). The model proposed by Callaghan *et al.* [28] was adopted here to estimate w at $U_{10} > 9.25 m \cdot s^{-1}$. Furthermore, w increases with increasing U_{10} following a power law, and then takes a constant value of 0.04 at $U_{10} > 24$ $m \cdot s^{-1}$ [3]. Equation (13) was thus used within the wind speed range of 9.25 to 24 $m \cdot s^{-1}$.

In 2000, Moore used a radiometric system to study remote sensing augmented reflectance (RSAR) in the presence of breaking waves [19]. The instrument consisted of a six-channel radiometer centered at 410, 440, 510, 550, 670, and 860 nm. Moore *et al.* used the law expressed in the Monahan and Muircheartaigh [34] model (for stable $\Delta T = 2$) to relate RSAR to U using the measurements. Then they established a model, $RSAR(412) = 3.4 \times 10^{-6} U^{2.55}$, with results given in Figure 10. They found that RSAR was essentially independent of wavelength and in the range 0.001–0.002 for wind speeds of 9–12 $m \cdot s^{-1}$. Furthermore, the Koepke whitecap reflectance model converted to augmented reflectance, $RSAR = 4.57 \times 10^{-7} U^{3.52}$, was given in solid line in Figure 11. In this study, RSAR(λ, U) is different from A(λ, U) but they are related as:

$$\mathrm{RSAR}(\lambda, \mathrm{U}) = \mathrm{A}(\lambda, \mathrm{U}) R_b(\lambda) \tag{14}$$

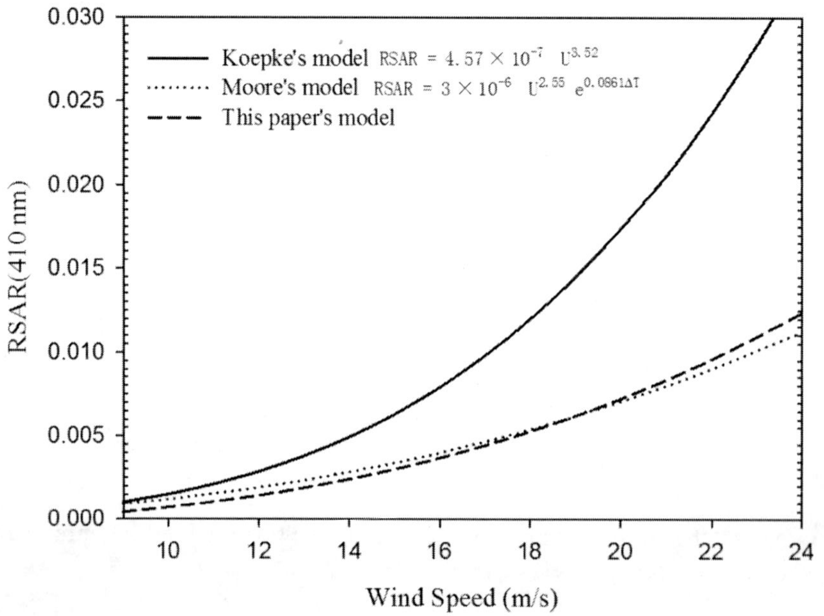

Figure 10. Predicted remote sensing augmented reflectance (RSAR) at 410 nm using three kinds of models. The solid line is the Koepke whitecap reflectance model [25] converted to augmented reflectance, RSAR = $4.57 \times 10^{-7} U^{3.52}$. The dotted line is the Moore model [19], RSAR $=3.4 \times 10^{-6} U^{2.55}$. The broken line is the product of A(λ)obtained using proposed Equation (13) and 0.065 (the value of $R_b(410)$) obtained by Moore *et al.* [19].

In the measurements reported by Moore *et al.* [19], $R_b(\lambda)$ was approximately 0.065 at 410 nm. Therefore, the value of RSAR(410) was calculated from Equations (13) and (14). Figure 10 shows the values of RSAR(410) estimated by the different models. At U_{10} =14 m·s^{-1}, RSAR(410) was 0.0015 obtained from Koepke's model, 0.0012 from Moore *et al.*'s model, and 0.0007 in the present proposed model. As the wind speed increased, the difference in RSAR(410) between Koepke's model and our model increased, whereas the discrepancy between Moore *et al.*'s model and our model decreased with increasing wind speed at U_{10}<19.5 m·s^{-1}. The uncertainty in the values of effective reflectance and the efficiency factor are relatively high in Koepke's model as a result of their use of data obtained under different observed conditions, including differences in the seawater components, sky condition, sea wave and solar glitter. The uncertainty in the value of w adopted in Koepke's paper was also relatively high since it referred to different authors, and because of the wide variation of the data and the subjective judgement in the analysis. The errors in both w and reflectance introduced a large overall uncertainty in Koepke's estimated RSAR(410). The effective reflectance errors did not appear in Moore *et al.*'s experiment since all the data was directly observed *in situ*. The dotted line in Figure 10 shows their result. The discrepancy between Moore *et al.*'s model and the present model increases either side of a wind speed of about 19 $m \cdot s^{-1}$. The RSAR calculated using our model is greater than that calculated from Moore's model, which was developed from *in situ* wind speed measurements in the range 9 to 12 $m \cdot s^{-1}$, whereas our experiments were conducted at wind speeds from 8.6 to 17.6 $m \cdot s^{-1}$. As pointed out above, ρ(λ) increases sharply wind speed increases from 9.25 to 14 $m \cdot s^{-1}$, then the rate of increase begins to slow and approaches a constant value at higher wind speeds, such that the proposed ρ(λ) caused the RSAR(410) discrepancy to decrease before U10<19.5 $m \cdot s^{-1}$, then increase after U_{10}>19.5 $m \cdot s^{-1}$. However, the present proposed model is reasonably consistent with Moore *et al.*'s result. Because of the parameter ρ(λ) introduced in this work, it is suggested that the model proposed here is more suitable than existing models for estimating A(λ) when U_{10}>14 $m \cdot s^{-1}$.

As shown in Figure 11, at the visible wavelengths, the averaged A(λ, U) given by (A(400, U)+A(401, U)+...+A(700, U))/301) is approximately 1.3% when U = 9 $m \cdot s^{-1}$ and w = 0.006; 2.2% when U = 10 $m \cdot s^{-1}$ and w = 0.008; 4.4% when U = 12 $m \cdot s^{-1}$ and w = 0.013; 7.4% when U = 14 $m \cdot s^{-1}$ and w = 0.020; 19% when U = 19 $m \cdot s^{-1}$ and w = 0.045; and 37.9% when U = 24 $m \cdot s^{-1}$ and w = 0.085 $m \cdot s^{-1}$. It is then necessary to consider the augmented reflectance ratio for remote sensing applications. Note that RSAR is not the best parameter for describing this when whitecaps appear on the sea surface. Generally, seawater is optically assumed to comprise pure seawater, suspended particulate, and colored, dissolved organic matter. These components tend to be temporally and spatially variable. This occurred in the present case, as Figure 3 shows. Previous models had assumed that the reflectance of background seawater was constant, as discussed above. Using RSAR to describe the augmented reflectance ratio introduces a large error, since background seawater reflectance varies considerably. Thus, to describe the augmented reflectance ratio for ocean seawater, the parameter $R_b(\lambda)$ was introduced into the model. In the present study, ρ(λ,U) is a significant parameter that establishes a relationship between $R_b(\lambda)$ and $R_w(\lambda)$. It is suggested that the proposed model A(λ, U) = w(U)ρ(λ,U)rigorously takes account of the reflectance error for ocean seawater when whitecaps appear.

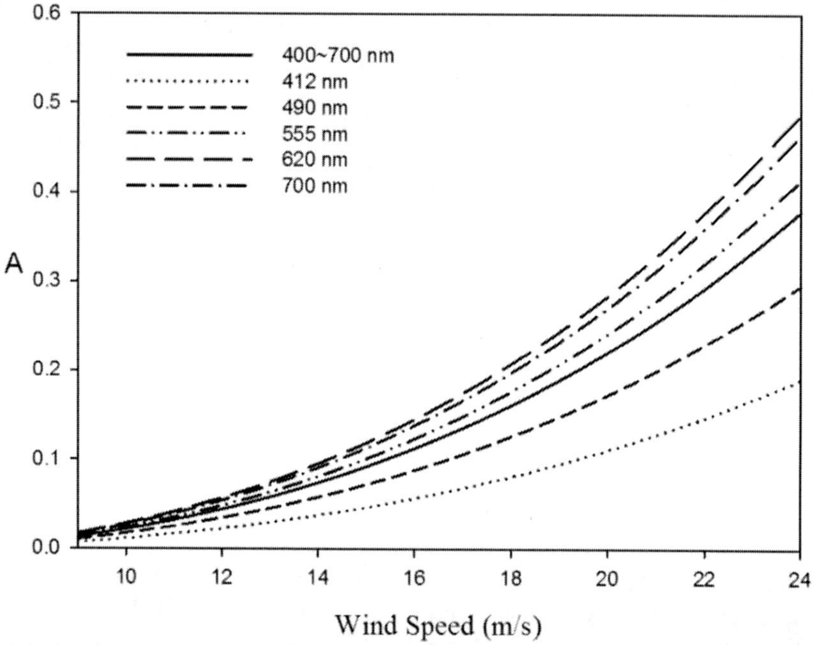

Figure 11. Change in A(λ, U) with increasing wind speed at wavelengths 412, 490, 555, 620 and 700 nm. The average value for the range 400–700 nm is also shown.

5. CONCLUSIONS

The earlier models for estimating the augmented reflectance on the sea surface when whitecaps appear had assumed that the reflectance of background seawater is constant. In fact, it tends to be more temporally and spatially variable than was first thought. As a result, the earlier models were highly uncertain when the background seawater reflectance changed. In the present work, a model is proposed for estimating the augmented reflectance ratio (A(λ, U)) for different fractional coverage of whitecaps (w). The augmented ratio of the reflectance of background water ρ(λ,U) and w are the crucial parameters in the proposed model. Using a specifically designed whitecap measurement system, whitecap reflectance measurements were conducted in the South China Sea and East China Sea, and the derived data was analyzed by a band ratio method to obtain w. The results showed that the Callaghan *et al.* [28] empirical equation was suitable for obtaining win the study region ρ(λ,U) was derived after determining the probability distribution function for whitecap reflectance.

An exponential equation was established to describe the relationship between ρ(λ,U) and U. Finally, A(λ, U) was obtained and given as

$$A(\lambda, U) = 4.82 \times 10^{-6} \frac{\rho(\lambda)}{\rho(500)} (U_{10} + 1.98)^3 (-283.3 U_{10}^{-2.2} + 4)$$

Estimating RSAR using A(λ, U) and $R_b(\lambda)$ showed that RSAR(λ, U) was in reasonably good agreement with earlier studies at wind speeds from 9 to 12 $m \cdot s^{-1}$, which confirmed that the model is credible. Generally, our results showed that A(400~700, U) is approximately 1.3% when U = 9 $m \cdot s^{-1}$ and w = 0.006; 2.2% when U = 10 $m \cdot s^{-1}$ and w = 0.008; 4.4% when U = 12 $m \cdot s^{-1}$ and w = 0.013; 7.4% when U = 14 $m \cdot s^{-1}$ and w = 0.020; 19% when U = 19 $m \cdot s^{-1}$ and w = 0.045; and 37.9% when U = 24 $m \cdot s^{-1}$ and w = 0.085 $m \cdot s^{-1}$. Because the reflectance of background seawater has been taken into account in this model, A(λ, U) = w(U)ρ(λ,U) gives a more scientifically rigorous method of determining the reflectance error when whitecaps appear.

ACKNOWLEDGEMENTS

This study was supported by the Youth Foundation of the State Key Laboratory of Tropical Oceanography (Grant No. SQ201102) and the Natural Science for Youth Foundation (Grant Nos. 41206029, 41406205, and 41206165). We express our appreciation to many colleagues in our group, who supported this field campaign. The data was provided by Zhantang Xu (email: xujiantang@163.com).

AUTHOR CONTRIBUTIONS

Zhantang Xu and Zhaohua Sun proposed the core idea of the observation method and designed the framework of the paper; Zhantang Xu performed the data analysis and manuscript writing; Yuezhong Yang and Wenxi Cao designed and provided the observation instrument; Wen Zhou, Junfang Lin, Guifen Wang, Qian Yang and Zhantang Xu observed the data *in situ*.

REFERENCES

1. Gordon, H.R.; Wang, M. Influence of oceanic whitecaps on atmospheric correction of ocean-color sensors. *Appl. Opt.* **1994**, *33*, 7754–7763.

2. Frouin, R.; Schwindling, M.; Deschamps, P.Y. Spectral reflectance of sea foam in the visible and near-infrared: In situ measurements and remote sensing implications. *J. Geophys. Res. Ocean.* **1996**, *101*, 14361–14371.
3. Holthuijsen, L.H.; Powell, M.D.; Pietrzak, J.D. Wind and waves in extreme hurricanes. *J. Geophys. Res. Ocean.* **2012**, *117*, 45–57.
4. Frouin, R.; Iacobellis, S.; Deschamps, P.Y. Influence of oceanic whitecaps on the global radiation budget. *Geophys. Res. Lett.* **2001**, *28*, 1523–1526.
5. Kokhanovsky, A.A. Spectral reflectance of whitecaps. *J. Geophys. Res. Ocean.* **2004**, *109*, 285–288.
6. Salisbury, D.J.; Anguelova, M.D.; Brooks, I.M. Global distribution and seasonal dependence of satellite-based whitecap fraction. *Geophys. Res. Lett.* **2014**, *41*, 1616–1623.
7. Gordon, H.R.; Wang, M. Retrieval of water-leaving radiance and aerosol optical thickness over the oceans with seawifs: A preliminary algorithm. *Appl. Opt.* **1994**, *33*, 443–452.
8. Moore, K.D.; Voss, K.J.; Gordon, H.R. Spectral reflectance of whitecaps: Instrumentation, calibration, and performance in coastal waters. *J. Atmos. Ocean. Technol.* **1998**, *15*, 496–509.
9. Cunningham, A.; Boyle, J.C.; Wood, P. Radiative transfer modelling of the relationship between seawater composition and remote sensing reflectance in sea lochs and fjords. *Int. J. Remote Sens.* **2002**, *23*, 3713–3724.
10. Payne, R.E. Albedo of the sea surface. *J. Atmos. Sci.* **1972**, *29*, 959–970.
11. Maul, G.A.; Gordon, H.R. On the use of the earth resources technology satellite (Landsat-1) in optical oceanography. *Remote Sens. Environ.* **1976**, *4*, 95–128.
12. Gordon, H.R.; Jacobs, M.M. Albedo of the ocean-atmosphere system: Influence of sea foam. *Appl. Opt.* **1977**, *16*, 2257–2260.
13. Whitlock, C.H.; Bartlett, D.S.; Gurganus, E.A. Sea foam reflectance and influence on optimum wavelength for remote sensing of ocean aerosols. *Geophys. Res. Lett.* **1982**, *9*, 719–722.
14. Allison, D.B.; Stramski, D.; Mitchell, B.G. Empirical ocean color algorithms for estimating particulate organic ca$_{Rb}$on in the southern ocean. *J. Geophys. Res. Ocean.* **2010**, *115*, 1971–1982.
15. Hu, C.M.; Chen, Z.Q.; Clayton, T.D.; Swarzenski, P.; Brock, J.C.; Muller-Karger, F.E. Assessment of estuarine water-quality indicators using MODIS medium-resolution bands: Initial results from Tampa Bay, FL. *Remote Sens. Environ.***2005**, *94*, 425–427.
16. Coble, P.; Hu, C.; Gould, R.W., Jr; Chang, G.; Wood, A.M. *Colored Dissolved Organic Matter in the Coastal Ocean An Optical Tool for Coastal Zone Environmental Assessment & Management*; Naval Research Lab, Stennis Space Center, MS. Oceanography Div.: Arlington, VA, USA, 2004.
17. Zhao, J.; Barnes, B.; Melo, N.; English, D.; Lapointe, B.; Muller-Karger, F.; Schaeffer, B.; Hu, C. Assessment of satellite-derived diffuse attenuation coefficients and euphotic depths in South Florida coastal waters. *Remote Sens. Environ.* **2013**, *131*, 38–50.
18. Wang, M.; Son, S.; Harding, L.W. Retrieval of diffuse attenuation coefficient in the Chesapeake Bay and tu$_{Rb}$id ocean regions for satellite ocean color applications. *J.Geophys. Res. Ocean.* **2009**, *114*, 244–254.
19. Moore, K.D.; Voss, K.J.; Gordon, H.R. Spectral reflectance of whitecaps: Their contribution to water-leaving radiance. *J. Geophys. Res. Ocean.* **2000**, *105*, 6493–6499.
20. Carratelli, E.P.; Dentale, F.; Reale, F. Numerical simulation of whitecaps and foam effects on satellite altimeter response. *Remote Sens.* **2014**, *6*, 3681–3692.
21. Reul, N.; Chapron, B. A model of sea-foam thickness distribution for passive microwave remote sensing applications. *J. Geophys. Res. Ocean.* **2003**, *108*, C10.
22. Mobley, C.D. Estimation of the remote-sensing reflectance from above-surface measurements. *Appl. Opt.* **1999**, *38*, 7442–7455.
23. Mobley, C.D. *Light and Water: Radiative Transfer in Natural Waters*; Academic Press: Manhattan, NY, USA, 1994.
24. Lee, Z.; Hu, C. Global distribution of Case-1 waters: An analysis from seawifs measurements. *Remote Sens. Environ.***2006**, *101*, 270–276.
25. Koepke, P. Effective reflectance of oceanic whitecaps. *Appl. Opt.* **1984**, *23*, 1816–1824.

26. Bortkovskii, R.; Novak, V. Statistical dependencies of sea state characteristics on water temperature and wind-wave age. *J. Mar. Syst.* **1993**, *4*, 161–169.
27. Ross, D.B.; Cardone, V. Observations of oceanic whitecaps and their relation to remote measurements of surface wind speed. *J. Geophys. Res.* **1974**, *79*, 444–452.
28. Callaghan, A.; de Leeuw, G.; Cohen, L.; O'Dowd, C.D. Relationship of oceanic whitecap coverage to wind speed and wind history. *Geophys. Res. Lett.* **2008**, *35*, L23609.
29. Moat, B.I.; Yelland, M.; Pascal, R. Oceanic whitecap coverage measured during UK-SOLAS cruises. In Proceedigs of 16th Conference on Air-Sea Interaction, Phoenix, AZ, USA, 10–15 January 2009.
30. Zhao, D.; Toba, Y. Dependence of whitecap coverage on wind and wind-wave properties. *J. Oceanogr.* **2001**, *57*, 603–616.
31. Goddijn-Murphy, L.; Woolf, D.K.; Callaghan, A.H. Parameterizations and algorithms for oceanic whitecap coverage. *J. Phys. Oceanogr.* **2011**, *41*, 742–756.
32. Anguelova, M.D.; Webster, F. Whitecap coverage from satellite measurements: A first step toward modeling the variability of oceanic whitecaps. *J. Geophys. Res.Ocean.* **2006**, *111*, C3.
33. Sugihara, Y.; Tsumori, H.; Ohga, T.; Yoshioka, H.; Serizawa, S. Variation of whitecap coverage with wave-field conditions. *J. Mar. Syst.* **2007**, *66*, 47–60.
34. Monahan, E.C.; Muircheartaigh, I. Optimal power-law description of oceanic whitecap coverage dependence on wind speed. *J. Phys. Oceanogr.* **1980**, *10*, 2094–2099.

CHAPTER 10

Four-dimensional variational ocean reanalysis: a 30-year high-resolution dataset in the western North Pacific (FORA-WNP30)

Norihisa Usui[1], Tsuyoshi Wakamatsu[2], Yusuke Tanaka[2], Nariaki Hirose[1], Takahiro Toyoda[1], Shiro Nishikawa[2], Yosuke Fujii[1], Yasushi Takatsuki[1], Hiromichi Igarashi[2], Haruka Nishikawa[2], Yoichi Ishikawa[2], Tsurane Kuragano[1], Masafumi Kamachi[2]

1 Oceanography and Geochemistry Research Department, Meteorological Research Institute, 1-1 Nagamine, Tsukuba, Ibaraki 305-0052, Japan

2 Center for Earth Information Science and Technology, Japan Agency for Marine-Earth Science and Technology, 3173-25, Showa-machi, Kanazawa-ku, Yokohama, Kanagawa 236-0001, Japan

ABSTRACT

We produced a four-dimensional variational ocean re-analysis for the Western North Pacific over 30 years (FORA-WNP30). It is the first-ever dataset covering the western North Pacific over 3 decades at eddy-resolving resolution. The four-dimensional variational analysis scheme version of the Meteorological Research Institute Multivariate Ocean Variational Estimation system (MOVE-4DVAR) is employed to conduct a long-term reanalysis experiment during 1982–2012. After evaluating the basic performance of FORA-WNP30, the interannual to decadal variability is analyzed. Overall, FORA-WNP30 reproduces basic features in the western North Pacific well. One of outstanding features in FORA-WNP30 is that anomalous events such as the Kuroshio large meander and anomalous intrusion of the Oyashio in the 1980s, when there were no altimeter data, are successfully reproduced. FORA-WNP30 is therefore a valuable dataset for a variety of oceanographic research topics and potentially for related fields such as climate study, meteorology and fisheries.

Keywords: FORA-WNP30, Ocean reanalysis, Data assimilation, 4DVAR, Western boundary current, Ocean model, Long-term, variability

1. INTRODUCTION

Data assimilation, which combines observations and numerical models to obtain the statistically best state estimate, has long been recognized as a useful tool for wide-ranging purposes in geophysics such as meteorology and oceanography. One of the major targets of data assimilation is "reanalysis," which aims at reconstructing a long-term historical dataset for the atmosphere, ocean and so on. In meteorology, various atmospheric reanalysis datasets have been produced (e.g., Kalnay et al. 1996; Uppala et al. 2005; Onogi et al. 2007) and have been extensively utilized in a variety of research fields of not only meteorology but also climate study and oceanography.

Ocean reanalyses have also become available along with advancements of ocean observing systems such as remotely sensed and in situ observing platforms in recent decades. One of the most important attempts in ocean reanalysis study is the production of global ocean reanalysis that covers a period of multiple decades for climate research, and such datasets are available from operational centers and some research groups (e.g., Scott et al. 2007; Carton and Giese 2008; Köhl and Stammer 2008; Saha et al. 2010; Storto et al. 2011; Balmaseda et al. 2013; Toyoda et al. 2013; Fujii et al. 2015; Osafune et al. 2015). Due partly to the size of the problem and technical limitations, relatively low-resolution ocean models are usually employed in these global reanalyses. Furthermore, a multi-model ensemble approach has been proposed recently in order to construct a more accurate dataset (Toyoda et al. 2015b; Han et al. 2015). State-of-the-art global reanalysis datasets are summarized in Balmaseda et al. (2015).

In addition to the enhancement of ocean observing systems, recent advances in computing capability enable us to perform long-term basin-scale reanalyses at eddy-resolving resolutions [e.g., (Miyazawa et al. 2009;

Moore et al. 2011; Oke et al. 2013)], which give a realistic representation of mesoscale variability as well as variations in large-scale circulation. The reanalysis period for the high-resolution datasets is however limited to the satellite altimeter era (since 1993) because realistic reproduction of actual mesoscale features such as meanders in a jet and mesoscale eddies strongly depends on satellite altimeter data (Oke and Schiller 2007). Therefore, to produce long-term high-resolution reanalysis data over a period longer than the altimeter era is a challenging issue, and at the present time there is no such dataset for the western North Pacific, which is the main target region in this study.

The western North Pacific is characterized by the existence of strong western boundary currents, the Kuroshio and Oyashio. Their variations accompany energetic variabilities in the associated mesoscale features such as eddies and meanders. Another prominent feature in this region is the existence of low-frequency variability on interannual to decadal time scales (e.g., Qiu 2003; Ohshima et al. 2006; Taguchi et al. 2007; Na et al. 2012; Usui et al. 2013; Kuroda et al. 2015), which has a strong connection with changes in atmospheric forcing. Accumulated evidence shows that the low-frequency variability is accompanied by modulation of the mesoscale variability. This region is also characterized by its marginal seas such as the East China Sea, the Japan Sea and the Okhotsk Sea, which connect with adjacent seas through narrow straits on the order of tens of kilometers in width. Therefore, eddy-resolving resolution with a typical horizontal grid spacing of about 1/10° is indispensable for realistic representation of ocean processes in the western North Pacific.

Strong sea surface temperature (SST) fronts associated with the western boundary currents exhibit an extremely large heat exchange between the ocean and atmosphere in the western North Pacific (Kida et al. 2015). This area, hence, is believed to be a key area for understanding mid-latitude air-sea interaction (e.g., Hirose et al. 2009; Taguchi et al. 2009; Kelly et al. 2010; D'Asaro et al. 2011; Nakamura et al. 2012; Sasaki et al. 2012). Moreover, the area is well known to contain some of the most productive fishing grounds supported by their high primary production, and variations in the western boundary currents at both the large- and meso-scales have a large impact on the area's marine ecosystems and fisheries (e.g., Miller et al. 2004; Nishikawa et al. 2011). Therefore, a long-term high-resolution reanalysis dataset for the western North Pacific would be a great benefit for various research fields such as oceanography, meteorology, climate study and fisheries. In this study, we produce a long-term ocean reanalysis dataset for the western North Pacific using an advanced ocean data assimilation system.

The Meteorological Research Institute (MRI) of the Japan Meteorological Agency (JMA) developed a data assimilation system with a multivariate three-dimensional variational (3DVAR) analysis scheme, namely the MRI Multivariate Ocean Variational Estimation system, MOVE-3DVAR (Usui et al. 2006a). The western North Pacific version of MOVE-3DVAR with an eddy-resolving ocean model was implemented at JMA in 2008 and has been used for monitoring and forecasting the ocean state around Japan. The four-dimensional variational (4DVAR) analysis scheme version of the MOVE system (MOVE-4DVAR) was recently developed, and it was demonstrated that MOVE-4DVAR provides better representation of mesoscale features compared to MOVE-3DVAR (Usui et al. 2015).

Running a 4DVAR assimilation system for producing a long-term ocean reanalysis with the eddy-resolving configuration is still challenging even on the current high-performance computer systems because of its high computational cost. Owing to the recent update of the Earth Simulator in Japan Agency for Marine-Earth Science and Technology (JAMSTEC) made in 2015 Spring, we now have enough computing resources, which enables us to perform a high-resolution, long-term reanalysis experiment with the 4DVAR system. Using this updated Earth Simulator, we conducted the reanalysis experiment and produced the new regional ocean reanalysis dataset, FORA-WNP30: Four-dimensional variational Ocean Re-Analysis for the Western North Pacific over 30 years.

FORA-WNP30 is produced for use in various research fields not only in oceanography but also in meteorology, climate study and fisheries. The purpose of this article is, therefore, to provide detailed descriptions and the basic performance of the FORA-WNP30 dataset. Configurations of the data assimilation system and reanalysis settings are described in Sects. 2 and 3, respectively. In Sect. 4, the FORA data are compared with independent observations, and the basic performance is discussed. The interannual to decadal variability represented by FORA-WNP30 is shown in Sect. 5. Section 6 gives a summary and future works.

2. DATA ASSIMILATION SYSTEM

The data assimilation system used in FORA-WNP30 is MOVE-4DVAR (Usui et al. 2015), which was developed as an extension of the 3DVAR version of the MOVE system (MOVE-3DVAR) for the western North Pacific. MOVE-4DVAR consists of an eddy-resolving ocean general circulation model (OGCM) and the 4DVAR assimilation scheme. The following subsections give descriptions of the OGCM and the assimilation scheme used in FORA-WNP30.

2.1 Ocean model

The OGCM used in MOVE-4DVAR is the western North Pacific version of the MRI Community Ocean Model version 2.4 (MRI.COM-WNP; Tsujino et al. 2006). MRI.COM is a depth-coordinate model that solves the primitive equations under the hydrostatic and the Boussinesq approximations. The vertical coordinate near the surface follows the surface topography like σ coordinates. The model configuration is basically the same as that used in Tsujino et al. (2006) except for vertical grid spacing near the bottom and for the sea ice model incorporated. A biharmonic operator is applied to the horizontal mixing of tracers. For the horizontal momentum mixing of a biharmonic Smagorinsky viscosity (Griffies and Hallberg 2000) is used. In order to determine the vertical viscosity and diffusivity we adopt the turbulent closure scheme of Noh and Kim (1999). The model is driven by wind stress and heat fluxes. Latent heat and sensible heat fluxes are calculated by the bulk formula of Kondo (1975) using the model sea surface temperature (SST). More detailed descriptions of the framework of MRI.COM and the setting of MRI.COM-WNP are given by Tsujino et al. (2006, 2010), respectively.

The model domain of MRI.COM-WNP covers the western North Pacific, spanning from 117°E to 160°W zonally and from 15°N to 65°N meridionally. The horizontal resolution is variable: it is 1/10° from 117°E to 160°E and 1/6° from 160°E to 160°W zonally, and it is 1/10° from 15°N to 50°N and 1/6° from 50°N to 65°N meridionally. There are 54 vertical levels with thickness increasing from 1 m at the surface to 600 m at 6300 m depth. Ocean states at the open boundaries are obtained from a North Pacific model with a horizontal resolution of 1/2∘×1/2∘using a one-way nesting method. Daily outputs from the North Pacific model are linearly interpolated in both time and space to update the open boundaries of MRI.COM-WNP at every time step.

A multicategory sea ice model with a single layer sea ice and snow cover is coupled to MRI.COM-WNP. In this model, five categories of sea ice are considered according to the thickness of the sea ice (0.0–0.6, 0.6–1.4, 1.4–2.4, 2.4–3.6 and 3.6–30.0 m). The sea ice model includes a thermodynamic and a dynamic component. The thermodynamical part is based on Mellor and Kantha (1989), and the dynamical part is based on the Los Alamos sea ice model (Hunke and Ducowicz 1997; Hunke and Ducowicz 2002). The sea ice model was recently introduced into MOVE-4DVAR although it was not included in the original version (Usui et al. 2015). It should however be noted that propagation of adjoint sensitivity through the sea ice model is ignored in the adjoint model.

The tangent linear and adjoint codes of MRI.COM-WNP were generated manually. We made some approximations to the codes. The vertical mixing coefficients in the turbulent closure scheme are calculated from the background state, and the dependency of the coefficients on the deviation from the background state is ignored. We also apply the same approximation to the bottom friction. In addition, some modifications are applied in order to avoid singularities. We have confirmed that the tangent linear and adjoint codes satisfy $(\mathbf{La})^T(\mathbf{La})=\mathbf{a}^T\mathbf{L}^T(\mathbf{La})$ for any vector a, where L denotes the tangent linear operator for MRI.COM-WNP. The left-hand side, the square of **La**, is obtained by integrating **a** using the tangent linear model. On the other hand, the right-hand side, the inner product of **a** and $\mathbf{L}^T\mathbf{La}$, is calculated by integrating **La** using the adjoint model $\mathbf{L}^T$ The tangent linear and adjoint codes of MRI.COM-WNP have already been used for various analyses. Fujii et al. (2008) conducted a singular vector analysis to estimate the optimal perturbations for a Kuroshio large meander. Furthermore, Fujii et al. (2013) identified the pathway of the North Pacific Intermediate Water from its origin in the subarctic North Pacific to the subtropical region.

2.2 Assimilation scheme

The assimilation scheme used for FORA-WNP30 includes two parts. One is ocean data assimilation, and the other is sea ice assimilation. The ocean data assimilation scheme is based on MOVE-4DVAR (Usui et al. 2015), but some modifications have been made compared to the original version. The largest difference from the original scheme in Usui et al. (2015) is the use of MOVE-3DVAR analysis fields as the first guess in the 4DVAR calculation, which is effective to reduce the number of iterations in the 4DVAR calculation (Usui and Fujii 2015). A schematic diagram of the FORA assimilation cycle is shown in Fig. 1.

Both MOVE-3DVAR and MOVE-4DVAR estimate temperature and salinity fields above 1500 m using a multivariate variational method with vertically coupled temperature-salinity (T-S) empirical orthogonal dunction (EOF) modal decomposition for the background error covariance matrix (Fujii and Kamachi 2003). The model domain of MRI.COM-WNP is divided into 13 subregions, and monthly T-S EOF modes are calculated in each subregion. In order to prevent discontinuities in the analysis fields, the subregions overlap in the boundary areas.

MOVE-4DVAR attempts to optimize the ocean state over the assimilation window $[t_I,t_F]$by controlling temperature and salinity increments to the initial condition at t_I. The cost function is defined by

$$J(\mathbf{z})=\frac{1}{2}\sum_{l}\mathbf{z}_l^T(\mathbf{B}_H)_l^{-1}\mathbf{z}_l+\sum_{t=t_I}^{t_F}\left[\frac{1}{2}\left(\mathbf{H}\mathbf{x}_t-\mathbf{y}_t^{\mathrm{TS}}\right)^T\mathbf{R}_t^{-1}\left(\mathbf{H}\mathbf{x}_t-\mathbf{y}_t^{\mathrm{TS}}\right)+\frac{1}{2\sigma_h^2}\left(\mathcal{H}(\mathbf{x}_t)-\mathbf{y}_t^{\mathrm{SLA}}-\mathbf{H}_h\bar{\mathbf{h}}\right)^T\right.$$
$$\left.\times\left(\mathcal{H}(\mathbf{x}_t)-\mathbf{y}_t^{\mathrm{SLA}}-\mathbf{H}_h\bar{\mathbf{h}}\right)\right]+J^c \quad (1)$$

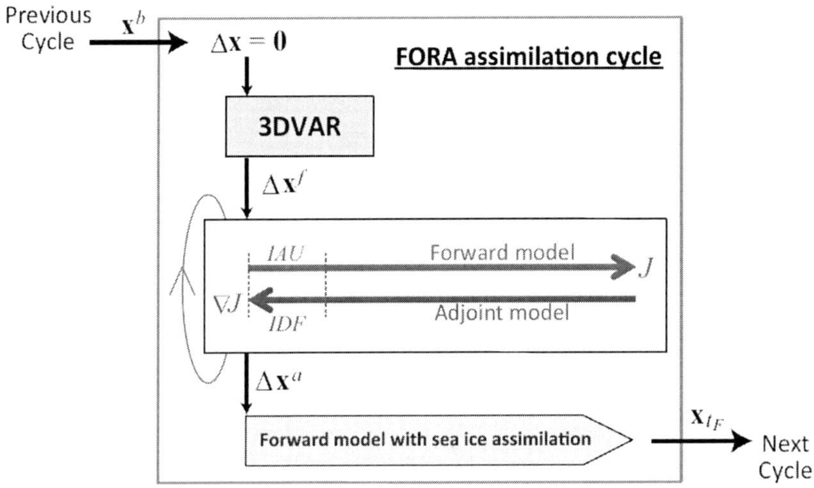

Figure. 1 Schematic diagram of the FORA assimilation cycle. The 3DVAR calculation is first carried out to estimate temperature and salinity increments Δx^f using the final state in the previous assimilation window as the background state x^b. Then, 4DVAR is performed to estimate the increment Δx^a using Δx^f as the first guess of the increment. In the 4DVAR calculation, the forward and adjoint models are integrated iteratively. The forward model with the initialization scheme of the incremental analysis update (IAU) is integrated to calculate the cost function J, and the adjoint model is integrated backward in time to calculate the gradient of the cost function ∇J. The IAU scheme works as the incremental digital filtering (IDF) during the adjoin model integration. Finally, the forward model is again integrated with the increment Δx^a and sea ice assimilation. The final state of the forward model x_{tF} is transferred to the background state in the next assimilation window

where $\mathbf{z}_l$ is the control variable composed of amplitudes for the vertical coupled T-S EOF modes, and $\mathbf{x}$ is the state vector of temperature and salinity fields, which is a function of the control variable $\mathbf{z}$. Matrices $(\mathbf{B}_H)_l$ and $\mathbf{R}_t$ are the horizontal correlation matrix of background errors and the observation error covariance matrix. The horizontal correlation matrix is modeled by using a to Kuragano and Kamachi (2000). Vectors y^{TS} and y^{SLA} are T-S profile observations and altimeter-derived sea level anomalies (SLAs), matrices $\mathbf{H}$ and $\mathbf{H}_h$ denote spatial interpolation from the model grid Gaussian-like function with area-dependent decorrelation scales, which are determined according to observation locations for T-S and SLA data, respectively, σ_h is the observation error for the altimeter-derived SLA data, and $\bar{h}$ is the mean sea surface dynamic height (SDH). The last term on the right-hand side, J^c, represents additional constraints, which are used for various purposes such as for avoiding density inversion (Fujii et al. 2005) and excessively cold temperature less than the freezing point (Usui et al. 2011). Note that in this study the constraint to prevent low temperature is set to work only in the region off the east of Japan. The subscripts l and t denote the l-th subregion and the time index. If we replace $\mathbf{x}_t$ with $\mathbf{x}_{tl}$, (1) becomes the cost function for MOVE-3DVAR.

The operator H calculates SDH at each observation point from the T-S field of the state vector $\mathbf{x}$ as follows:

$$\mathcal{H}(\mathbf{x}) = \mathbf{H}_h\mathbf{h}, \tag{2}$$

$$h_i = -\frac{1}{\rho_s}\int_0^{z_m} \rho_i'(T, S, p)dz \tag{3}$$

where $\mathbf{h}$ is the vector consisting of SDH at each horizontal grid point, h_i is the ith component of $\mathbf{h}$, i.e., SDH at the ith horizontal grid point, z and z_m denote the vertical coordinate and the reference depth for the SDH calculation, T and S are temperature and salinity, p is the pressure, ρ_s is the surface density, and ρ' is the density deviation from the reference state (T=0∘C and S=35 psu). Note that the z-axis is positive upward, and z=0 is the sea surface. This analysis scheme enables us to estimate variations in salinity accompanying temperature changes by using the T-S EOF modes, even if few salinity data are available.

The analysis increment for the temperature and salinity fields with respect to the background state at the initial time is calculated by

$$\Delta\mathbf{x} = \mathbf{x}(\mathbf{z}) - \mathbf{x}^b = \mathbf{S}\sum_l \mathbf{W}_l\mathbf{U}_l\mathbf{\Lambda}_l\mathbf{z}_l \tag{4}$$

where $\mathbf{x}^b$ represents the background state, $\mathbf{S}$ is a diagonal matrix for standard deviation of background errors, $\mathbf{U}_l$ is a matrix composed of dominant T-S EOF modes, $\mathbf{\Lambda}_l$ is a diagonal matrix for the singular values of T-S EOF modes, and $\mathbf{W}_l$ is a diagonal weight matrix for the lth subregion whose mth diagonal element, w_{lm}, needs to satisfy $\sum_l w^2_{lm}=1$ (Fukumori 2002).

In order to filter out high-frequency noises MOVE-4DVAR employs an initialization scheme of Incremental Analysis Updates (IAU; Bloom et al. 1996) in contrast to the conventional 4DVAR scheme, in which the analysis increment is added to the first-guess field at once at the initial time. The time evolution of the state vector during [tl, t] is thus expressed by

$$\mathbf{x}_t = \mathcal{M}_{t_I,t}\left(\mathbf{x}_{t_I}\right) + \Delta\mathbf{x}_t^{\mathrm{cor}} \tag{5}$$

where

$$\Delta x_t^{\mathrm{cor}} \equiv g_t\left(\mathbf{x}(\mathbf{z}) - \mathbf{x}_{t_I}\right) = g_t\left(\Delta\mathbf{x} + \mathbf{x}^b - \mathbf{x}_{t_I}\right) \tag{6}$$

Operator $M_{tI,t}$ denotes a nonlinear model, $\mathbf{x}_{tI}$ is the model state at the initial time step corresponding to the final state of the previous assimilation window, and $\mathbf{\Delta x}_t^{\mathbf{cor}}$ represents the correction term of IAU, which works during the initialization period from t_I to t_I+N. It should be noted that an Incremental Digital Filtering (IDF; Polavarapu et al. 2000) works on adjoint variables during the backward integration of the adjoint model, because IDF is the adjoint of IAU (Usui et al. 2015).

To reduce model bias the background state $\mathbf{x}^b$ is determined by

$$\mathbf{x}^b = r_c\,\mathbf{x}_{\mathrm{clim}} + (1 - r_c)\,\mathbf{x}_{t_I}, \tag{7}$$

where $\mathbf{x}_{\mathrm{clim}}$ is the climatology and r_c is the weight for the climatology. The weight is set to be 0.02, which is equivalent to a restoring time scale of about 16 months. Note that if r_c is zero x^b becomes $\mathbf{x}_{tI}$.

The ocean data assimilation for FORA-WNP30 is carried out according to the following steps (see Fig. 1): (1) MOVE-3DVAR is conducted to estimate the increment $\mathbf{\Delta x}^f$; (2) MOVE-4DVAR is performed by using $\mathbf{\Delta x}^f$ as the first guess of the increment. In both MOVE-3DVAR and MOVE-4DVAR, the same observation data within the assimilation window are used. In order to minimize the cost function, the preconditioned optimizing utility for large-dimensional analyses (POpULar; Fujii 2005) is adopted in the MOVE system. POpULar is based on a nonlinear preconditioned quasi-Newton method and can minimize a nonlinear cost function without inversion of a nondiagonal background error covariance matrix.

After the step (2), the forward model is again integrated using the increment estimated by MOVE-4DVAR. During this model integration, ice concentration data are assimilated. The sea ice assimilation scheme is based on a simple least squares analysis and nudging. The least squares analysis follows a similar method of Lindsay and Zhang (2006). The analysis value of the ice concentration at the *i*th horizontal grid C_i^a is written by

$$C_i^a = C_i^f + K_i\left(C_i^o - C_i^f\right), \tag{8}$$

where C_i^f is the first guess obtained from the model results, C_i^o is the observation interpolated into the model grid, and $\boldsymbol{K_i}$ is the weighting factor, which determines the strength of the analysis increment. The main target of the sea ice assimilation in FORA-WNP30 is changes in sea ice distribution in the Okhotsk Sea north of Hokkaido Japan, where a large seasonal change in ice area is apparent. In order to assign a large value to the weighting factor around the sea ice edge, we assume that the weight depends on mean ice thickness for each grid (ice volume divided by grid area). To be specific, the weight is set to be 1.0 (0.0) when the mean thickness is less than 0.3 m (greater than 2.0 m) and decreases as the thickness becomes larger when the thickness is between the two values.

The model ice concentration is then nudged to the analysis result with a restoring time scale of 1 day. During the nudging procedure, ice thickness is also corrected so that the volume of sea ice is conserved except for cases of adding (production) or removing (melting) ice (Usui et al. 2010). When changing the ice volume because of the nudging correction, water fluxes corresponding to the changes in ice volume are transferred to the ocean model after conversion to salt fluxes.

3. REANALYSIS SETTINGS

This section describes experimental settings for FORA-WNP30, which are summarized in Table 1. The FORA-WNP30 dataset covers the period 1 January 1982–31 Decmber 2012. The assimilation window and the IAU period in MOVE-4DVAR are set to 10 and 3 days, respectively. We assimilate in-situ temperature and salinity profiles above 1500 m, gridded SST, altimeter-derived sea surface height (SSH) and sea ice concentration. The temperature and salinity profiles are collected from the World Ocean Database 2013 (WOD13; Boyer et al. 2013) and the Global Temperature Salinity Profile Program (GTSPP; Hamilton 1994). The gridded SST data are Merged satellite and in situ Global Daily Sea Surface Temperature (MGDSST, Kurihara et al. 2006) and have a horizontal resolution of 1/4∘×1/4The altimeter-derived SSH observations are AVISO along-track multimission products for TOPEX/Poseidon, Jason-1/2, ERS-1/2, Envisat, GFO and Cryosat-2 (AVISO 2015). The sea ice concentration observations used for the ice assimilation are gridded data with a horizontal resolution of 1/4∘×1/4 derived from the Defense Meteorological Satellite Program (DMSP) special sensor microwave imager (SSM/I).

Table 1 Overview of the FORA-WNP30 dataset

Ocean ice model	MRI.COM version 2.4	
Model domain	Western North Pacific (117°°E–160°°W, 15°°–65°°N)	
Horizontal resolution	1/10° × 1/10°° × 1/10°	
Vertical levels	54 levels	
Data assimilation	MOVE-4DVAR with sea ice assimilation	
Method (ocean)	4DVAR	
Method (sea ice)	Least squares analysis and nudging	
Assimilation window	10 days (ocean), daily nudging (sea ice)	
Observations		
T-S profiles	WOD2013, GTSPP	(1982–2012)
SST	MGDSST	(1982–2012)
SSH	Along-track multimission altimeter products	(1993–2012)
Sea ice	SSM/I ice concentration	(1982–2012)
Forcing	JRA-55	
Period	1982–2012	

Figure 2 shows the time series of the number of assimilated observations in each assimilation window. The SSH data have been available since 1993, and the number is closely linked to the number of satellites (Fig. 2b). The temperature and salinity data (sum of satellite-based MGDSST and in-situ profile observations) are almost constant before 2003. Despite the slight increase of the number of observation points (blue line in Fig. 2c), the number of temperature and salinity data becomes about double in 2008 compared to the level of 2003, indicating the large contribution of Argo floats. The total number of observations shows the gap between 1992 and 1993 due to the introduction of a satellite altimeter and continuously increases after the 2000s (Fig. 2a).

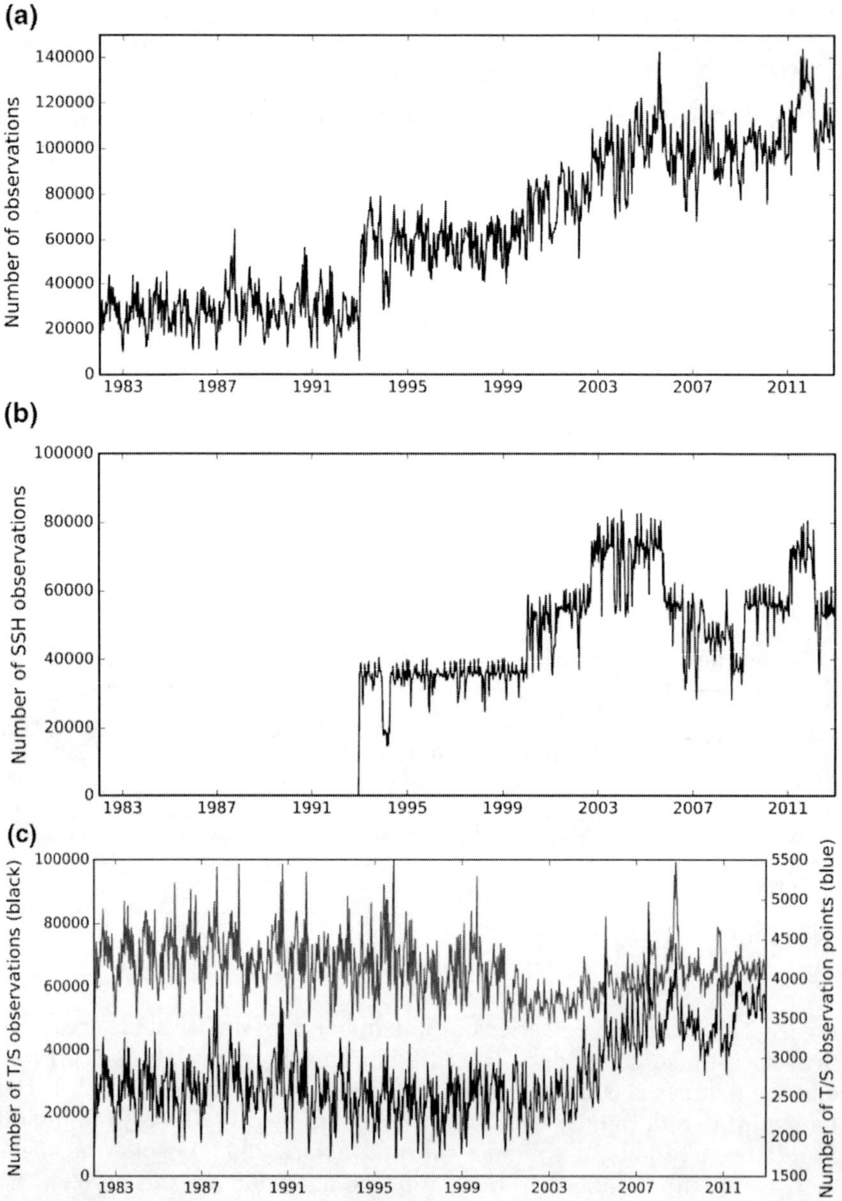

Figure. 2 The time series of the number of assimilated observations in each window. **A** Total, **B** SSH, **C** sum of SST and profile data (*black line*) and number of observation points (*blue line*)

It should be noted that in-situ temperature and salinity profiles are assimilated after vertical interpolation from observed depth levels to model vertical levels (37 levels in the top 1500 m). The number of vertical levels for each profile is taken into consideration in the number of observations shown by the solid black line in Fig. 2. For example, it is counted as 37(1) for a temperature profile from the surface to 1500 m (SST, i.e., available data are only at the top layer). In contrast, the number of observation points shown by the solid blue line in Fig. 2c does not count the number of vertical levels and hence corresponds to the number of observation stations. To check the horizontal distribution of observation points, in Fig. 3 we plot the density distribution of observation points in each 5o×5ogrid, which are calculated every 5 years. The figure indicates that there are relatively dense in-situ observations around Japan in the 1980s although for many of them the observation depths are limited to the surface layers. On the other hand, in-situ observation points in the open ocean, which are sparse in the 1980s and 1990s, have become much denser since the 2000s because of the deployment of Argo floats.

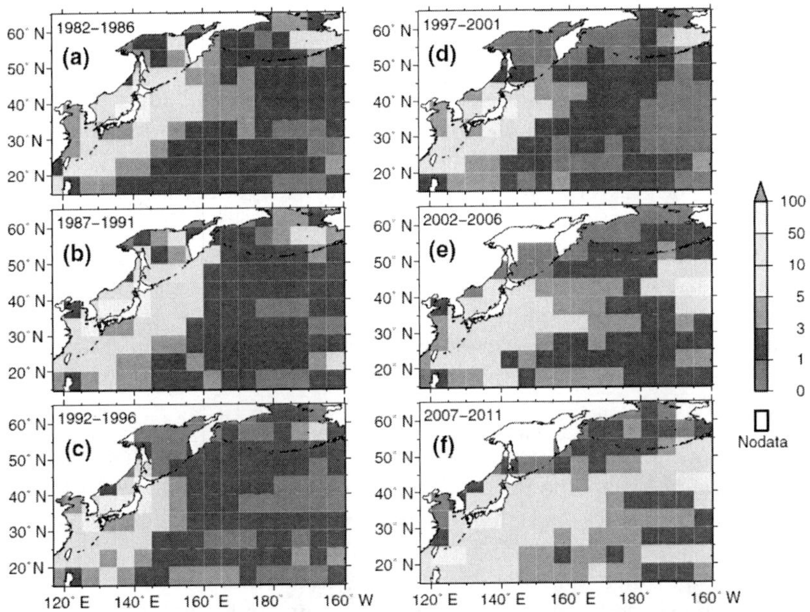

Figure. 3 Geographical distribution of the average station number of in-situ observations per 10 days in each 5° grid. Means of every 5 years are presented in each panel

The monthly T-S EOF modes and their related statistics such as standard deviations of background errors and singular values of the T-S EOF modes are calculated from historical T-S profiles. We use 12 dominant modes for calculating the background term of the cost function in both MOVE-3DVAR and MOVE-4DVAR, which explain more than 90_% of the total variance. The mean SDH h in (1) is calculated by the following procedure. First, we produce monthly mean temperature and salinity fields during 1993–2012, corresponding to the reference period for the AVISO SLA observations. The monthly fields are obtained from a 3DVAR analysis using in-situ T-S profiles and monthly mean climatology from World Ocean Atlas 2013 (WOA13; Locarnini et al. 2013; Zweng et al. 2013) as a first guess. Second, monthly SDH fields are calculated from the monthly T-S fields using (3). Then, we obtain the mean SDH by averaging the monthly SDH fields over the period 1993–2012.

The MOVE-4DVAR assimilation scheme described in the previous section is based on the so-called strong constraint 4DVAR, in which it is assumed that model errors originate from the initial condition of its 10-day assimilation window, and the model physics and forcing field do not give rise to any errors. Our former studies have shown that MRI.COM-WNP, the forward model in MOVE-4DVAR, has a good performance in reproducing mesoscale features such as Kuroshio meanders and mesoscale eddies as well as large-scale circulation patterns (e.g., Tsujino et al. 2006; Usui et al. 2006). In order to reduce the model error, we also performed 3DVAR analysis over the North Pacific model from which the boundary values are provided to the western North Pacific model. The model resolution (~~10 km), however, is not sufficient to represent the coastal phenomena, and it does not include tidal forcing. We therefore apply a special treatment to in-situ observations in coastal regions with bathymetry less than about 1000 m. To be specific, the observation time of the profile data within an assimilation window in the coastal region is set to the initial time of the assimilation window, indicating that those observations are assimilated through 3DVAR.

The JRA-55 atmospheric reanalysis product (Kobayashi et al. 2015) is used for the forcing ocean model. Surface atmospheric parameters originally defined on a reduced Gaussian grid (TL319) are interpolated to the regular longitude grid of about 0.56° resolution. Further, 6 hourly data are averaged to produce daily data. In order to avoid unphysical atmospheric forcing in the coastal zone due to the discrepancy in land mask definitions between JRA55 and FORA-WNP, JRA55 ocean grid data are saturated to the land grid by successive applications of the Laplacian operator. Since in general atmospheric reanalysis products have relatively large errors in precipitation, the freshwater flux is corrected by restoring sea surface salinity toward the monthly mean climatology with a time scale of 1 day to reduce a model bias.

The initial condition on 1 January 1982 was obtained from a preliminary experiment, in which the ocean model is spun up from an initial state of no motion with climatological temperature and salinity for about 20 years using climatological forcing, and subsequently data assimilation of in-situ profile data is conducted with MOVE-3DVAR using the forcing of JRA-55 during the period 1971–1981.

One of the biggest technical challenges in deriving a long-term reanalysis product using an advanced data assimilation system is its high computational cost. In the FORA system, the original data assimilation package, MOVE, is tuned to give the best performance on the specific high performance computational platform, NEC SX-ACE. About 12 % reduction of computational time relative to the original MOVE system is marked. Further, the FORA analysis system adapts the hybrid approach of 3DVAR and 4DVAR in optimizing its control parameters for computational efficiency as mentioned before. The upper limit in the number of iterations in the descent algorithm is set to 10 in 3DAR and 30 in 4DVAR. The upper limit of 3DVAR iterations is reduced to 9 when excessive reduction of the cost function due to a local minimum is observed. On average, the first 3DVAR analyses mark about 20 % reduction of the cost function, and the following 4DAR analyses mark about 40 % reduction, implying that 4DVAR analysis fields have higher accuracy than 3DVAR.

4. GENERAL FEATURES OF THE REANALYSIS DATA

4.1 Comparison with independent observations

4.1.1 Current field

To examine the performance of the FORA data, the current field of FORA is compared to the observations by the JMA research vessels. The current data are not assimilated in the system. We use the current observations during 2001–2012 obtained by the acoustic Doppler current profiler (ADCP) installed in two JMA research vessels, R/V Ryofu Maru and R/V Keifu Maru. To correct misalignment of the transducer and/or gyro settings and reduce other uncertainties, the ADCP data were processed by the CODAS software system, which was developed at the University of Hawaii. The ensemble averaging interval is 300 s. To exclude tidal current in the observations, we estimate barotropic tidal currents using the OSU tidal data inversion global model of "TXPO7.2" (Egbert and Erofeeva 2002) and subtract it from the ADCP data.

In addition to the FORA data, we evaluate another reanalysis product, which was recently produced at JMA using the MOVE-3DVAR system. As mentioned before, MOVE-3DVAR employs the same ocean model (MRI.COM-WNP) as that used in FORA, and hence the 3DVAR product has the same horizontal resolution. The atmospheric forcing used is JRA-55, which is the same as in FORA except for 3-hourly temporal resolution. Assimilated observations in MOVE-3DVAR are also basically the same as those used in FORA except for in-situ temperature and salinity profiles. In the 3DVAR product, in addition to in-situ T-S observations obtained from WOD13 and GTSPP, T-S observations uniquely collected at JMA are assimilated. Thus, differences in assimilated fields between FORA and MOVE-3DVAR mainly originate from the difference in the analysis schemes used.

All ADCP data at 100 m depth during the cruise (ship speed exceeds 10 knot) are compared to the same position of FORA and MOVE-3DVAR, which are linearly interpolated from four nearest data grid. Table 2 shows the current field statistics in the six areas shown in Fig. 4, i.e., the averaged current velocity and its standard deviation, correlation coefficient of current velocity, slope of the regression line, standard deviation of the reanalysis data from regression line, and standard deviation of the current direction difference between the reanalysis data and ADCP data. Both FORA and MOVE-3DVAR show high correlations with the ADCP current velocity in South of Honshu, the East China Sea and Kuroshio extension region and relatively low in the Japan Sea and Oyashio region. In each region, the correlation coefficient of the current velocity of FORA is higher than that of MOVE-3DVAR, and those differences are statistically significant at the 99 % confidence level. As for standard deviations of difference from the regression line and the difference of the current direction, those of FORA are smaller than those of the 3DVAR product in most regions. No statistic of FORA is significantly worse than that of MOVE-3DVAR.

In April 2011, a warm eddy exists off Sanriku at around 39°N and 143°E, and R/V Keifu Maru had cruised around this area to deploy and recover an ocean bottom seismograph. Figure 5shows the comparison of reanalysis data with the ADCP data of R/V Keifu Maru during 11–20 April 2011. In Fig. 5, locations of assimilated observation data during April 11–20 are also plotted. The data density seems to resolve the mesoscale eddies in the area well. The FORA data reproduced the current field around the warm eddy well, especially in the southwest part. The 3DVAR product also reproduced the warm eddy field; however, the detailed shape of the eddy (for example, around 38.5°N and 143.3°E) is not well matched with the ADCP data. Figure 6shows the scatter diagram of the current velocity and the histogram of difference of the current direction between the reanalysis data and ADCP data. The results show the current velocity and direction of the FORA data are reproduced much better than those of MOVE-3DVAR.

Table 2 Current field statistics of reanalysis data (FORA and 3DVAR system) at 100 m depth in each area in Fig. 4 compared to the observed data in terms of the averaged current velocity and its standard deviation, correlation coefficients, slope of regression line, standard deviation of the reanalysis data from regression line and standard deviation of the current direction difference

Area (# of data)	Current velocity				Diff. of direction
	Average[2] SD (m s^{-1})	Correlation	Regression	SD (m s^{-1})	SD (°)
South of Honshu ($N = 41{,}122$)					
FORA	0.479 ± 0.397	0.855	0.816	0.206	46.4
3DVAR	0.438 ± 0.375	0.802	0.722	0.224	51.0
ADCP	0.550 ± 0.416				
East China Sea ($N = 2632$)					
FORA	0.256 ± 0.301	0.849	0.853	0.159	68.8
3DVAR	0.238 ± 0.298	0.837	0.833	0.163	73.1
ADCP	0.369 ± 0.300				
Kuroshio extension ($N = 24{,}141$)					
FORA	0.496 ± 0.372	0.730	0.687	0.254	54.9
3DVAR	0.462 ± 0.337	0.623	0.531	0.264	59.4
ADCP	0.546 ± 0.396				
Mixed water region ($N = 15{,}297$)					
FORA	0.265 ± 0.218	0.580	0.549	0.178	64.7
3DVAR	0.246 ± 0.197	0.453	0.387	0.176	73.0
ADCP	0.314 ± 0.230				
Oyashio region ($N = 6535$)					
FORA	0.240 ± 0.152	0.430	0.301	0.137	67.0
3DVAR	0.243 ± 0.162	0.315	0.234	0.153	76.1
ADCP	0.335 ± 0.217				
Japan Sea ($N = 3380$)					
FORA	0.115 ± 0.080	0.328	0.220	0.076	103.5
3DVAR	0.128 ± 0.095	0.111	0.089	0.095	100.6
ADCP	0.172 ± 0.119				

Observed current data are obtained by the shipboard acoustic Doppler current profiler of the JMA research vessels during 2001–2012. The data are processed by CODAS software developed at the University of Hawaii (averaging time: 300 s) and subtracted tidal current estimated by the OSU tidal data inversion (Egbert and Erofeeva 2002). Asterisk (*) indicates vthe difference between the FORA and 3DVAR system is statistically significant at the 99 % confidence level

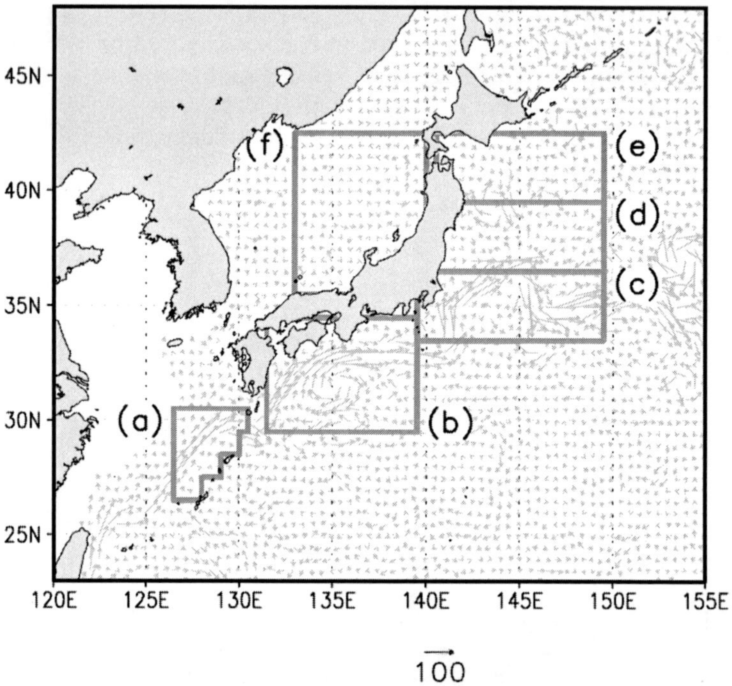

Figure. 4 Map of areas to calculate current field statistics shown in Table 2. A snapshot of the FORA daily current field (unit in cm s^{-1}) at 100 m depth is also shown. Area names are as follows: **A** East China Sea, **B** South of Honshu, **C** Kuroshio extension region, **D** mixed water region, **E** Oyashio region and **F** Japan Sea

4.1.2 Sea level

We next evaluate sea-level variability in FORA using tide gauge observations, which are also not used in data assimilation. Daily mean sea-level data during the period of 2001–2010 at 63 tide gauge stations (see Fig. 7 for their locations) are used for the evaluation. Tidal variations consisting of diurnal and semi-diurnal signals are eliminated by a tide-killer filter with a cutoff period of 48 h designed by Hanawa and Mitsudera (1985). In addition, inverted barometer (IB) correction represented by

$$(\mathrm{IB}) = (P - Pref)/\rho g,$$

is applied to the daily mean sea level, where ρρ is the water density, g is the acceleration of gravity, P is sea level pressure (SLP), and P_{ref} is the mean SLP over the entire ocean. We use daily-mean SLP observed at a meteorological weather station close to each tide gauge station. The mean SLP over the entire ocean is calculated using SLP derived from JRA-55.

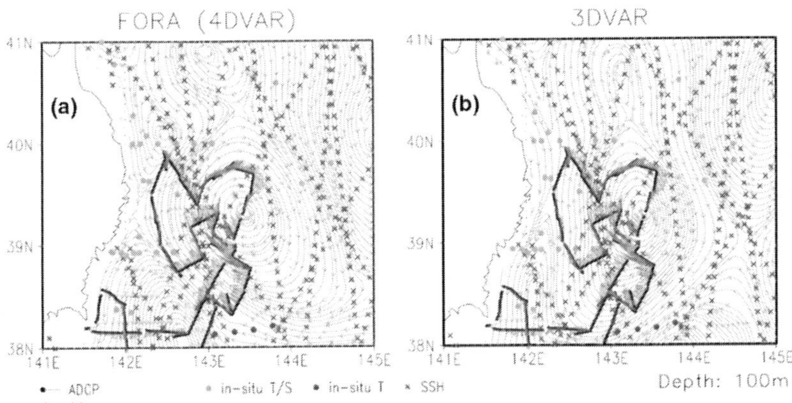

Figure. 5 Comparison of reanalysis data with ADCP data for the **A** FORA and **B** 3DVAR system. Streamlines (*light blue*) are the current field at 100 m depth of reanalysis for the daily mean of 16 April 2011. Sticks (*red*) are the current vector of ADCP data for 11–20 April 2011 observed by R/V Keifu Maru. *Black dot* indicates the origin of a vector and is plotted at the ship position of each datum. Assimilated observation during 11–20 April 2011 is also shown as an *orange circle* (in-situ temperature and salinity), *blue circle* (in-situ temperature) and *purple cross* (SSH)

Figure 7a shows the root-mean square difference (RMSD) between the FORA SSH field and tide gauge observations during the period of 2001–2010. The matchup between the tide gauge observations and FORA data is taken after linear interpolation of the FORA SSH field into the location of each tide gauge station. In order to minimize the ground deformation effect in the observed sea level, RMSD is calculated with sea-level anomalies from the annual mean for each year. The calculated RMSD is relatively large along the south coast of Japan, where natural sea-level variability is relatively high because of variations in the Kuroshio Current flowing along the coast (e.g., Kawabe 1980). In other regions, it is largely less than 4.5 cm except for some specific tide gauge stations. The largest RMSD occurs at Hachijojima Island (RMSD: 7.5 cm) located close to the Kuroshio Current axis, but it is sufficiently small compared to sea level variance there (35.9 cm), which is mainly caused by Kuroshio path variations.

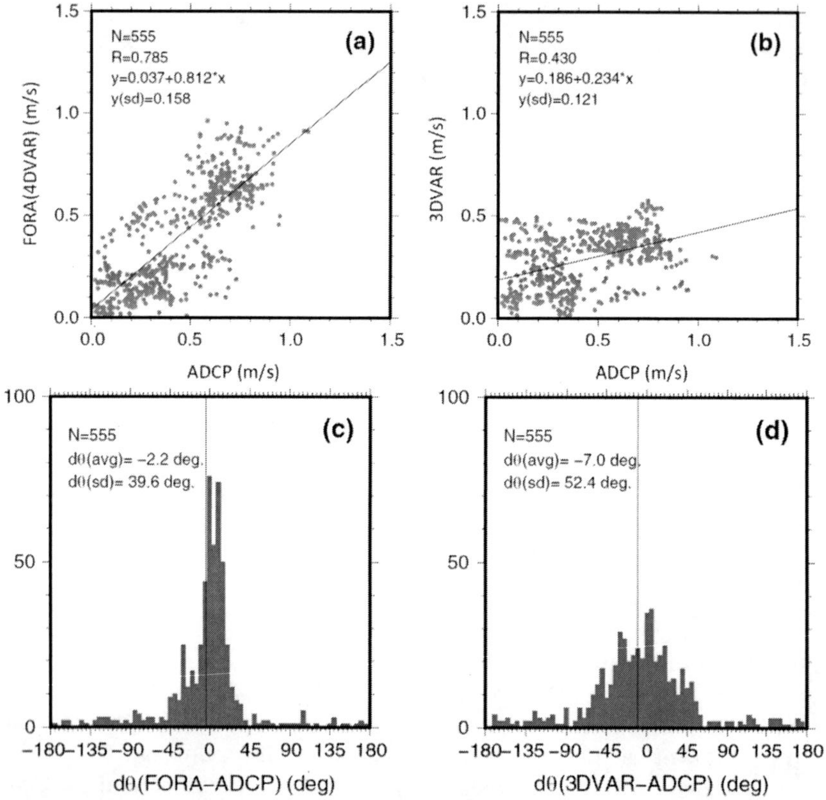

Figure. 6 Comparison of the current velocity of reanalysis data and ADCP data during 11–20 April 2011 in the area shown in Fig. 5 for **A** the FORA and **B** 3DVAR system. A linear regression is plotted as a *black line* in each panel. Histograms of the direction differences between reanalysis and ADCP data in 5° bins for **C** the FORA and **D** 3DVAR system. The average difference is shown as a *black line* in each panel. Statistics are also shown in the each panel

We then calculate the RMSD for MOVE-3DVAR during 2001–2010 based on the same manner as that for FORA and compare two RMSDs. To make it easy to evaluate how much the 4DVAR scheme used in FORA improves the sea level variability compared to the 3DVAR product, the improve ratio (IR) defined by

$$(\mathrm{IR}) = \frac{(\mathrm{RMSD})_{3\mathrm{DVAR}} - (\mathrm{RMSD})_{\mathrm{FORA}}}{(\mathrm{RMSD})_{3\mathrm{DVAR}}} \times 100,$$

is plotted in Fig. 7b. The calculated IR is positive at most tide gauge stations, indicating that the FORA result improves sea-level variability in most areas compared to the 3DVAR results. At tide gauge stations located in the open ocean such as Hachijojima and Chichijima, the IR is relatively high, and some of them exceed 10%. There are however several stations showing negative IR although their values are relatively small. The IR at tide gauge stations surrounded by complex coastal topography (e.g., the inner bay) or near the tip of a peninsula tends to be small. This is probably because representation of the coastal topography of the 10 km model used in both FORA and 3DVAR is insufficient.

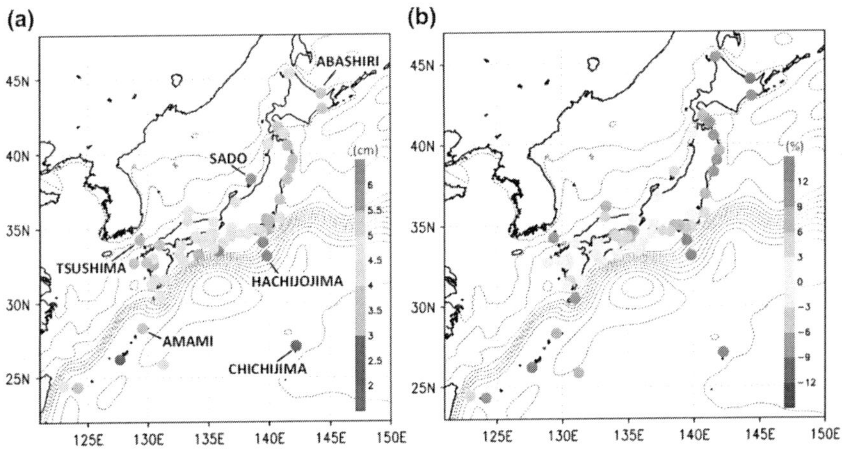

Figure. 7 A RMSD between the FORA SSH field and tide gauge records at 63 stations (unit in cm). **B** Improve the ration (%) of sea-level variability in FORA defined as the relative difference in SSH RMSD for FORA and MOVE-3DVAR (see text for detailed definition). *Black dotted lines* indicate the climatological annual mean SSH field with an interval of 10 cm

To further look into the accuracy of the FORA result, 1-year time series of sea level in 2010 (2009 only for Tsushima) at six tide gauge stations are shown in Fig. 8. For comparison, we also plot the sea level derived from the AVISO daily gridded multi-satellite product with a horizontal resolution of 0.25° (AVISO 2015). Note that the AVISO data are not shown for the tide gauge station of Abashiri because grid point values adjacent to the station are all missing values. RMSDs of FORA, MOVE-3DVAR and AVISO data for the corresponding year are depicted in the figure.

The calculated RMSDs for FORA are smaller than not only those for MOVE-3DVAR, but also those for AVISO at all tide gauge stations except for

Tsushima where FORA has the largest RMSD among the three datasets. Although RMSDs in Fig. 8 are calculated for the specific year (2010 or 2009), we confirmed that the results are robust by calculating the RMSD for 2001–2010. Comparison of the time series indicates that the FORA result reproduces features of sea level variability observed by the tide gauges well. At Hachijojima Island, where the variance of the sea level is the largest among the six stations, short-term variability on the time scale of about a month is enhanced. This variability, mainly due to short-term fluctuations of the Kuroshio path propagating to the downstream, is well represented in FORA. In contrast, temporal variability in AVISO is too small, and consequently it fails to represent the short-term sea level variations (for example, September–October 2010 in Fig. 8b).

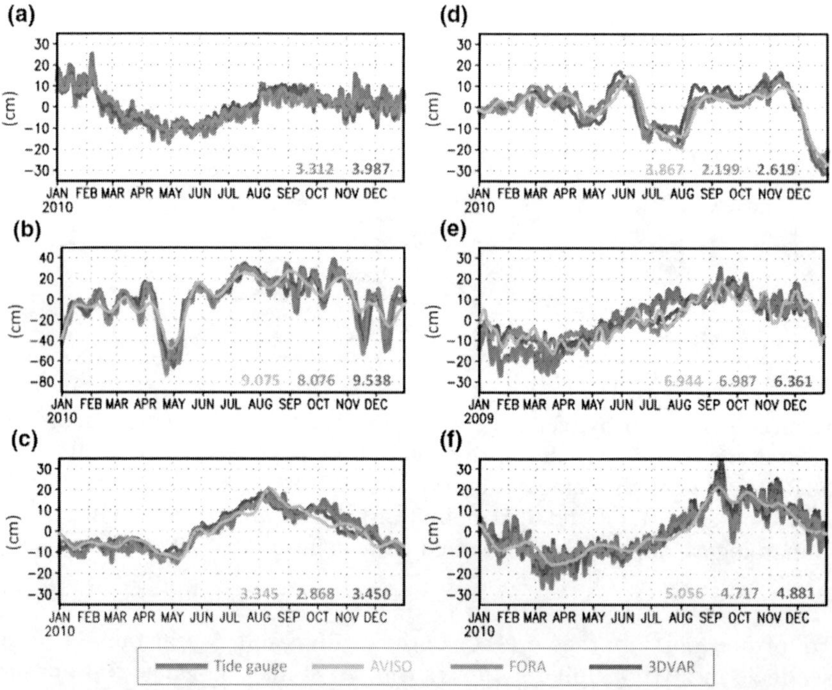

Figure. 8 Time series of SLAs at six tide gauge stations: **A** Abashiri, **B** Hachijojima, **C** Amami, **D** Chichijima, **E** Tsushima and **F** Sado (see Fig. 7a for their locations). The SLAs here denote anomalies from annual mean sea level for each year. The time series except for Tsushima are from 1 January through 31 December 2010. That at Tsushima is for 2009 because available data in 2010 are limited. *Four lines* indicate time series for (*gray*) the tide gauge, (*green*) AVISO, (*red*) FORA and (*blue*) 3DVAR. Note that time series of AVISO data are not shown for Abashiri because *grid point*

values adjacent to the tide gauge station are all missing values. RMSDs of SLAs for AVISO, FORA and 3DVAR calculated with tide gauge observations for 2010 (or 2009 at Tsushima) are shown by *green*, *red* and *blue* numbers (unit in cm)

At Amami and Chichijima, which are located in the open ocean, variations in sea level on the time scale of a few months are predominantly caused by westward propagating mesoscale eddies. Such variations of sea level in FORA compare very well with the tide gauge observations although MOVE-3DVAR and AVISO also capture basic features of the observed sea level variability, implying that mesoscale eddies are properly assimilated in FORA. Those features are consistent to Usui et al. (2015), in which it was shown that the 4DVAR scheme provides better representation of mesoscale variability compared to 3DVAR.

Sea levels at Abashiri, Tsushima and Sado, which are located in marginal seas such as the Japan Sea and the Okhotsk Sea, exhibit high-frequency variability on a time scale shorter than 1 month, which is known to be induced by atmospheric forcing (Fukumori et al. 1998). Sea levels for FORA and MOVE-3DVAR show similar high-frequency variability, but it is not represented in AVISO. At the tide gauge station of Tsushima, which is located at Tsushima Island, the IR is the lowest among the 63 tide gauge stations. Sea levels at Tsushima not only for FORA but also for MOVE-3DVAR and AVISO show systematic bias compared to the tide gauge observation (Fig. 8e). Tsushima Island, which is not resolved in the AVISO data, is situated at the Tsushima Strait, and the bottom topography around the island is complicated. Thus, the tide gauge record at Tsushima might be affected by local processes. In addition, the bathymetric data are uncertain around the island (Hirose 2005). These might lead to the systematic bias.

4.2 SST and SSH fields

In this section, we evaluate SST and SSH fields in FORA using satellite-derived observations. Figure 9a, b compares mean and standard deviation of FORA SST during the period of 1982–2012 with MGDSST. The annual mean SST in FORA compares well with that for MGDSST. Taking a difference of the mean SST between FORA and MGDSST, we however recognize a weak cold bias over a wide area. We have confirmed that a similar bias pattern appears in MOVE-3DVAR as well as in a free-simulation experiment using the same model and the same atmospheric forcing (not shown), implying that this cold bias originates from the atmospheric forcing or errors in model physics. It is also worth noting that a warm bias is recognizable along the Kuroshio in the East China Sea and south of Japan. In SST fields, the Kuroshio is recognized as a warm tongue, which especially emerges in winter (Kida et al. 2015). The FORA SST field shows a realistic representation of the warm tongue compared to satellite images (not shown). In contrast, it is modest in MGDSST because the horizontal resolution is coarse compared to FORA, and the dataset is produced by an objective analysis without a dynamic model. Therefore, this warm bias is caused by the difference in representation of SST fronts around the Kuroshio between FORA and MGDSST.

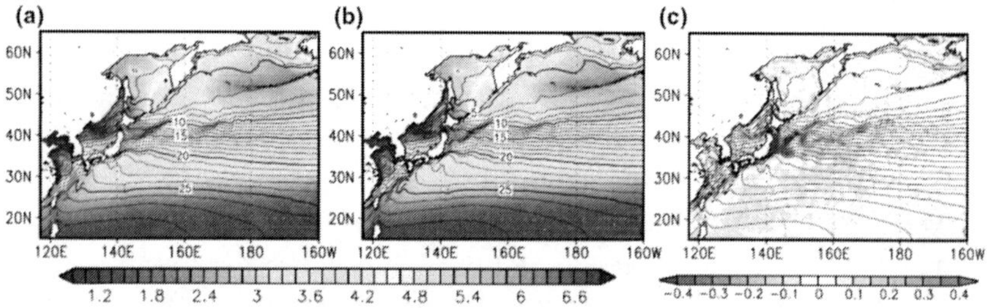

Figure. 9 (Contours) annual mean and (*shades*) standard deviation of SST for **A** MGDSST and **B** FORA. **C** Mean difference in SST between FORA and MGDSST. These values are calculated using the SST fields during the period from 1982 to 2012. Units in °C

The observed feature of the variance field is also well captured in FORA. The SST fields for both FORA and MGDSST show large variability in higher latitudes, shelf regions and frontal regions (Fig. 9). The variability in higher latitudes and shelf regions is due to large seasonal changes (e.g., Levitus 1987). In frontal regions such as the Kuroshio-Oyashio confluence region east of Japan, on the other hand, ocean current variability is considered to have a large impact on variations in SST. The current variability is caused by energetic mesoscale eddy activity associated with meanders of the Kuroshio Extension Current and pinch-off eddies. SST variability and its relation to current variability around the frontal regions in the western North Pacific are well summarized in Kida et al. (2015).

In order to evaluate mesoscale eddy variability with a time scale longer than 1 month, the SSH field in FORA is then compared to the AVISO SSH field. We take correlation between FORA and AVISO using monthly mean SSH fields in 1993–2012. For comparison, correlation between MOVE-3DVAR and AVISO is also calculated. Two correlation maps are compared in Fig. 10. For both FORA and MOVE-3DVAR, correlation exceeds 0.7 in most areas in the western North Pacific except for some seasonal sea ice areas. Focusing on an area with correlation greater than 0.9, it is widely distributed in the subtropical gyre for FORA, while it is limitted in the western part of the subtropical gyre in MOVE-3DVAR. Compared between the two maps, the correlation for FORA is much improved in the Kuroshio Extension and mixed water region off the east of Japan, where mesoscale eddy activity is the most enegetic in the western North Pacific. Therefore, the higher correlation implies better representation of mesoscale variability in FORA.

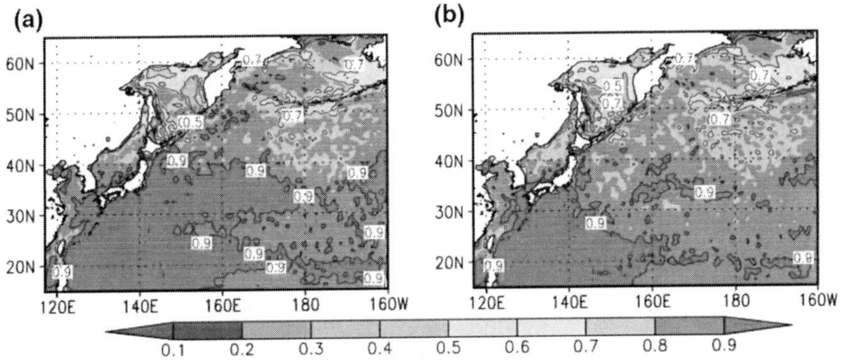

Figure. 10 Correlation maps of SSH fields between AVISO and the reanalysis product: **A** FORA and **B** 3DVAR. Monthly mean SSH fields from January 1993 to December 2012 are used for calculation of the correlation

In the subarctic region of the North Pacific where barotropic responses to atmospheric forcing are enhanced (Kuragano et al. 2014), the correlation is not so high compared to that in the lower latitudes. This is probably because the observation operator *H* for altimeter-derived SLA observations in (2) does not take into consideration the barotropic component of sea level changes. Therefore, a refinement of the observation operator for SLA observations could lead to further improvement of the SSH field.

4.3 Mixed layer depth and water mass properties

The ocean surface mixed layer directly communicates with the atmosphere. It provides the boundary condition (e.g., sea surface temperature) to the atmosphere, and, on the other hand, water masses are generated in the mixed layer under the atmospheric forcing and subducted to the interior ocean. Therefore, spatial and temporal variability of the surface mixed layer depth (MLD) and water masses is one of the important metrics for assessing the reproduction of the surface layer variability (e.g., Toyoda et al. 2015a). In addition, it has been indicated that smaller scale (eddy) processes are essential for the reproduction of MLD and surface-intermediate water masses (e.g., Ishizaki and Ishikawa 2004; Tsujino and Yasuda 2004; Nishikawa et al. 2010). Thus, the present product (FORA-WNP30) is expected to effectively reproduce the MLD-water mass variability by adopting a high-resolution OGCM.

Figure 11 presents the distributions of the winter- and summertime MLDs. MLD is defined in this study as the depth where potential density exceeds the 10-m depth value by 0.125 kg m^{-3}(e.g., Levitus 1982). The mean MLD distribution from FORA (Fig. 11a, c) captures the gross features of the observational MLD climatologies (Fig. 11b, d), although the wintertime MLDs are slightly deeper in the former case. At least in terms of the climatology, the seasonal cycle of the MLD distribution in the western North Pacific is, therefore, well reproduced in FORA. The horizontal patterns of the wintertime MLD differences between the composites in the positive and negative PDO (Pacific decadal oscillation; Mantua et al. 1997) phases are shown in Fig. 11e, f. Our result indicates large positive MLD anomalies in the central subtropics. Figure 11f shows that this pattern is supported by the observational estimate by using the recent Argo observations (MILA-GPV; Hosoda et al. 2010). Therefore, FORA represents the overall MLD variability on the interannual-decadal scale as well as the seasonal scale.

Figure 12 compares cross sections of the 31-year (1982–2012) mean climatological temperature and salinity along 160.5°E and 179.5°W of the FORA-WNP30 with those of the latest observational climatology (WOA13). The temperature section along 160.5°E (Fig. 12a–c) contains the central-eastern part of Subtropical mode water (STMW; Masuzawa 1969) (about 16–19 °C), which originates from the Kuroshio water. The FORA result reproduces the subsurface temperature structure including STMW, although a small warm bias in the upper thermocline region (upper 200–300 m) is found (Fig. 12c). The salinity section along 179.5°W (Fig. 12d–f) includes the salinity minimum structure of North Pacific intermediate water (NPIW; Talley 1993) (green shades in Fig. 12d, e), which is generated in the Okhotsk Sea and Kuroshio-Oyashio interfrontal zone (e.g., Yasuda et al. 1996). Note that the realistic reproduction of the NPIW distribution is known to be a difficult issue in model simulation (e.g., Ishizaki and Ishikawa 2004). The FORA result reproduces this structure well.

Figure 13 presents the time series of annual mean area and temperature of STMW in the vertical cross section along the 137°E line from the FORA results (red lines) and observations (black lines). Here, the observations are based on the long-term ship observation dataset (temperature and salinity, from 1971 to 2011) along 137°E by the Japan Meteorological Agency (JMA). STMW is defined by the following criteria: potential vorticity (PV) value is less than $2.1\times10^{-10}m^{-1}\ s^{-1}$ with potential density range from 24.8 to 25.8 (see the figure caption for details).

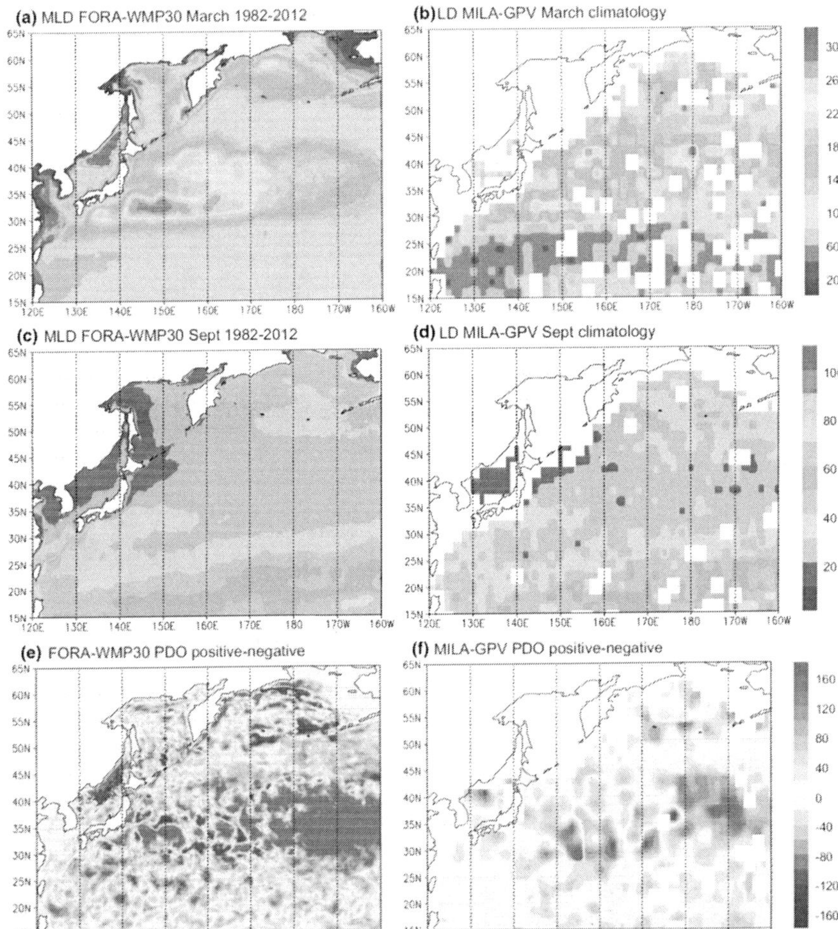

Figure. 11 A–D MLD distributions in March (**A**, **B**) and September (**C**, **D**) from FORA averaged over the 1982–2012 period (**A**, **C**) and from MILA-GPV climatology data. (**E**, **F**) Differences between the March MLD composites in the positive (2003, 2005 and 2010) and negative (2006, 2007, 2009) PDO phases from FORA (**E**) and from MILA-GPV monthly data (**F**)

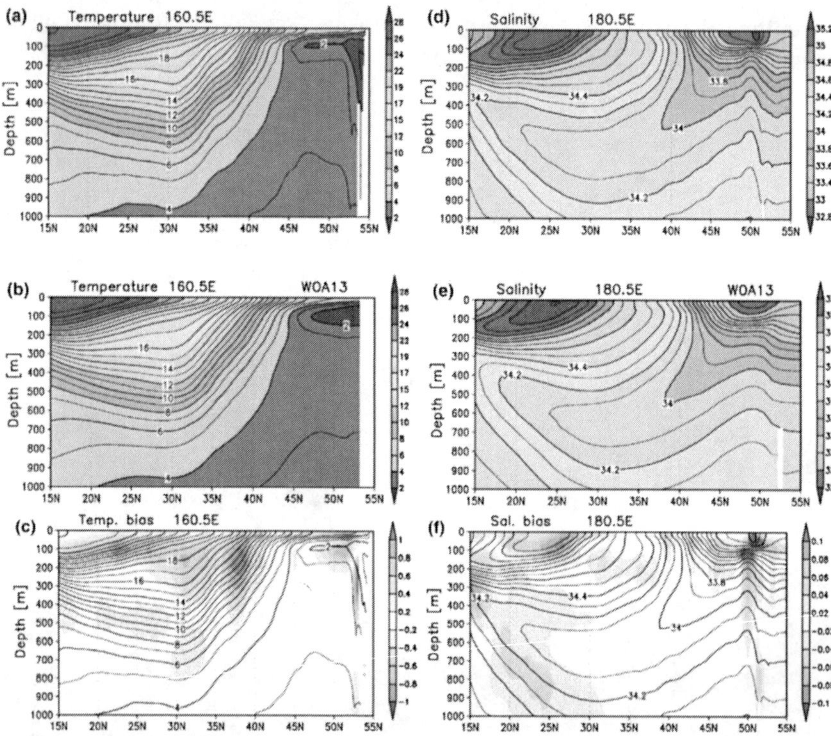

Figure. 12 Climatological mean temperature [°C] along 160.5°E for **A** FORA and **B** WOA13. **C** Difference (*shades*) between **A** and **B**. Contours in **C** are the same as in **A**. **D–F** Same as in **A–C** but for salinity [psu] along 179.5°W. The FORA values are calculated using the fields during the period from 1982 to 2012

The observational results show decadal and interannual variability of the STMW area and temperature along 137°E. The FORA results reproduce these variabilities fairly well, although there are some differences (e.g., in 2003). The differences may partly come from temporal and spatial inhomogeneity of the observational data. Previous studies indicated that the decadal variability of STMW is closely related to that of the Kuroshio Extension (KE) state (e.g., Oka et al. 2015). Therefore, the good reproduction of the KE state in the present high-resolution system as described later (Sect. 5.2) would contribute to the above-mentioned reproduction of the STMW features.

Overall, FORA-WNP30 reproduces the climatological structures and decadal–interannual variability of MLD and major water masses in the North Pacific well.

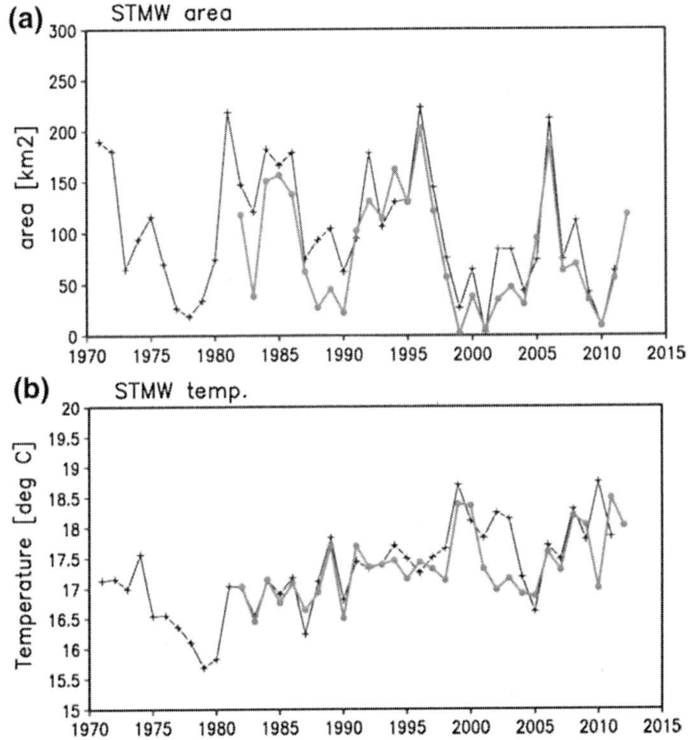

Figure. 13 Time series of annual STMW **A** area [km^2] and **B** area-mean temperature [°C] along 137°E. Here, STMW is defined as the water with potential vorticity (PV) less than 2.1×10^{-10} m^{-1} s^{-1} and potential density range from 24.8 to 25.8 kg m^{-3} in the latitudes 23°–33°N and the depths 91–570 m. PV here is simply calculated from the vertical density gradient and the Coriolis parameter. *Red lines* are from the FORA results, and *black lines* are from the JMA's 137°E line ship observation dataset (temperature and salinity)

5. INTERANNUAL TO DECADAL VARIABILITY AND ANOMALOUS EVENTS IN THE SEAS AROUND JAPAN

5.1 Kuroshio south of Japan

It is well known that the Kuroshio south of Japan takes different paths: the straight path and the large meander (LM) path (Kawabe 1995). The straight path is sometimes called the non-large meander (NLM) path. This bimodal feature is unique to the Kuroshio and is not observed in other western boundary currents such as the Gulf Stream and Agulhas Current.

In order to evaluate the Kuroshio path variations south of Japan, we plot in Fig. 14 the time series of the Kuroshio southernmost latitude defined between 136°–141°E. The Kuroshio latitude of JMA data in Fig. 14 is subjectively determined on the basis of various information such as satellite SST images, tide gauge data, objectively analyzed subsurface temperature distribution using hydrographic observations and recent MOVE-WNP assimilation model results. The southernmost latitude for FORA is identified by tracking the maximum velocity at 100 m. During the reanalysis period, four LM events occurred, and their periods are depicted by gray shades.

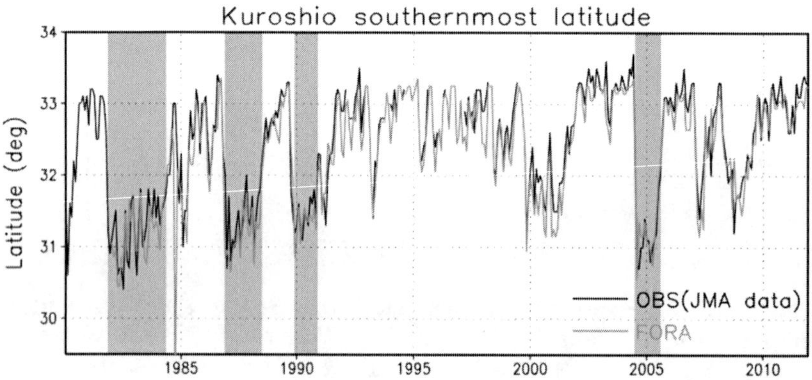

Figure. 14 Time series of the southernmost latitude of the Kuroshio path south of Japan defined between 136°E and 140°E. *Black line* is based on JMA data obtained from the web site at http://www.data.jma.go.jp/gmd/kaiyou/data/shindan/b_2/kuroshio_stream/kuro_slat.txt, and *red line* represents the time series calculated from FORA monthly fields. *Gray shading* indicates large meander periods

The Kuroshio path variation represented by FORA compares quantitatively well with the JMA data. It is worth emphasizing that FORA captures the observed features well, even in the 1980s, when there is no altimeter data. One reason for the precise representation of the Kuroshio path variation in the 1980s would be the relatively dense in-situ observations around the Japanese coasts as described in Sect. 3.

Formation of past LMs are also well represented in FORA. As an example, the time evolution of the Kuroshio path during formation of the LM in

1986–1988 is shown in Fig. 15. A small-scale meander first appears off the southeast of Kyushu (Fig. 15a). Then, the meander moves eastward and finally develops into a LM path. This feature is consistent with that indicated by previous studies addressing the LM formation (e.g., Usui et al. 2008). In addition, a sizable anticyclonic eddy appears in the deep layer around 3000 m during the formation of this LM (not shown). Previous modeling studies have indicated that the LM is caused by baroclinic instability, and a deep-anticyclonic eddy is formed below the Kuroshio because of the baroclinic instability (Endoh and Hibiya 2001; Tsujino et al. 2006). Here it should be noted that the 4DVAR system used in FORA controls the temperature and salinity fields above 1500 m as described in Sect. 2, indicating that the deep-anticyclonic eddy represented in FORA is generated through the model dynamics.

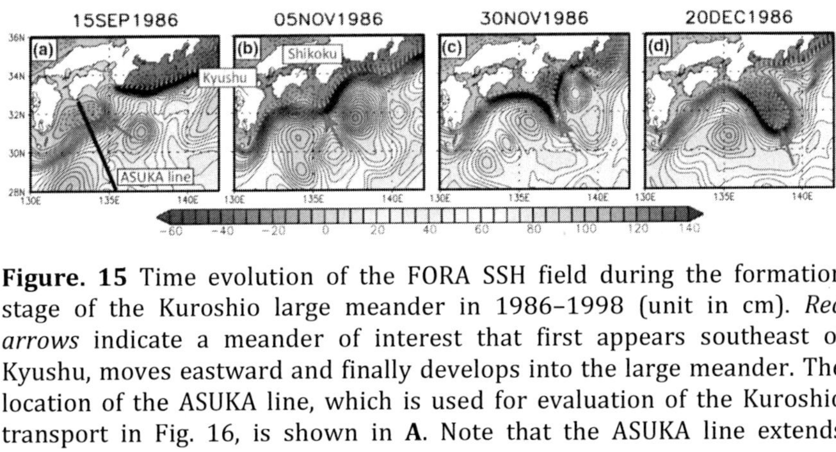

Figure. 15 Time evolution of the FORA SSH field during the formation stage of the Kuroshio large meander in 1986–1998 (unit in cm). *Red arrows* indicate a meander of interest that first appears southeast of Kyushu, moves eastward and finally develops into the large meander. The location of the ASUKA line, which is used for evaluation of the Kuroshio transport in Fig. 16, is shown in **A**. Note that the ASUKA line extends southward to 26°N

We next look at the Kuroshio transport across the ASUKA (Affiliated Surveys of the Kuroshio off Cape Ashizuri) line off Shikoku, which coincides with a subsatellite track of TOPEX/Poseidon and Jason-1/2 (see Fig. 15a for geographical location). Figure 16 compares the time series of the Kuroshio transport above 1000 m across the ASUKA line between FORA and an altimeter-based transport estimate, which is deduced from sea-level difference across the Kuroshio observed by TOPEX/Poseidon and Jason-1 (Imawaki et al. 2001).

We calculated an eastward and throughflow transport of the Kuroshio in a similar way to the method of Imawaki et al. (2001). As can be found in Fig. 15, a stationary anticyclonic circulation exists to the south of Shikoku. The eastward transport includes the contribution of the stationary anticyclonic circulation. In contrast, the throughflow transport is calculated by subtracting a westward transport associated with the anticyclonic circulation from the eastward transport. It can thus be regarded as the net Kuroshio transport.

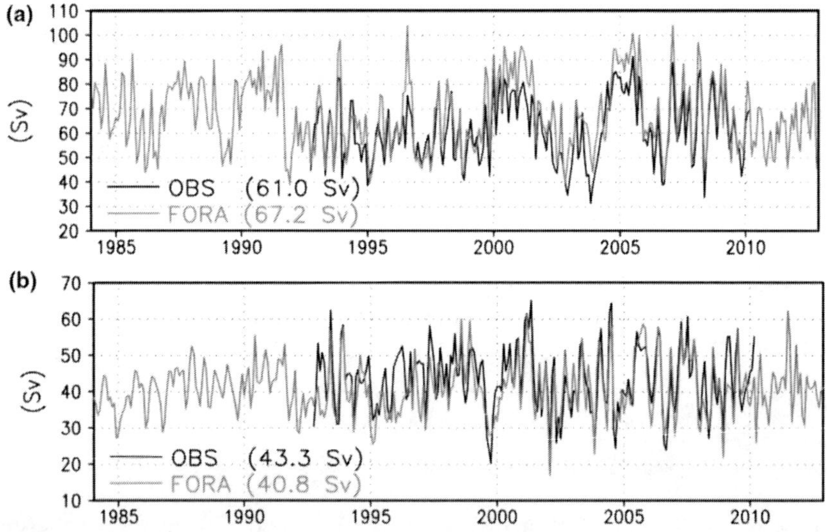

Figure. 16 Time series of the Kuroshio **A** eastward and **B** throughflow transport above 1000 m across the ASUKA line. See text for definition of the two transports. *Black lines* indicate transports estimated from satellite altimeter data (Imawaki et al. 2001), and *red lines* are FORA results. Mean transport for each time series averaged over October 1992–March 2010 is also shown

Both time series of the eastward and the throughflow transport for FORA are in good agreement with the observation. The eastward transport, however, has a somewhat large bias. The mean eastward transports averaged from October 1992 to March 2010 are 61.0 and 67.2 Sv for observation and FORA, respectively. This bias reduces for the throughflow transport, implying that the bias originates in a strong anticyclonic circulation off Shikoku. The assimilation scheme for altimeter-derived sea-level anomalies could be one possible cause for this bias. As described in Sect. 2, the observation operator for altimeter data in MOVE-4DVAR takes model-data misfits in terms of surface dynamic height, which is calculated with a reference depth of 1500 m. The reference depth might not be appropriate for the stationary anticyclonic circulation, which has a deep vertical structure (e.g., Kobashi and Hanawa 2004). At the same time, the observation operator does not take into consideration ocean mass-related sea-level changes. Thus, improving the assimilation scheme for altimeter-derived sea-level anomalies might lead to a reduction of the bias for the anticyclonic circulation off Shikoku.

5.2 Kuroshio extension

The Kuroshio extension is an eastward-flowing free jet after the Kuroshio separates from the Japanese coast. The KE region exhibits the highest mesoscale eddy variability in the North Pacific. In addition, low-frequency changes on a decadal time scale are also one of prominent features of the KE jet variability. The KE decadal variability is considered to be remotely forced by the large scale wind stress field in the North Pacific (Qiu 2003; Taguchi et al. 2007).

In order to look into the KE variability represented by FORA, we plot in Fig. 17 the time series of three indices for the KE jet. For comparison, indices calculated from the AVISO altimetry data are also shown. All indices for both FORA and AVISO are calculated on the basis of the monthly mean fields. The latitudinal position of the KE jet in Fig. 17a is determined by zonally (141°–153°E) averaging latitudinal positions defined by the maximum surface velocity. The path length of the KE jet, which characterizes the KE dynamic state, is calculated by tracking the maximum surface velocity (141°–153°E). The eddy kinetic energy (EKE) is calculated using the surface velocity anomaly and is averaged over 141°–153°E and 32°–38°N. The surface velocity anomaly used for the EKE calculation is defined as a deviation from the climatological mean state from 1993 to 2012 for both FORA and AVISO.

The KE variability on interannual to decadal time scales in FORA is in good agreement with that in the AVISO data. The meridional shift of the KE jet in Fig. 17a on decadal time scales is synchronized with the decadal changes in the strength of the KE southern recirculation gyre characterized by the area-averaged SSH anomaly (see the gray dotted line in Fig. 17a), suggesting that the northward (southward) shift of the KE jet is accompanied by a strong (weak) recirculation gyre. The time series of the path length of the KE jet shown in Fig. 17b also exhibits distinct low-frequency changes, which largely correlate with the meridional shift of the KE jet. When the KE jet migrates southward, the path length tends to be large, meaning a spatially convoluted KE path (unstable state). In contrast, the KE path length tends to be small, that is, KE is in a stable state when the KE jet migrates northward. These features of the KE variability in FORA are consistent with findings in a number of previous studies (e.g., Qiu and Chen 2005; Nakano and Ishikawa 2010). Furthermore, the time series of the KE latitude and path length in the 1980s, when there are no altimeter data, also seems to be consistent with a time series by Seo et al. (2014), who inferred the KE dynamic state using the wintertime SST gradient and constructed the time series of the KE latitude and the path length from 1982 to 2011 (see Fig. 4 in Seo et al. 2014). We, however, need further evaluation of the KE variability especially in the 1980s, which remains for future studies.

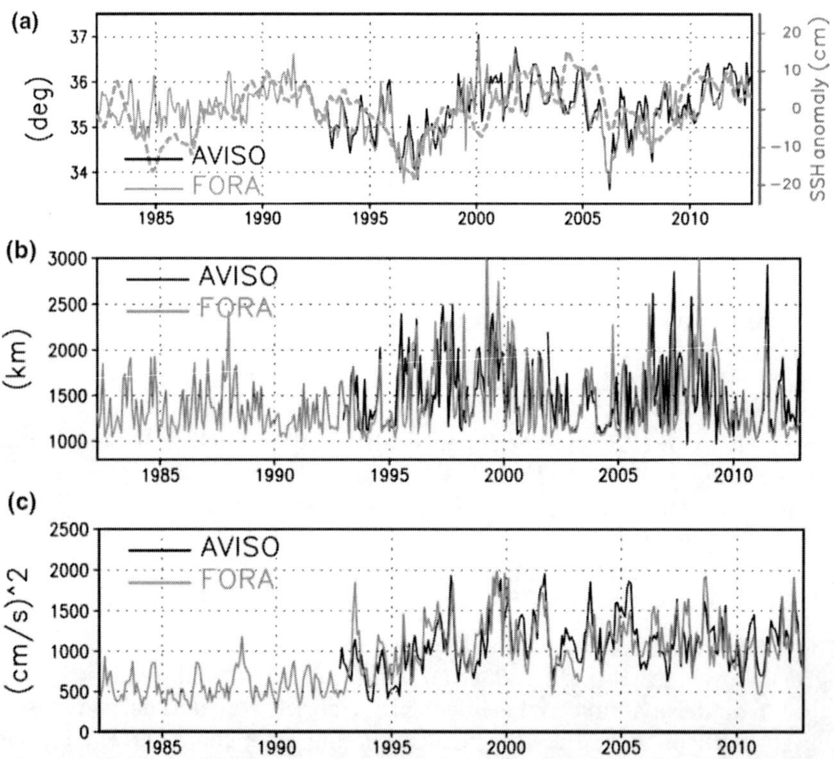

Figure. 17 Time series of three indices for the KE state. **A** KE latitude defined as a zonal average of the latitudinal position of the KE jet between 141°E and 153°E, **B** KE path length integrated from 142°E to 153°E and **C** area-averaged EKE within 141°–153°E and 32°–38°N. *Black* and *red lines* denote indices calculated from AVISO data and FORA, respectively. Time series of SSH anomaly averaged over the KE southern recirculation region (140°–155°E, 30°–35°N) with a 5-month running mean is also shown by the *gray dotted line* in (**A**)

We finally mention the continuity of the FORA data in the KE region. Figure 17c shows the time series of EKE in the KE region, which reflects mesoscale eddy activity. A distinct gap is recognizable in 1993, when the satellite altimeter data begin to be assimilated. The EKE level before 1993 is obviously low compared to that during the altimeter era. A similar EKE gap can be seen in the Subtropical Countercurrent region (not shown), where mesoscale eddy activity is high as well as the KE region. We therefore need to pay special attention to the energy gap when looking into the energy field, and reduction of the gap should be the subject for future studies.

5.3 Oyashio

The Oyashio Current, the western boundary current of the wind-driven subarctic gyre in the North Pacific Ocean, flows southwestward along the Kuril Islands and the southeast coast of Hokkaido (see Fig. 25b for geographical location) and then splits into two branches. One branch continues southward along the east coast of Honshu and is known as the coastal branch or the first branch. The other, called the offshore branch or the second branch, turns direction and flows eastward. The Oyashio water is characterized by cold and fresh water, which is formed as a result of merging waters from the East Kamchatska Current (EKC) and Okhotsk Sea (e.g., Kono and Kawasaki 1997). It meets with warm and saline Kuroshio water off the east of Honshu to form complicated frontal structures, and thus this region is often called the mixed water region (MWR).

The first branch of the Oyashio has significant seasonal variation, which is well characterized by the horizontal extent of the Oyashio water in MWR. The horizontal extent increases from winter to early spring and reaches the maximum in March–April. It then gradually decreases, and the minimum extent occurs in November (Yasuda 2003). Moreover, the Oyashio Current exhibits prominent interannual variability, which is considered to have a relation to changes in large-scale atmospheric forcing in the North Pacific (Sekine 1988; Sekine 1999) although its mechanism is still controversial.

Figure 18 presents year-to-year changes of the Oyashio area in MWR averaged over March–May, which is defined by the cold water area at 100 m with temperature less than 5 °C (Kawai 1972) within a rectangular region (141°–148°E, 30°–43°N). The time series of the Oyashio area represented by FORA indicates that the maximum and minimum during 1982–2012 appeared in 1984 and 2000, respectively. The maximum extent in 1984 represented by FORA is consistent with the fact that an anomalous southward intrusion of the Oyashio occurred in 1984 (Sekine 1988).

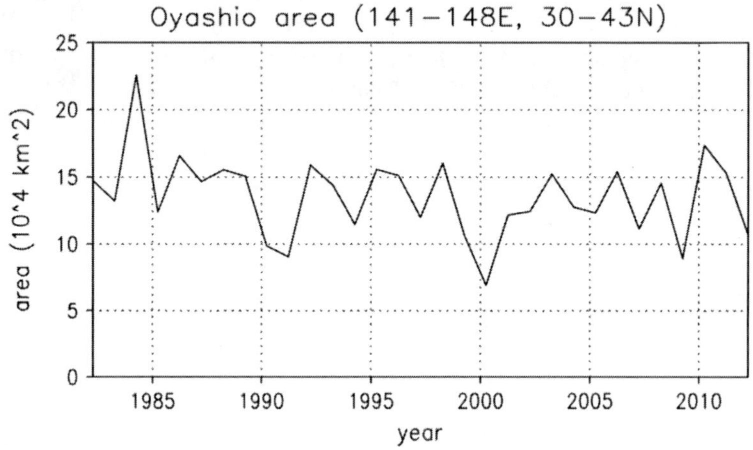

Figure. 18 Time series of the Oyashio area in spring (March–May) during the period from 1982 to 2012 (unit in 1044 km22). The Oyashio area is defined by the area of cold water at 100 m with temperature less than 5 °C and is calculated within 141°–148°E and 30°–43°N

In order to take a look at the difference between the two anomalous years, Fig. 19 compares the temperature and horizontal velocity distributions at 100 m in April 1984 and April 2000. In April 1984, the Oyashio water characterized by temperature less than 5 °C is widely distributed in MWR. It intrudes southward along the east coast of Honshu, and the southern tip of the Oyashio water extends to around 36°N. These features shown by FORA are in good agreement with observed features reported by former studies (Okuda 1986; Sekine 1988).

In contrast, the Oyashio area is hardly found in MWR in April 2000. It is noticeable that a warm-anticyclonic eddy around 146°E and 41.5°N exists, which seems to block the southward intrusion of the Oyashio water. The red solid line in Fig. 19 denotes a trajectory of the eddy during the period of 3 November 1998 through 23 September 2000, which is detected using daily SSH fields in FORA. This long-lived mesoscale eddy is first detected at the KE region in November 1998 and then migrates northward along the Japan Trench located there; it reaches southeast of Hokkaido in the winter to spring of 2000. Similar northward-moving mesoscale eddies are often observed in this area (Itoh and Yasuda 2010), and their relation to variability in the Oyashio Current has also been argued (Isoguchi and Kawamura 2006; Kuroda et al. 2015).

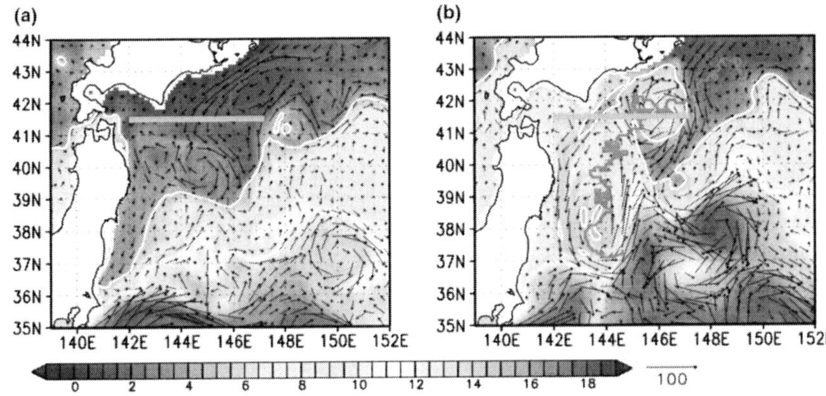

Figure. 19 Comparison of monthly mean temperature (unit in °C) and velocity (unit in cm s^{-1}) fields at 100 m in **A** April 1984 and **B** 2000. The *solid white line* indicates a temperature isoline of 5 °C, which is often used for an indicator of the Oyashio area. The *red solid line* represents a trajectory of the anticyclonic warm-core eddy located around 146°E and 41.5°N in April 2000. The eddy's trajectory during the period from November 1998 to September 2000 is plotted. The *solid purple line* along 41.5°N denotes a repeat *hydrographic line* observed by R/V Kofu-maru

Figure 20 shows vertical sections of temperature and salinity along 41.5°N (see Fig. 19) for hydrographic observations and FORA in April 1984 and April 2000. The hydrographic data in Fig. 20 were obtained by the research vessel Kofu Maru, which had been operated by Hakodate Marine Observatory in JMA. There are some discrepancies between the observation and FORA. For example, the subsurface temperature inversion in 1984, known as the mesothermal structure (Ueno and Yasuda 2000), is somewhat weak, and exists temperature near the surface in 2000 is warmer compared to the observation. The FORA results, nevertheless, capture the observed differences in the temperature and salinity fields between 1984 and 2000 well.

In April 1984, the surface layers are occupied by cold and fresh Oyashio water. On the other hand, warm and saline Kuroshio water is widely distributed in April 2000. A distinct feature of the temperature and salinity fields in 2000 is the existence of the long-lived warm-core eddy located around 146°E. The vertical structure of the eddy extends to about 600 m. The above results imply that the southward intrusion of the Oyashio water characterized by the Oyashio area in the MWR region should be considered together with mesoscale eddies in the region as well as the strength of the Oyashio Current. In addition, it is worth noting that the KE jet was in an unstable state in 2000 (see Fig. 17b); hence, there were many mesoscale eddies in the KE region. This suggests a possibility that the KE state is also a possible factor to control the variations in the Oyashio intrusion through the mesoscale eddy activity. The FORA data, a long-term high-resolution dataset, could be useful to evaluate this hypothesis, which remains for future studies.

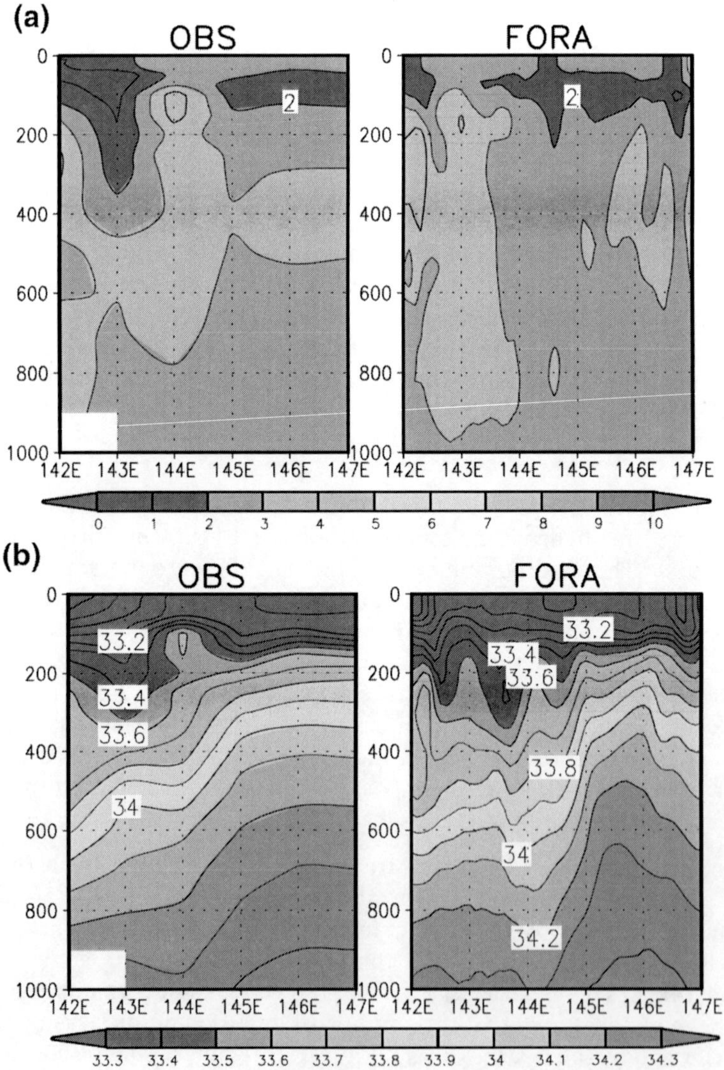

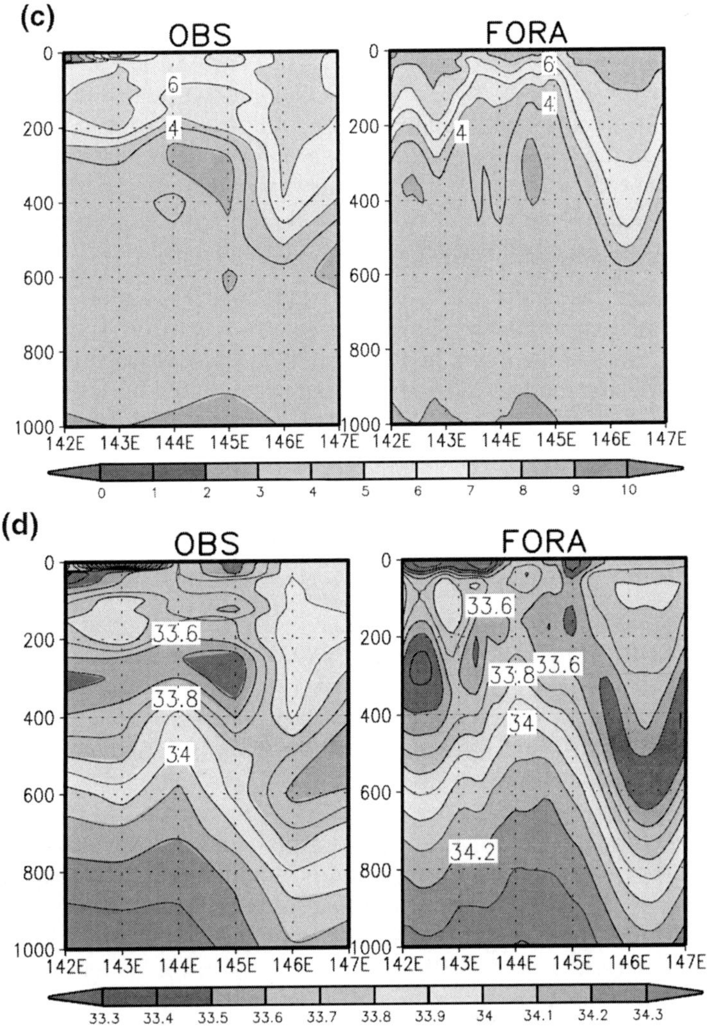

Figure. 20 A, **C** Temperature (unit in °C) and **B**, **D** salinity (unit in psu) sections from the surface down to 1000 m along 41.5°N (see also Fig. 19) in (**A**, **B**) April 1984 and (**C**, **D**) April 2000. Observation and FORA results are compared

5.4 Japan Sea

The Japan Sea is a semi-enclosed marginal sea of the North Pacific Ocean. It is connected to adjacent seas through four narrow and shallow straits. The straits of Tsushima (TSM) and Tsugaru (TGR) connect the Japan Sea to the

East China Sea and the North Pacific, respectively. There are two straits between the Japan Sea and Okhotsk Sea: the Soya Strait (SOY) and Tatar Strait (TTR). Because of the semi-enclosed character, volume transports through the straits are thought to be important for mean state and variability in the Japan Sea. This is why a number of observational studies have attempted to estimate an accurate volume transport through each strait (e.g., Fukudome et al. 2010; Ito et al. 2003; Fukamachi et al. 2010). Recent observational efforts are well summarized in Na et al. (2009).

Annual mean transport over 31 years (1982–2012) and its standard deviation at each strait are shown in Table 3. A positive transport value represents northward flows in TSM and TTR and eastward flows in TGR and SOY. A water budget in the Japan Sea represented by FORA is that an inflow transport at TSM (2.32 Sv) almost balances with the outflow ones at TGR and SOY (1.63 and 0.67 Sv). A contribution of the transport at TTR (0.016 Sv) is very small. The ratios of the outflow transports to the inflow through TSM are thus estimated at 70 and 29 % for TGR and SOY, respectively, which are consistent with former estimates (Cho et al. 2009; Na et al. 2009).

Observational transport estimates at the straits are also given in Table 3 to be compared with the FORA results. Fukudome et al. (2010) showed 10-year time series of volume transport at TSM obtained from the acoustic Doppler current profiler (ADCP) data from February 1997 to February 2007, and the 10-year average of the transport is estimated at 2.65 Sv. They also showed volume transports through the east and west channels of TSM as 1.20 and 1.45 Sv. Ito et al. (2003) estimated the mean volume transport through TGR as 1.5 Sv using ADCP data from November 1999 to March 2000. Although this observation is conducted only in a short term, this value is almost the same as other observational results (Shikama 1994; Nishida et al. 2003). Fukamachi et al. (2010) showed the annual volume transport through SOY was 0.62–0.67 Sv from September 2006 to July to 2008 using bottom-mounted ADCP data; however, the annual mean from 2004 to 2005 was 0.94–1.04 Sv (Fukamachi et al. 2008). They suggested that the interannual variation in the transport through SOY is large. We also calculate the volume transport in FORA at each strait for the same period as the reference observation, which is summarized in Table 3. Note that the calculated transports for FORA get closer to the observational estimates at almost straits compared to the annual mean transports for 1982–2012.

Table 3 Comparison of volume transport at each strait in the Japan Sea between FORA and observations

	Tsushima	Tsugaru	Soya	Tatar
FORA-WNP30 (1982–2012)	2.32 ± 0.19	1.63 ± 0.07	0.67 ± 0.11	0.016 ± 0.014
Observation/FORA-WNP30	2.65 / 2.40	1.5 / 1.61	0.62–0.67 / 0.70 0.94–1.04 / 0.82	
(Period)	(Feb 1997–2006)	(Nov 1999–Mar 2000)	(Sep 2006–Jul 2008, May 2004–2005)	
(Reference)	(Fukudome et al. 2010)	(Ito et al. 2003)	(Fukamachi et al. 2008, 2010)	
Trend (Sv/year)	1.62×10^{-2}	6.12×10^{-3}	9.28×10^{-3}	8.62×10^{-4}

The volume transports for FORA are estimated for 1982–2012 and for the same periods as reference observations. Standard deviations and linear trends for the annual mean transport in FORA are also calculated

The mean transports represented by FORA are largely consistent with the observations, while there are some minor discrepancies compared to the observational estimates. The volume transport in FORA is slightly lower (higher) at TSM (TGR). The lower transport at TSM seems to be attributed to underestimating at the western channel of TSM (1.09 Sv). It should however be noted that there is an imbalance of the observation-based transport estimates among the straits (Han et al. 2015); thus, it would also be one cause for the mismatch of the transports between FORA and the observations.

Figure 21a shows seasonal variation of volume transport calculated as a 31-year average at each strait. Annual cycles of volume transports at TSM and SOY are almost similar with their maximum in summer and their minimum in winter. In contrast, volume transport at TGR has the maximum in November and minimum in February. These features in FORA are consistent with seasonal variations indicated by previous studies (e.g., Fukudome et al. 2010). Figure 21b shows year-to-year variations in annual mean volume transports through the four straits. In addition to the interannual variations, the time series of the annual mean transports exhibit positive linear trends, which are estimated at 1.62×10^{-2}, 9.28×10^{-3} and 6.12×10^{-3} Sv/year for TSM, SOY and TGR, respectively. These trends are significant on 99 % confidence level.

The upper ocean heat content in the Japan Sea also shows a significant increasing trend as shown in Fig. 22. The spatial pattern of the trend compares well with that in Na et al. (2012). Further, in Fig. 23 we compare time-depth sections of temperature between FORA and an observation at a JMA long-term monitoring station PM-5 (134.72°E, 37.72°N) located around the Yamato Basin, where the upper ocean heat content shows a significant increasing trend (see Fig. 22). The isotherms are deepened with time for both the data. This feature is consistent with the increasing trend in the upper ocean heat content.

Na et al. (2012) also suggested that low-frequency variability in the upper ocean heat content on a decadal time scale is apparent in the Japan Sea. To examine the long-term variation of the heat content in FORA, we calculated the detrended heat content anomaly averaged over the Japan Sea (130°E–140°E, 36°N–43°N) and a 13-month running-mean filter was applied (Fig. 24). The time series of the heat content anomaly in the Japan Sea exhibits significant variations on interannual to decadal time scales. The decadal variations are in phase with those in Na et al. (2012). Although the previous studies are limited to temperature analysis, the FORA data including both the temperature and velocity fields enable us to investigate the variations of the Japan Sea in more detail, such as the relationship between the ocean heat content and volume transport through TSM.

Therefore, this dataset has the potential to clarify the variations in the Japan Sea on the decadal time scales using not only the temperature but also the velocity field.

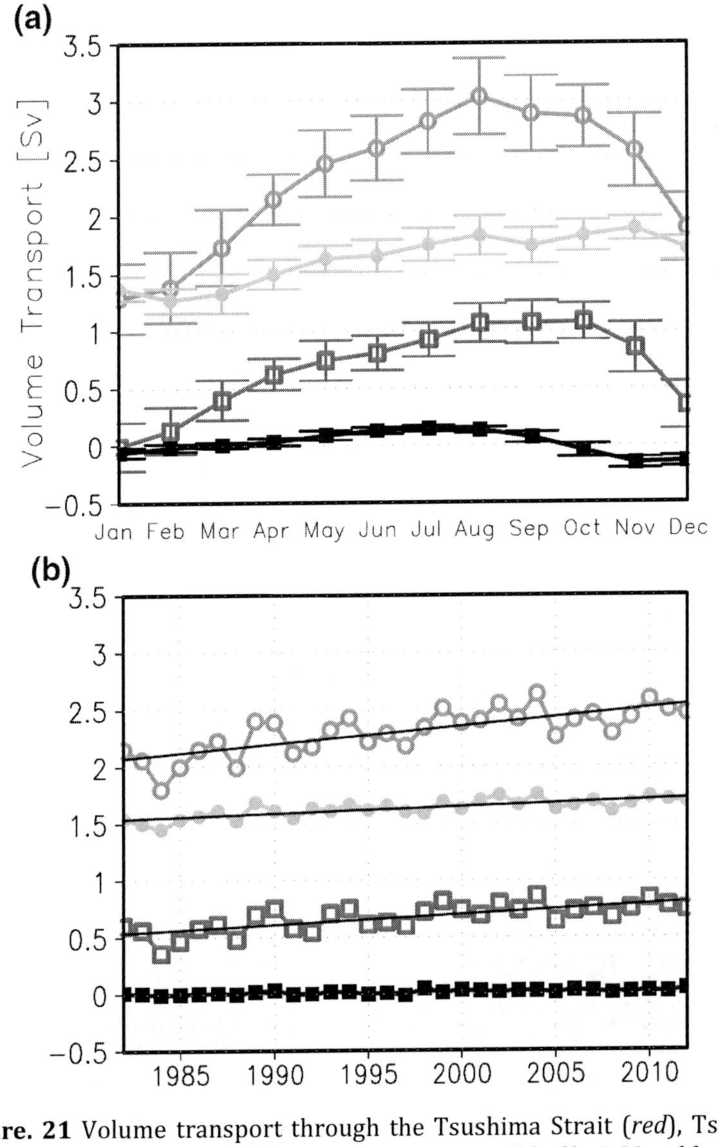

Figure. 21 Volume transport through the Tsushima Strait (*red*), Tsugaru Strait (*green*), Soya Strait (*blue*) and Tatar Strait (*black*). **A** Monthly mean averaged over 31 years with standard deviations for each month (*error bar*) and **B** annual mean with linear trend (*black line*)

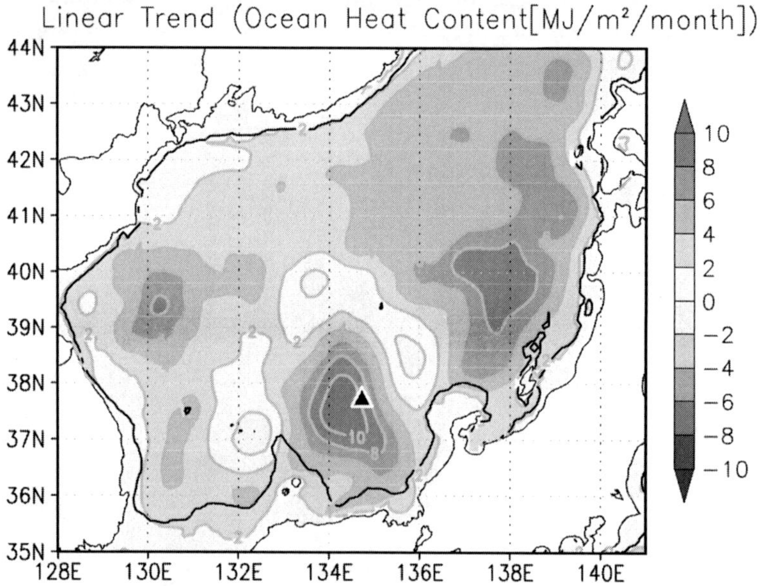

Figure. 22 Linear trend over 31 year of the upper 300 m heat content in the Japan Sea. *Black thick line* is the contour of 300 m depth. *Triangle mark* is we compare observing point (PM-5)

5.5 Sea ice in the Okhotsk Sea

One of the characteristics of FORA-WNP30 is that the sea ice model and ice assimilation scheme are incorporated. In this subsection, we therefore present sea ice variability in the Okhotsk Sea and its relation to oceanic variations represented by FORA. The Okhotsk Sea, located to the northeast of Hokkaido, is known to be one of the southernmost seasonal sea ice zones in the Northern Hemisphere. A large amount of sea ice is produced at coastal polynyas in the northwest shelf region (Kimura and Wakatsuchi 2004), which are open water areas caused by strong offshore wind (Martin et al. 1998). In the northwestern coastal polynya, dense shelf water is formed because of brine rejection associated with the active ice production (Shcherbina et al. 2003), and it is thought to be the ventilation source of NPIW (Talley 1991).

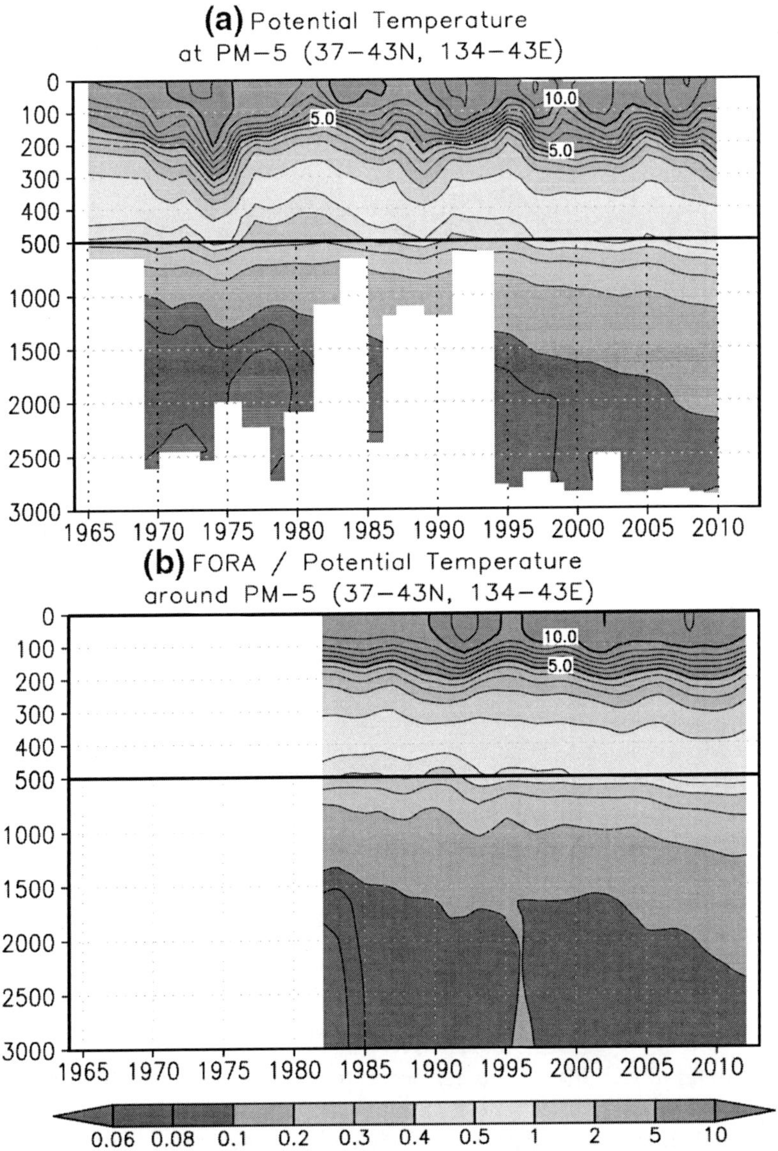

Figure. 23 Time series in *vertical* profiles of potential temperature in winter at the station PM-5 (*left*) observation and (*right*) re-analysis

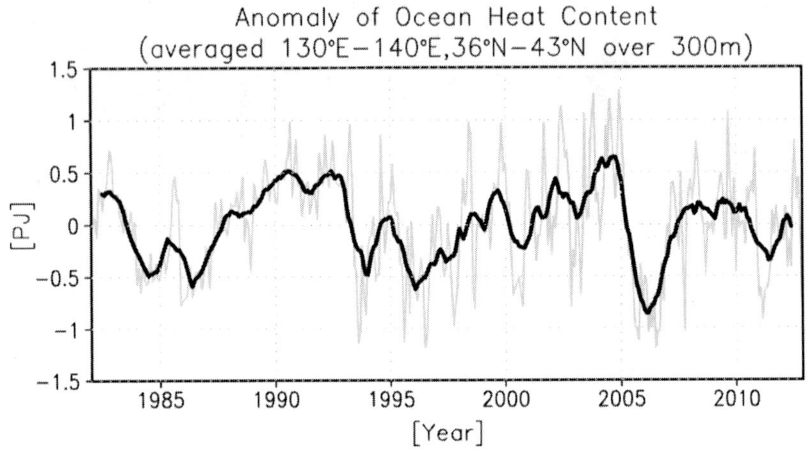

Figure. 24 Time series of the monthly detrended heat content anomaly over 300 m in the Japan Sea (*gray line*) and applied 13-month running-mean filter (*black line*)

In autumn sea ice begins to cover the northwestern shelf region of the Okhotsk Sea and then spreads southward along the east coast of Sakhalin. The southernmost tip of sea ice reaches the north coast of Hokkaido in mid to late winter, and the maximum ice extent occurs in March. Subsequently, the sea ice extent starts decreasing, and most of the sea ice disappears by June. The sea ice extent in the Okhotsk Sea shows large interannual variability (Ohshima et al. 2006). In Fig. 25a, we compare interannual variations of the sea ice extent in March in an observation and FORA. The observation data used for the comparison are calculated from quasi-five-daily ice charts produced by JMA, which are created through a manual analysis of in situ, satellite and aircraft observations. The ice extent for FORA is defined here by the cumulative area of grid cells in the Okhotsk Sea that have more than 15 % sea ice concentration.

Interannual variability of the sea ice extent in FORA compares well with the observed variability. The maximum sea ice extent during the period from 1982 to 2012 occurred in 2001, and the minimum was in 2009 in both datasets. It should be noted that the lowest sea ice extent since 1970 was updated recently in the 2014/2015 winter. Figure 25b displays the year-to-year change of the sea ice edge in March, which is defined by the 15 % ice concentration isoline. The figure indicates that the year-to-year variation depends on how much sea ice expands to the southeastern part of the Okhotsk Sea.

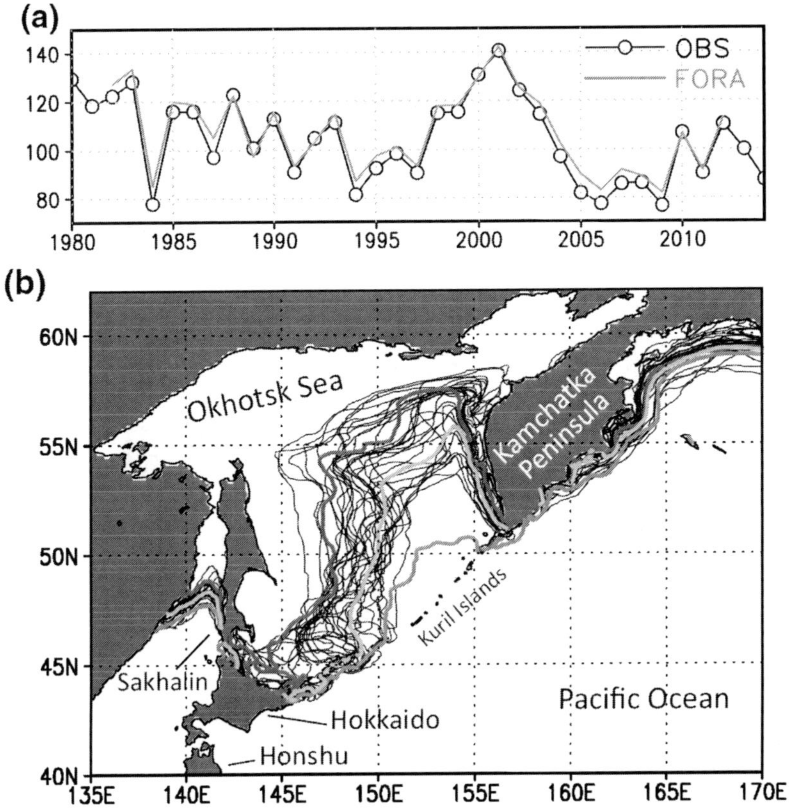

Figure. 25 A Time series of sea ice extent in the Okhotsk Sea in March for (*black*) observation and (*red*) FORA. **B** Year-to-year variations of the sea ice edge in March defined by the 15 % isoline of monthly mean sea ice concentration in FORA. The climatological ice edge in March and those in the maximum and minimum ice extent years (2001 and 2009) are depicted by *green*, *red* and *blue solid lines*, respectively

A recent study by Nakanowatari et al. (2010) pointed out that the maximum sea ice extent in the Okhotsk Sea is highly correlated with SST around the EKC region in the preceding autumn. To confirm this relation using the FORA data, in Fig. 26 we plot the lag correlation maps between the sea ice extent of the Okhotsk Sea in March and SST in the preceding October to December. As shown by Nakanowatari et al. (2010), the correlation maps indicate that SST off the east coast of Kamchatka Peninsula in autumn has a large negative correlation with the sea ice extent in the subsequent winter. It is also worth emphasizing that the area with high correlation coefficients in FORA, which is confined along EKC compared to the maps in Nakanowatari et al. (2010), moves southwestward along EKC and a part of the correlation signal enters the southeastern part of the Okhotsk Sea through straits along the Kuril Islands, where there is large interannual variability in the sea ice extent. These features in FORA support a hypothesis proposed by Nakanowatari et al. (2010) that the interannual variability of the sea ice extent in the Okhotsk Sea is influenced by advection of temperature anomalies by EKC.

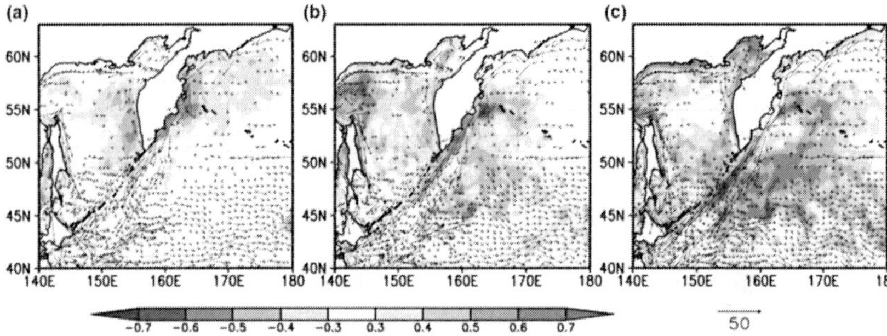

Figure. 26 Lag correlation maps between sea ice extent of the Okhotsk Sea in March and monthly mean SST in the preceding **A** October, **B** November and **C** December. Climatological monthly mean fields of FORA surface velocity (unit in cm s^{-1}) are also shown. Note that the climatological surface velocities with magnitude less than 3 cm s^{-1} are masked

6. SUMMARY AND FUTURE WORK

In this study, we have produced a four-dimensional variational ocean re-analysis for the Western North Pacific over 30 years (FORA-WNP30), which is the first-ever dataset covering the western North Pacific over 3 decades at eddy-resolving resolution. MOVE-4DVAR, which consists of an ocean ice model for the western North Pacific and a 4DVAR assimilation scheme, is employed to assimilate in-situ profiles, gridded SST and altimeter-derived SSH observations. In addition, ice concentration observations are assimilated through a least squares analysis and nudging. A long-term assimilation experiment is conducted using the JRA-55 atmospheric forcing to produce the reanalysis dataset during 1982 to 2012.

We first compare the FORA data with the MOVE-3DVAR reanalysis dataset and independent observations such as ADCP current measurements and tide-gauge sea-level records. The comparison indicates that the FORA data reproduce the observed variability in ocean current and sea level. The comparison between FORA and MOVE-3DVAR also shows that FORA has higher accuracy than the 3DVAR product. We then evaluate the mean field represented by FORA focusing on the SST, SSH and MLD fields. Overall, the FORA dataset reproduces basic features in the western North Pacific well, although there are some biases.

The interannual to decadal variability and anomalous events in the seas around Japan are then analyzed. One of the outstanding features in FORA is that anomalous events such as the Kuroshio large meander and anomalous intrusion of the Oyashio in the 1980s, when there are no altimeter data, are successfully reproduced. In the Japan Sea, FORA reproduces variations in the upper ocean heat content as well as inflow/outflow transports on interannual to decadal time scales well, implying that the FORA dataset is useful for further analysis of the climate variability in the sea. Variations of sea ice extent in the Okhotsk Sea and their relation to SST changes in the preceding autumn are also properly represented by FORA compared to former observational studies. To summarize the above, we conclude that FORA is a valuable dataset for a variety of oceanographic research topics and potentially for related fields such as climate study, meteorology and fisheries.

There are, however, some issues found in FORA that need to be resolved in future works. As mentioned in Sect. 5.2, the EKE field in FORA has a distinct gap in 1993, when the satellite altimeter data begin to be assimilated. The mesoscale eddy activity in FORA before 1993 is at a similar level as that in a free simulation, but is relatively low compared to that in the altimeter era, resulting in the EKE gap. It is assumed that one cause for the EKE gap is a model bias derived from atmospheric forcing used, and thus using an atmospheric forcing adjusted for ocean models would be a useful approach to reduce the gap. Use of such adjusted forcing is becoming common in the ocean modeling community (Brodeau et al. 2010; Griffies et al. 2009). We also pointed out a cold bias in the FORA SST and a potential issue in the design of the observation operator for the SLA data assimilation. Another issue found in FORA but not explicitly mentioned in the previous sections is that excessively cold water with temperature less than the freezing point is sometimes analyzed near the surface layer in the southern part of the Okhotsk Sea. This issue appears in summer and arises from the analysis increment crossing over a strong temperature front between cold Okhotsk Sea water and warm water brought by the Soya Warm Current. As described in Sect. 2, we added a constraint term to the cost function in MOVE-4DVAR to prevent such an unrealistically cold temperature, but it was set to work only in MWR off the east of Honshu, where the same issue can occur (Usui et al. 2011). It is therefore possible to resolve this issue by setting the constraint to work in the Okhotsk Sea as well as in MWR. These improvements should be made when updating the dataset in the future.

The FORA-WNP30 dataset is available for research use. Instruction on the acquisition of the dataset can be found at http://synthesis.jamstec. go. jp/FORA/.

ACKNOWLEDGEMENTS

The authors thank Naohiro Kosugi and Daisuke Sasano for kindly providing the repeat hydrographic data along 137°E. Thanks are extended to Akane Kishi for collection of tide gauge observations and SLP data and subsequent data processing. Comments from two anonymous reviewers greatly helped improve an early version of the manuscript. The tide gauge data and SLA data were obtained from the web sites of the Japan Oceanographic Data Center and JMA, respectively. The reanalysis experiment was carried out on the Earth Simulator with the support of JAMSTEC. This work was supported by JSPS KAKENHI grant nos. 26400472 and 16K17805 and by the Social Implementation Program on Climate Change Adaptation Technology(SI-CAT) of the Ministry of Education, Culture, Sports, Science and Technology (MEXT).

REFERENCES

1. AVISO (2015) MSLA and (M)ADT near-real time and delayed time products. CLS-DOS-NT-06-034, Issue 4.4, Date:2015/06/30, Nomenclature: SALP-MU-P-EA-21065-CLS
2. Balmaseda M, Mogensen K, Weaver A (2013) Evaluation of the ecmwf ocean reanalysis system oras4. Q J R Meteorol Soc 139:1132–1161
3. Balmaseda M, Hernandez F, Storto A, Palmer M, Alves O, Shi L, Smith G, Toyoda T, Valdivieso M, Barnier B, Behringer D, Boyer T, Chang YS, Chepurin G, Ferry N, Forget G, Fujii Y, Good S, Guinehut S, Haines K, Ishikawa Y, Keeley S, Köhl A, Lee T, Martin M, Masina S, Masuda S, Meyssignac B, Mogensen K, Parent L, Peterson K, Tang Y, Yin Y, Vernieres G, Wang X, Waters J, Wedd R, Wang O, Xue Y, Chevallier M, Lemieux JF, Dupont F, Kuragano T, Kamachi M, Awaji T, Caltabiano A, Wilmer-Becker K, Gaillard F (2015) The ocean reanalyses intercomparison project (ora-ip). J Oper Oceanogr 8(S1):s127–s146
4. Bloom SC, Takacs LL, Silva AMD, Ledvina D (1996) Data assimilation using incremental analysis updates. Mon Wea Rev 124:1256–1271
5. Boyer TP, Antonov JI, Baranova OK, Coleman C, Garcia HE, Grodsky A, Johnson DR, Locarnini RA, Mishonov AV, O'Brien TD, Paver CR, Reagan JR, Seidov D, Smolyar IV, Zweng MM (2013) World ocean database 2013. In: Levitus S, Mishonov A (eds) NOAA Atlas NESDIS 72, pp 209
6. Brodeau L, Barnier B, Treguier AM, Penduff T, Gulev S (2010) An era40-based atmospheric forcing for global ocean circulation models. Ocean Modell 31:88–104
7. Carton JA, Giese BS (2008) A reanalysis of ocean climate using simple ocean data assimilation (soda). Mon Wea Rev 136:2999–3017
8. Cho YK, Seo GH, Choi BJ, Kim S, Kim Y, Dever EP (2009) Connectivity among straits of the northwest pacific marginal seas. J Geophys Res 114:C06018. doi:10.1029/2008JC005218

9. D'Asaro E, Black P, Centurioni L, Harr P, Jayne S, Lin II, Lee C, Morzel J, Mrvaljevic R, Niiler P, Rainville L, Sanford T, Tang T (2011) Typhoon-ocean interaction in the western north pacific: Part 1. Oceanography 24:24–31
10. Egbert G, Erofeeva S (2002) Efficient inverse modeling of barotropic ocean tides. J Atmos Oceanic Technol 19:183–204
11. Endoh T, Hibiya T (2001) Numerical simulation of the transient reponse of the kuroshio leading to the large meander formation south of japan. J Geophys Res 106:26,833–26,850
12. Fujii Y (2005) Preconditioned optimizing utility for large-dimensional analyses (popular). J Oceanogr 61:167–181
13. Fujii Y, Kamachi M (2003) Three-dimensional analysis of temperature and salinity in the equatorial pacific using a variational method with vertical coupled temperature-salinity eof modes. J Geophys Res 108:3297. doi:10.1029/2002JC001745
14. Fujii Y, Ishizaki S, Kamachi M (2005) Application of nonlinear constraints in a three-dimensional variational ocean analysis. J Oceanogr 61:655–662
15. Fujii Y, Tsujino H, Usui N, Nakano H, Kamachi M (2008) Application of singular vector analysis to the kuroshio large meander. J Geophys Res. doi:10.1029/2007JC004476
16. Fujii Y, Nakano T, Usui N, Matsumoto S, Tsujino H, Kamachi M (2013) Pathways of the north pacific intermediate water identified through the tangent linear and adjoint models of an ocean general circulation model. J Geophys Res 118:2035–2051
17. Fujii Y, Tsujino H, Toyoda T, Nakano H (2015) Enhancement of the southward return flow of the atlantic meridional overturning circulation by data assimilation and its influence in an assimilative ocean simulation forced by core-ii atmospheric forcing. Clim Dyn **(in press)**
18. Fukamachi Y, Tanaka I, Ohshima KI, Ebuchi N, Mizuta G, Yoshida H, Takayanagi S, Wakatsuchi M (2008) Volume transport of the soya warm current revealed by bottom-mounted adcp and ocean-radar measurement. J Oceanogr 64:385–392
19. Fukamachi Y, Ohshima KI, Ebuchi N, Bando T, Ono K, Sano M (2010) Volume transport in the soya strait during 2006–2008. J Oceanogr 66:685–696
20. Fukudome K, Yoon JH, Alexander O, Takikawa T, Hans IS (2010) Seasonal volume transport variation in the tsushima warm current through the tsushima straits from 10 years of adcp observations. J Oceanogr 66:539–551
21. Fukumori I (2002) A partitioned kalman filter and smoother. Mon Wea Rev 130:1370–1383
22. Fukumori I, Raghunath R, Fu LL (1998) Nature of global large-scale sea level variability in relation to atmospheric forcing: a modelling study. J Geophys Res 103:5493–5512
23. Griffies SM, Hallberg RW (2000) Biharmonic friction with a smagrinsky- like viscosity for use in large-scale eddy-permitting ocean models. Mon Wea Rev 128:2935–2946
24. Griffies SM, Biastoch A, Böning C, Bryan F, Danabasoglu G, Chassignet EP, England MH, Gerdes R, Haak H, Hallberg RW, Hazeleger W, Jungclaus J, Large WG, Madec G, Pirani A, Samuels BL, Scheinert M, Gupta AS, Severijns CA, Simmons HL, Treguier AM, Winton M, Yeager S, Yin J (2009) Coordinated ocean-ice reference experiments (cores). Ocean Modell 26:1–46
25. Hamilton D (1994) Gtspp builds an ocean temperature-salinity database. Earth Sys Monitor 4:4–5
26. Han A, Hirose N, Usui N, Miyazawa Y (2015) Multi-model ensemble estimation of volume transport through the straits of the east/japan sea. Ocean Dyn 66:59–76
27. Hanawa K, Mitsudera F (1985) On the data processings of daily mean values of oceanographical data: note on the daily mean sealevel data (in japanese). Bull Coast Oceanogr 23:4–5
28. Hirose N (2005) Least-squares estimation of bottom topography using horizontal velocity measurements in the Tsushima/Korea Straits. J Oceanogr 61:789–794
29. Hirose N, Nishimura K, Yamamoto M (2009) Observational evidence of a warm ocean current preceding a winter teleconnection pattern in the northwestern pacific. Geophys Res Lett 36:L09705. doi:10.1029/2009GL037448
30. Hosoda S, Ohira T, Sato K, Suga T (2010) Improved description of global mixed-layer depth using argo profiling floats. J Oceanogr 66:773–787

31. Hunke EC, Ducowicz JK (1997) An elastic-viscous-plastic model for sea ice dynamics. J Phys Oceanogr 27:1849–1867
32. Hunke EC, Ducowicz JK (2002) The elastic-viscous-plastic sea ice dynamics model in general orthogonal curvilinear coordinates on a sphere: Incorporation of metric terms. Mon Weather Rev 130:1848–1865
33. Imawaki S, Uchida H, Ichikawa H, Fukasawa M, Umatani S, the ASUKA Group (2001) Satellite altimeter monitoring the kuroshio transport south of Japan. Geophys Res Lett 28:17–20
34. Ishizaki H, Ishikawa I (2004) Simulation of formation and spreading of salinity minimum associated with npiw using a high-resolution model. J Oceanogr 60:463–485
35. Isoguchi O, Kawamura H (2006) Oyashio seasonal intensification and its effect on subsurface temperature variation off the Sanriku coast. J Geophys Res. doi:10.1029/2006JC003628
36. Ito T, Togawa O, Ohnishi M, Isoda Y, Nakayama T, Shima S, Kuroda H, Iwahashi M, Sato C (2003) Variation of velocity of volume transport of the Tsugaru warm current in the winters of 1999–2000. Geophys Res Lett 30:1678
37. Itoh S, Yasuda I (2010) Characteristics of mesoscale eddies in the Kuroshio-Oyashio extension region detected from the distribution of the sea surface height anomaly. J Phys Oceanogr 40:1018–1034
38. Kalnay E, Kanamitsu M, Kistler R, Collins W, Deaven D, Gandin L, Iredell M, Saha S, White G, Woollen J, Zhu Y, Chelliah M, Ebisuzaki W, Higgins W, Janowiak J, Mo KC, Ropelewski C, Wang J, Leetmaa A, Reynolds R, Jenne R, Joseph D (1996) The NCEP/NCAR 40-year reanalysis project. Bull Amer Meteor Soc 77:437–470
39. Kawabe M (1980) Sea level variations along the south coast of Japan and the large meander in the Kuroshio. J Oceanogr Soc Japan 36:97–104
40. Kawabe M (1995) Variations of current path, velocity, and volume transport of the Kuroshio in relation with the large meander. J Phys Oceanogr 25:3103–3117
41. Kawai H (1972) Hydrography of the Kuroshio extension. In: Stommel H, Yoshida K (eds) Kuroshio—its physical aspects. University of Tokyo Press, Tokyo, pp 235–352
42. Kelly KA, Small RJ, Samelson RM, Qiu B, Joyce TM, Kwon YO, Cronin MF (2010) Western boundary currents and frontal airsea interaction: Gulf stream and Kuroshio extension. J Clim 23:5644–5667
43. Kida S, Mitsudera H, Aoki S, Guo X, Ito S, Kobashi F, Komori N, Kubokawa A, Miyama T, Morie R, Nakamura H, Nakamura T, Nakano H, Nishigaki H, Nonaka M, Sasaki H, Sasaki YN, Suga T, Sugimoto S, Taguchi B, Takaya K, Tozuka T, Tsujino H, Usui N (2015) Oceanic fronts and jets around Japan: a review. J Oceanogr 71:469–497
44. Kimura N, Wakatsuchi M (2004) Increase and decrease of sea ice area in the sea of okhotsk:ice production in coastal polynyas and dynamic thickening in convergence zones. J Geophys Res 109:C09S03. doi:10.1029/2003JC001864
45. Kobashi F, Hanawa K (2004) Hydrographic features off the southeast coast of Kyushu during the Kuroshio small meanders: a case study for small meanders that occurred in 1994 and 1995 spring. J Oceanogr 60:645–661
46. Kobayashi S, Ota Y, Harada Y, Ebita A, Moriya M, Onoda H, Onogi K, Kamahori H, Kobayashi C, Endo H, Miyaoka K, Takahashi K (2015) The JRA-55 reanalysis: general specifications and basic characteristics. J Meteor Soc Japan 93:5–48
47. Köhl A, Stammer D (2008) Decadal sea level changes in the 50-year GECCO ocean synthesis. J Clim 21:1876–1890
48. Kondo J (1975) Air-sea bulk transfer coefficients in diabatic conditions. Bound Layer Meteorol 9:91–112
49. Kono T, Kawasaki Y (1997) Modification of the western subarctic water by exchange with the Okhotsk sea. Deep Sea Res 44:689–711
50. Kuragano T, Kamachi M (2000) Global statistical space-time scales of oceanic variability estimated from the TOPEX/POSEIDON altimetry data. J Geophys Res 105:955–974
51. Kuragano T, Fujii Y, Toyoda T, Usui N, Ogawa K, Kamachi M (2014) Seasonal barotropic sea surface height fluctuation in relation to regional ocean mass variation. J Oceanogr 70:45–62

52. Kurihara Y, Sakurai T, Kuragano T (2006) Global daily sea surface temperature analysis using data from satellite microwave radiometer, satellite infrared radiometer and in-situ observations. Weath Bull 73:s1–s18 (in Japanese)
53. Kuroda H, Wagawa T, Shimizu Y, Ito S, Kakehi S, Okunishi T, Ohno S, Kusaka A (2015) Interdecadal decrease of the oyashio transport on the continental slope off the southeastern coast of Hokkaido, Japan. J Geophys Res 120:2504–2522. doi:10.1002/2014JC010402
54. Levitus S (1982) Climatological atlas of the world ocean. NOAA Professional Paper 13, NOAA/ERL GFDL, Princeton, New Jersey
55. Levitus S (1987) A comparison of the annual cycle of two sea surface temperature climatologies of the world ocean. J Phys Oceanogr 17:197–214
56. Lindsay RW, Zhang J (2006) Assimilation of ice concentration in an ice-ocean model. J Atmos Oceanic Technol 23:742–749
57. Locarnini RA, Mishonov AV, Antonov JI, Boyer TP, Garcia HE, Baranova OK, Zweng MM, Paver CR, Reagan JR, Johnson DR, Hamilton M, Seidov D (2013) World ocean atlas 2013, volume 1: temperature. In: Levitus S, Mishonov A (eds) NOAA Atlas NESDIS 73, pp 40
58. Mantua NJ, Hare SR, Zhang Y, Wallace JM, Francis RC (1997) A Pacific interdecadal climate oscillation with impacts on salmon production. Bull Amer Meteor Soc 78:1069–1079
59. Martin S, Drucker R, Yamashita K (1998) The production of ice and dense shelf water in the Okhotsk sea polynyas. J Geophys Res 103:27,771–27,782
60. Masuzawa J (1969) Subtropical mode water. Deep Sea Res 16:463–472
61. Mellor GL, Kantha L (1989) An ice-ocean coupled model. J Geophys Res 94:10,937–10,954
62. Miller AJ, Chai F, Chiba S, Moisan JR, Neilson DJ (2004) Decadalscale climate and ecosystem interactions in the North Pacific ocean. J Oceanogr 60:163–188
63. Miyazawa Y, Zhang T, Guo X, Tamura H, Ambe D, Lee JS, Okuno A, Yoshinari H, Setou T, Komatsu K (2009) Water mass variability in the western North Pacific detected in a 15-year eddy resolving ocean reanalysis. J Oceanogr 65:737–756
64. Moore AM, Arango HG, Broquet G, Edwards C, Veneziani M, Powell BS, Foley D, Doyle J, Costa D, Robinson P (2011) The regional ocean modeling system (roms) 4-dimensional variational data assimilation systems: Part II—performance and application to the California current system. Prog Oceanogr 91:50–73
65. Na H, Isoda Y, Kim K, Kim YH, Kim SJ (2009) Recent observations in the straits of the East/Japan Sea; a review of hydrography, currents and volume transports. J Mar Syst 78:200–205
66. Na H, Kim KY, Cheng KI, Park JJ, Kim K, Minobe S (2012) Decadal variability of the upper ocean heat content in the East/Japan Sea and its possible relationship to northwestern Oacific variability. J Geophys Res 117:C02917. doi:10.1029/2011JC007369
67. Nakamura H, Nishina A, Minobe S (2012) Response of storm tracks to bimodal Kuroshio path states south of Japan. J Clim 25:6881–6893
68. Nakano H, Ishikawa I (2010) Meridional shift of the Kuroshio extension induced by response of recirculation gyre to decadal wind variations. DeepNSea Res 57:1111–1126
69. Nakanowatari T, Ohshima KI, Nagai S (2010) What determines the maximum sea ice extent in the Sea of Okhotsk? Importance of ocean thermal condition from the Pacific. J Geophys Res 115:C12031. doi:10.1029/2009JC006070
70. Nishida Y, Kanomata I, Tanaka I, Sato S, Takahashi S, Matsubara H (2003) Seasonal and interannual variations of the volume transport through the Tsugaru strait. Umi no kenkyu 12:487–499 (in Japanese with English abstract)
71. Nishikawa H, Yasuda I, Itoh S (2011) Impact of winter-to-spring environmental variability along the Kuroshio jet on the recruitment of Japanese sardine. Fish Oceanogr 20:570–582
72. Nishikawa S, Tsujino H, Sakamoto K, Nakano H (2010) Effects of mesoscale eddies on subduction and distribution of subtropical mode water in an eddy-resolving OGCM of the western North Pacific. J Phys Oceanogr 40:1748–1765
73. Noh Y, Kim HJ (1999) Simulations of temperature and turbulence structure of the oceanic boundary layer with the improved nearsurface processes. J Geophys Res 104:15,621–15,634

74. Ohshima KI, Nihashi S, Hashiya E, Watanabe T (2006) Interannual variability of sea ice area in the sea of Okhotsk: importance of surface heat flux in fall. J Meteor Soc Japan 84:907–919
75. Oka E, Qiu B, Takatani Y, Enyo K, Sasano D, Kosugi N, Ishii M, Nakano T, Suga T (2015) Decadal variability of subtropical mode water subduction and its impact on biogeochemistry. J Oceanogr 71:389–400
76. Oke PR, Schiller A (2007) Impact of ARGO, SST, and altimeter data on an eddy-resolving ocean reanalysis. Geophys Res Lett 34:L19601. doi:10.1029/2007GL031549
77. Oke PR, Sakov P, Cahill ML, Dunn JR, Fiedler R, Griffin DA, Mansbridge JV, Ridgway KR, Schiller A (2013) Towards a dynamically balanced eddy-resolving ocean reanalysis: BRAN3. Ocean Modell 67:52–70
78. Okuda K (1986) Occurrence of extremely low temperature in the coastal region of the Tohoku area associated with interannual variation of the Oyashio. Bull Tohoku Reg Fish Res Lab 48:87–96 (in Japanese)
79. Onogi K, Tsutsui J, Koide H, Sakamoto M, Kobayashi S, Hatsushika H, Matsumoto T, Yamazaki N, Kamahori H, Takahashi K, Kadokura S, Wada K, Kato K, Oyama R, Ose T, Mannoji N, Taira R (2007) The JRA-25 reanalysis. J Meteor Soc Japan 85:369–432
80. Osafune S, Masuda S, Sugiura N, Doi T (2015) Evaluation of the applicability of the estimated state of the global ocean for climate research (ESTOC) data set. Geophys Res Lett 42:4903-4911
81. Polavarapu S, Tanguay M, Fillion L (2000) Four-dimensional variational data assimilation with digital filter initialization. Mon Wea Rev 128:2491–2510
82. Qiu B (2003) Kuroshio extension variability and forcing of the pacific decadal oscillations: responses and potential feedback. J Phys Oceanogr 33:2465–2482
83. Qiu B, Chen S (2005) Variability of the Kuroshio extension jet, recirculation gyre and mesoscale eddies on decadal timescales. J Phys Oceanogr 35:2090–2103
84. Saha S, Moorthi S, Pan HL, Wu X, Wang J, Nadiga S, Tripp P, Kistler R, Woollen J, Behringer D, Liu H, Stokes D, Grumbine R, Gayno G, Wang J, Hou YT, Chuang HY, Juang HMH, Sela J, Iredell M, Treadon R, Kleist D, Delst PV, Keyser D, Derber J, Ek M, Meng J, Wei H, Yang R, Lord S, Dool HVD, Kumar A, Wang W, Long C, Chelliah M, Xue Y, Huang B, Schemm JK, Ebisuzaki W, Lin R, Xie P, Chen M, Zhou S, Higgins W, Zou CZ, Liu Q, Chen Y, Han Y, Cucurull L, Reynolds RW, Rutledge G, Goldberg M (2010) The NCEP climate forecast system reanalysis. Bull Am Met Soc 91:1015–1057
85. Sasaki YN, Minobe S, Asai T, Inatsu M (2012) Influence of the Kuroshio in the east china sea on the early summer (baiu) rain. J Climate 25:6627–6645
86. Scott FA, Wang W, Stammer D (2007) Variability of pacific subtropical cells in the 50-year ecco assimilation. Geophys Res Lett 34:L05604. doi:10.1029/2006GL028478
87. Sekine Y (1988) Anomalous southward intrusion of the Oyashio east of Japan: 1. Iinfluence of the seasonal and interannual variations in the wind stress over the North Pacific. J Geophys Res 93:2247–2255
88. Sekine Y (1999) Anomalous southward intrusion of the Oyashio east of Japan 2. Two-layer numerical model. J Geophys Res 104:3049–3058
89. Seo Y, Sugimoto S, Hanawa K (2014) Long-term variations of the Kuroshio extension path in winter: meridional movement and path state change. J Clim 27:5929–5940
90. Shcherbina A, Talley LD, Rudnick DL (2003) Direct observations of brine rejection at the source of North Pacific intermediate water in the Okhotsk sea. Science 302:1952–1955
91. Shikama N (1994) Current measurement in the tsugaru strait using the bottom-mounted ADCPS. Kaiyo Mon 26:815–818 (in Japanese)
92. Storto A, Dobricic S, Masina S, Pietro PD (2011) Assimilating alongtrack altimetric observations through local hydrostatic adjustments in a global ocean reanalysis system. Mon Wea Rev 139:738–754
93. Taguchi B, Xie SP, Schneider N, Nonaka M, Sasaki H, Sasai Y (2007) Decadal variability of the Kuroshio extension: observations and an eddy-resolving model hindcast. J Clim 20:2357–2377

94. Taguchi B, Nakamura H, Nonaka M, Xie SP (2009) Influences of the Kuroshio/Oyashio extensions on air-sea heat exchanges and storm-track activity as revealed in regional atmospheric model simulations for the 2003/04 cold season. J Climate 24:6536–6560
95. Talley LD (1991) An okhotsk sea water anomaly: implications for ventilation in the North Pacific. Deep Sea Res 38(Suppl. 1):S171–S190
96. Talley LD (1993) Distribution and formation of North Pacific intermediate water. J Phys Oceanogr 23:517–537
97. Toyoda T, Fujii Y, Yasuda T, Usui N, Iwao T, Kuragano T, Kamachi M (2013) Improved analysis of the seasonal-interannual fields by a global ocean data assimilation system. Theor Appl Mech Jpn 61:31–48
98. Toyoda T, Fujii Y, Kuragano T, Kamachi M, Ishikawa Y, Masuda S, Sato K, Awaji T, Hernandez F, Ferry N, Guinehut S, Martin M, Peterson KA, Good S, Valdivieso M, Haines K, Storto A, Masina S, Köhl A, Zuo H, Balmaseda M, Yin Y, Shi L, Alves O, Smith G, Chang YS, Vernieres G, Wang X, Forget G, Heimbach P, Wang O, Fukumori I, Lee T (2015a) Intercomparison and validation of the mixed layer depth fields of global ocean syntheses. Clim Dyn. doi:10.1007/s00382-015-2637-7
99. Toyoda T, Fujii Y, Kuragano T, Kosugi N, Sasano D, Kamachi M, Ishikawa Y, Masuda S, Sato K, Awaji T, Hernandez F, Ferry N, Guinehut S, Martin M, Peterson KA, Good S, Valdivieso M, Haines K, Storto A, Masina S, Köhl A, Yin Y, Shi L, Alves O, Smith G, Chang YS, Vernieres G, Wang X, Forget G, Heimbach P, Wang O, Fukumori I, Lee T, Zuo H, Balmaseda M (2015b) Interannual-decadal variability of wintertime mixed layer depths in the north pacific detected by an ensemble of ocean syntheses. Clim Dyn. doi:10.1007/s00382-015-2762-3
100. Tsujino H, Yasuda T (2004) Formation and circulation of mode waters of the North Pacific in a high-resolution GCM. J Phys Oceanogr 34:399–417
101. Tsujino H, Usui N, Nakano H (2006) Dynamics of Kuroshio path variations in a high-resolution GCM. J Geophys Res 111:C11001. doi:10.1029/2005JC0031180
102. Tsujino H, Motoi T, Ishikawa I, Hirabara M, Nakano H, Yamanaka G, Yasuda T, Ishizaki H (2010) Reference manual for the meteorological research institute community ocean model (mri.com) version 3. Tech. Rep. 59, Meteorol. Res. Inst., Tsukuba, Japan
103. Ueno H, Yasuda I (2000) Distribution and formation of the mesothermal structure (temperature inversions) in the North Pacific subarctic region. J Geophys Res 105:16,885–16,897
104. Uppala SM, Kållberg PW, Simmons AJ, Andrae U, Bechtold VDC, Fiorino M, Gibson JK, Haseler J, Hernandez A, Kelly GA, Li X, Onogi K, Saarinen S, Sokka N, Allan RP, Andersson E, Arpe K, Balmaseda MA, Beljaars ACM, Berg LVD, Bidlot J, Bormann N, Caires S, Chevallier F, Dethof A, Dragosavac M, Fisher M, Fuentes M, Hagemann S, Hólm E, Hoskins BJ, Isaksen L, Janssen PAEM, Jenne R, Mcnally AP, Mahfouf JF, Morcrette JJ, Rayner NA, Saunders RW, Simon P, Sterl A, Trenberth KE, Untch A, Vasiljevic D, Viterbo P, Woollen J (2005) The ERA-40 re-analysis. Q J R Meteorol Soc 131:2961–3012
105. Usui N, Fujii Y (2015) Improvements of MOVE-4DVAR toward operational use. Weath Bull 82:s43–s53 (in Japanese)
106. Usui N, Ishizaki S, Fujii Y, Tsujino H, Yasuda T, Kamachi M (2006a) Meteorological research institute multivariate ocean variational estimation (move) system: some early results. Adv Space Res 37:806–822
107. Usui N, Tsujino H, Fujii Y, Kamachi M (2006b) Short-range prediction experiments of the Kuroshio path variabilities south of Japan. Ocean Dyn 56:607–623
108. Usui N, Tsujino H, Nakano H, Fujii Y (2008) Formation process of the Kuroshio large meander in (2004) J Geophys Res 113:C08047. doi:10.1029/2007JC004675
109. Usui N, Imaizumi T, Tsujino H (2010) Toward introduction of assimilation of ice concentration into move/mri.com: a model validation and a simple assimilation experiment in the sea of Okhotsk. Weath Bulletin 77:s72–s82 (in Japanese)
110. Usui N, Ishizaki S, Fujii Y, Kamachi M (2011) Improving strategies with constraints regarding non-gaussian statistics in a three-dimensional variational assimilation method. J Oceanogr 67:253–262

111. Usui N, Tsujino H, Nakano H, Matsumoto S (2013) Long-term variability of the Kuroshio path south of Japan. J Oceanogr 69:647–670
112. Usui N, Fujii Y, Sakamoto K, Kamachi M (2015) Development of a four-dimensional variational assimilation system for coastal data assimilation around Japan. Mon Wea Rev 143:3874–3892
113. Yasuda I (2003) Hydrographic structure and variability in the Kuroshio- Oyashio transition area. J Oceanogr 59:858–402
114. Yasuda I, Okuda K, Shimizu Y (1996) Distribution and modification of north pacific intermediate water in the Kuroshio-Oyashio interfrontal zone. J Phys Oceanogr 26:448–465
115. Zweng MM, Reagan JR, Antonov JI, Locarnini RA, Mishonov AV, Boyer TP, Garcia HE, Baranova OK, Johnson DR, DSeidov, Biddle MM (2013) World ocean atlas 2013, volume 2: salinity. NOAA Atlas NESDIS 74

CHAPTER 11

Eocene (Lutetian) Shark-Rich Coastal Paleoenvironments of the Southern North Sea Basin in Europe: Biodiversity of the Marine Fürstenau Formation Including Early White and Megatooth Sharks

C. G. Diedrich

PaleoLogic, Nansenstraße 8, D-33790 Halle, Germany

ABSTRACT

The Fürstenau Formation (Lutetian, Paleogene, Eocene) is based on type sections near Fürstenau in Germany (central Europe) and is built of 22 meter thick marine glauconitic and strongly bioturbated sands, clays, and a vertebrate-rich conglomerate bed. The conglomerate layer from the Early Lutetian transgression reworked Lower Cretaceous, and Paleogene marine sediments. It is dominated by pebbles from the locally mountains which must have been transported by an ancient river in a delta fan. Marine reworked Lower Cretaceous and Paleogen pebbles/fossils, were derived from the underlying deposits of northern Germany (= southern pre North Sea basin). The benthic macrofauna is cold upwelling water influenced and non-tropical, and medium divers. The vertebrate fish fauna is extremely rich in shark teeth, with about 5,000 teeth per cubic meter of gravel. The most dominant forms are teeth from sand shark ancestors Striatolamia macrota, followed by white shark ancestors Carcharodon auriculatus. Even teeth from the magatooth shark ancestor Carcharocles sokoloviare present in a moderately diverse and condensed Paleogene fish fauna that also includes rays, chimaeras, and more then 80 different bony fish. Fragmentary turtle remains are present, and few terrestrial vertebrates and even marine mammals with phocids, sirenians and possibly whales.

1. INTRODUCTION

Early research in the Dalum fossil locality in the Fürstenau region of north-west Germany (Figure 1(a)) concentrated largely on the stratigraphy and the fossil shark tooth content [1]. A preliminary analysis of the conglomerate components has also been published by Bartholomäus [2]. The site has only been excavated by private collectors, who have been active there since 1971 [2]. Whereas aragonite and calcite fossils at Dalum have all been dissolved away, they are still preserved in a sand pit at Osteroden (Figure 1(b)) from which 11,000 fish otoliths have previously been analysed by Schwarzhans [3]. The extremely rare teeth of terrestrial mammals from the Dalum and Osteroden sites were the focus of a paper by Franzen and Mörs [4].

The Paleogene (Middle Eocene) marine localities at Dalum and Osteroden (Figures 1(a) and 1(b)), in the southern pre-North Sea basin, represent an important environmental bridge between marine, coastal, deltaic swamp, and terrestrial faunas (Figure 1(c)). The marine-dominated fossils at Dalum also offer valuable information on the paleobiodiversity of the central European tropical terrestrial faunas during the Eocene.

Other important marine Eocene sites in Europe are at Helmsleben in Germany, and at Oosterzele and Balegem in Belgium (Figure 1(c)), which are also well known for their enormous quantities of shark teeth [5] and which also contain rare remains of terrestrial mammals [4]. The marine Paris Basin to the south of the Artois submarine swell (Figure 1(c)) is also one of the classic fossil regions of Europe in which many of the shark species presented herein were first found, as described in the monograph by Lériche [6], as well as by other authors. Finally, the Lower Eocene London Clay Formation in England, for example, on the Isle of Sheppey (Figure 1(c)), represents near-shore, fresh-water-influenced, coastal paleoenvironments that contain overlapping terrestrial and marine faunal remains [7].

The Eocene terrestrial fauna (Figure 1(c)) was made famous by the abundance of early mammal, bird, reptile, anuran, and fish skeletons, as well as botanical specimens and insects, from the German at UNESCO World Heritage Messel sites at Geiseltal (an Early to Middle Eocene lake site) [8] and at Messel (Middle Eocene) [9]. The Eckfeld Maar volcano lake site has also yielded important fossil material of terrestrial fauna and flora from the Middle Eocene [10], as have the Eocene coal mine at Stolzenbach [11] and the Swiss mammal site at Egerkingen, which is of Upper Eocene of age [12].

(b)

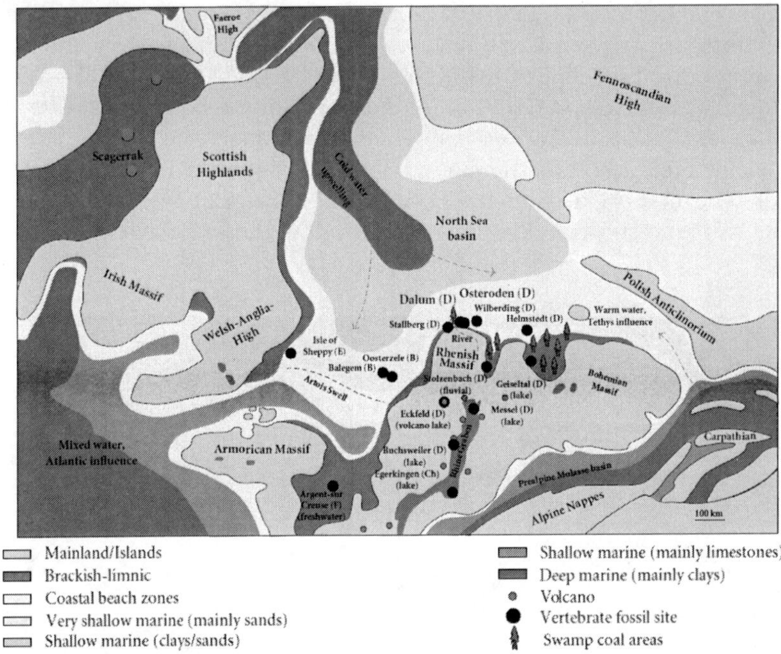

(c)

Figure 1: (a) The Dalum site near Fürstenau and (b) the Osteroden sand pit near Bippen, in north-western Germany. (c) Paleogeography of important Eocene marine and terrestrial vertebrate fossil sites of Europe (compiled from [4, 5, 9–14]).

This study presents preliminary stratigraphic and sedimentological analyses focussed on the marine, terrestrially influenced, vertebrate-rich gravels at Dalum, and includes a preliminary statistically based overview of the site with regard to the quality and quantity of its fossils. This information is used for a paleoenvironmental reconstruction resulting in the compilation of a new paleogeographical map, and to provide an overview of the (mainly marine) biodiversity, and marine palaeocurrents on the southern pre North Sea basin.

2. MATERIAL AND METHODS

In May and June of 2011 the UNESCO-supported "Geo and Naturpark TERRA.Vita" near Osnabrück (NW Germany) started a systematic excavation and research project (under the leadership of the PaleoLogic private research institute) at the Dalum forest site near Fürstenau (Figure 1(a)) and at a second locality in the Osteroden sand pit, Bippen (Figure 1(b)). The eventual objective, following the completion of the research program, was to develop and promote the Shark Center at Bippen, which is a visitor and research center within the UNESCO-supported "Geo and Naturpark TERRA.Vita" of north-west Germany.

In May 2011 a 4-day field program using heavy machinery was completed at the Dalum and Osteroden sites (Figures 1(a) and 1(b)). Stratigraphic cross-sections were prepared at both of these sites (making use of 5-meter drillholes at Osteroden) indicating a combined 22-meter thick marine section (Figure 3).

A total of 180 cubic meters of conglomeratic material was excavated from the Dalum site for sieving, of which only 250 buckets (10 litres each) were sieved during the following two weeks (i.e., only 0.1% of the recovered material). It was only possible to excavate 10 cubic meters of material at the Osteroden site, which has not yet been sieved (other than for a preliminary sample) because of the intense weathering and damage shown by all of the fossils, and the large degree of root dissolution in the teeth.

Attention has therefore been focussed on the Dalum gravel, which was sieved into two different fractions of +4 mm and −4 to +1 mm, in order to obtain a good variety of gravel types (Figures 4–5), invertebrate remains (Figures 6–8), and vertebrate remains (Figures 9–15), and to facilitate the gravel pebble analysis. Only material greater then 4 mm is presented herein, which is adequate for qualitative and quantitative gravel analyses and for analysis of the main vertebrate fauna. In total, 14,437 fossils were recovered and studied, of which 95% are shark teeth and only 5% (747 specimens) are teeth from other types of animals (Figure 5).

More than half of the 13,683 shark teeth recovered were identifiable, but lower proportions of the ray teeth and fin spines and the bone fish teeth and bones. The finer material (less than 4 mm) appears (after sieving a preliminary sample) to contain many additional species, which will be recovered and examined during the coming years from the retained fine fraction of the gravel. A separate, more detailed fish analysis (covering all material >1 mm) will also be the subject of future research, after much more material has been recovered to allow a more complete picture of the fish and invertebrate paleobiodiversity in the pre-North Sea basin during the Eocene.

All the material described and illustrated herein from the 2011 excavations (14,440 fossils in total) is now housed in the Shark Center, Bippen (SCB).

3. GEOLOGY AND SEDIMENTOLOGY

3.1. Glacial Tectonics

The Tertiary sediments in north-west Germany were deformed and compressed as a result of Middle Pleistocene glacial tectonics [13]. The Ankumer range in particular is the result of proglacial folding due to a Saalian glacier, as indicated by a U-shaped moraine that extends as far as the north-western part of the Wiehengebirge mountain range near Bramsche (Figure 2()) [15]. The Eocene sediments at the Dalum (Figure 2()) and Osteroden sites are therefore strongly folded. The fold axis at Dalum is oriented east to west (parallel to the main glacial fold axis, cf. Figure 2()), with a "saddle" that dips at about 10–15° towards the east. In a road cutting about 50 meters to the east of the new sampling trench and parallel to it (Figure 2()) the top of the saddle structure is missing in the road section [1] having been eroded during the Saalian and further the Weichselian Ice Age. At the new section in the forest instead the top of the saddle is preserved and the convex deformed sediment layers that would normally be expected are more or less horizontal (Figure 2()), indicating that the saddle has been compressed from above. At Osteroden the deformation is more complex, and in the recent pit the gravel layers dip at about 30–60° to the north-east. This is also parallel to the main axis of glacial deformation (Figure 2(a)).

3.2. Paleogene Stratigraphy, Sedimentology, and Dating in North-West Germany

No firmly dated early Tertiary (Paleocene) deposits are known in north-west Germany [13].

Early Eocene (Ypresian) marine sediments have, however, been dated by microfossils such as diatoms, radiolarians, and foraminifers [16, 17], and are known from a clay pit at Wilberding, near Damme in north-west Germany [18], and also from another clay pit at Stallberg near Emsbüren (Figure 1(c)) which contains shark remains and reworked iron-carbonate concretions from the Lower Cretaceous [19]. These gray-green, glauconitic marine clays of Wilberding and Stallberg containing phosphorite nodules (phosphorites), that appear to be of Lower Eocene age, represent the earliest known marine Tertiary sediments in north-west Germany. The Stalberg section also contained many mixed layers 1 mm to 40 mm thick of volcanic ash (tuff) and marine sediments, which are explained as being the result of volcanic ash clouds from the marine strait Skagerrak (possibly also from the Eifel volcanos, e.g., the Eckfelder Maar, cf. Figure 1(c)) [20]. This was a tectonically active period that was also responsible for the Rhine Graben structure of central Europe (Figure 1(c)) [20], and it was also the time when the Messel and Geiseltal fossil sites were formed in Germany, together with the Stolzenbach site and the coastal swamps of the Rhenish Massif (Figure 1(c)).

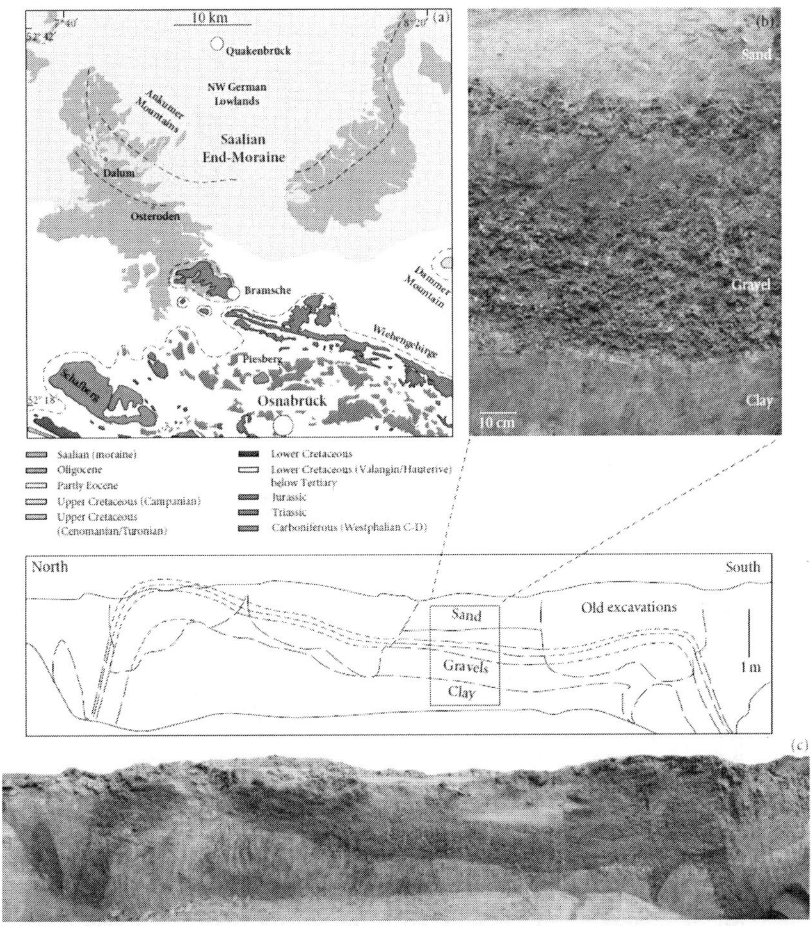

Figure 2: Glacial tectonics and distribution of Tertiary deposits as result of (a) the Saalian Ice Age end-moraine forming the Ankumer Mountains (geology redrawn and simplified from [15]), including the Eocene sites at Dalum and Osteroden; (b) north-south section through the gravels at Dalum; (c) photo of the same section at Dalum, showing the Eocene marine deposits, which where deformed into a saddle-like structure (well visible in the Eocene gravel layer) by glacial compression.

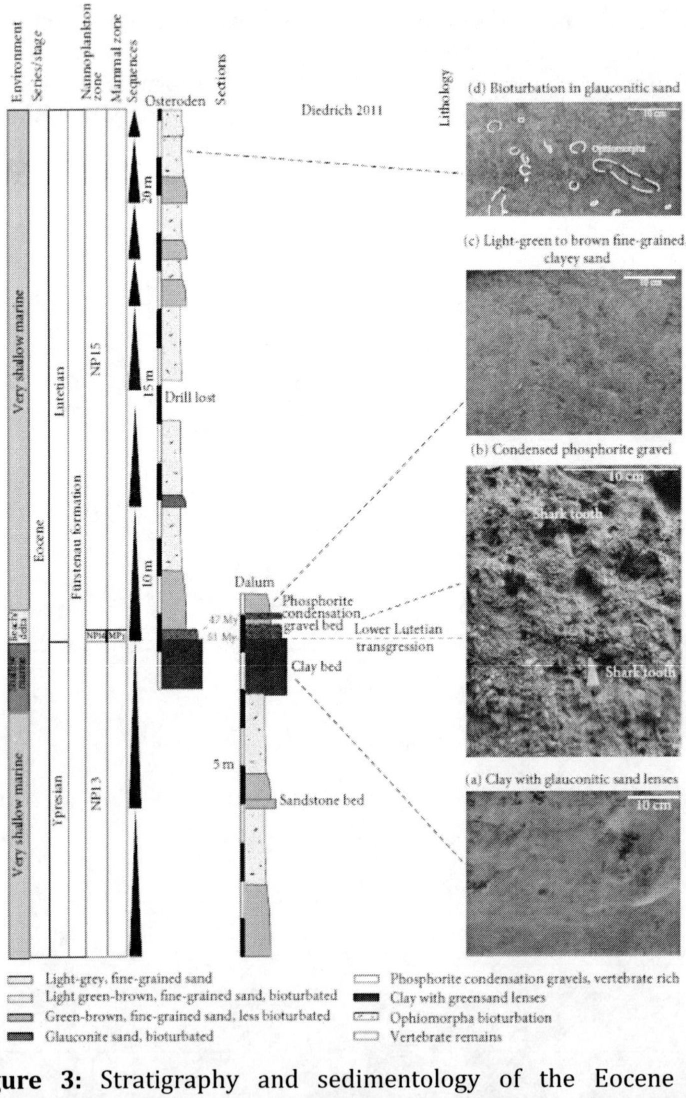

Figure 3: Stratigraphy and sedimentology of the Eocene Fürstenau Formation vertebrate sites at Osteroden and Dalum (NW Germany). (a) Clay with glauconitic sand lenses. (b) Condensed phosphorite gravel bed with abundant vertebrate remains, especially shark teeth. (c) Light-green to brown fine-grained, clayey sand, at the base of the sequence. (d) Bioturbated glauconitic sands, with Ophiomorpha and other burrows (top of sequence).

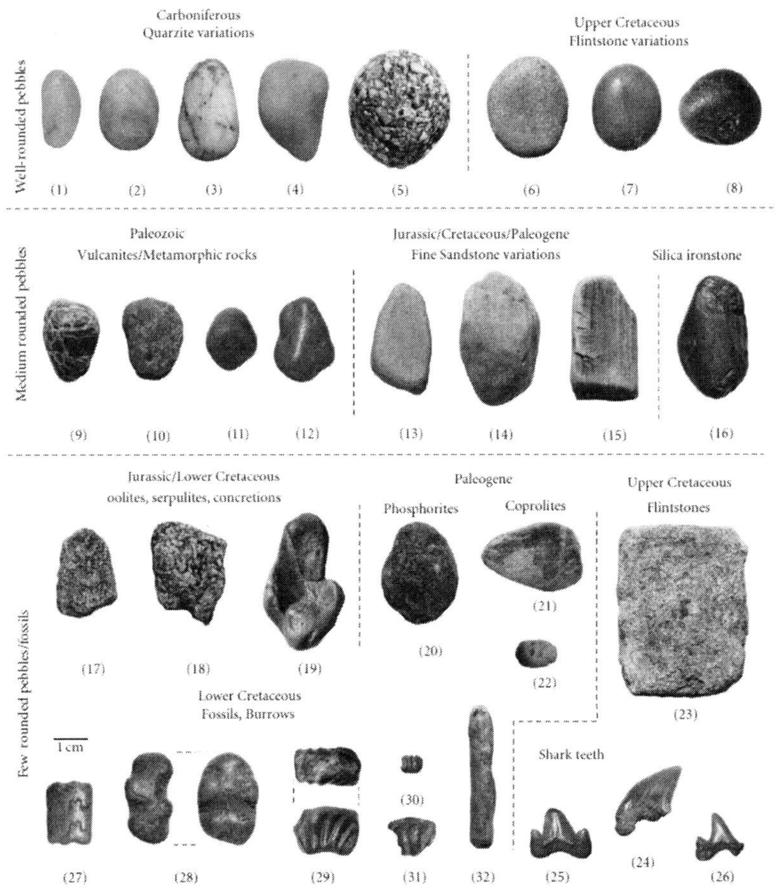

Figure 4: Middle Eocene (basal Lutetian) gravel types. (1–3(?4)) Carboniferous (Westphalian C-D, originating from the Schafberg and Piesberg mountains) quartzite variants which were already well rounded in the Paleozoic, and have only been transported a short distance since then (SCB no. Sedim-2). (5) Carboniferous conglomerate (Westphalian C-D), originating from the Schafberg and Piesberg mountains (SCB no. Sedim-3). (6–8) Upper Cretaceous (Campanian-Maastrichtian) flints (SCB no. Sedim-6), consisting of Campanian "Rijckholt" flint (6-7) from the Aachen region and northern Netherlands (SCB no. Sedim-6a) and "Maas-Egg," black flint (8) from the Maastrichtian of the northern Netherlands (SCB no. Sedim-6b). (9) Paleozoic metamorphite of uncertain origin (SCB no. Sedim-10), (10–12) red Paleozoic volcanic rocks of uncertain origin, possibly from Scandinavia (SCB no. Sedim-13), (13) Paleogene white fine-grained sandstone of uncertain origin (SCB no. Sedim-9), (14-15) Paleogen glauconitic fine-grained sandstones (SCB no. Sedim-11), and (16) Jurassic

silica/iron-carbonate concretion from the Osnabrücker Bergland (SCB no. Sedim-17). (18) Jurassic or Lower Cretaceous serpulid (SCB no. Sedim-4). (19) Iron-carbonate concretion fragment with orthicon (Bochianites) ammonite negatives from the Valanginian/Hauterivian, Lower Cretaceous (SCB no. Sedim-18). (20) Middle Eocene phosphorite nodule (SCB no. Sedim-7). (21-22) Eocene phosphatic shark coprolites (SCB no. Sel-890, 898). (23) Upper Cretaceous (Campanian) platy flint (SCB no. Sedim-8). (24) Shark tooth from Squalicorax kaupi, reworked from the Campanian, Upper Cretaceous (SCB Dal-Sel-901), labial. (25) Shark tooth from Cretolamna appendiculata, reworked from the Santonian-Campanian, Upper Cretaceous (SCB Dal-Sel-882), labial. (26) Shark tooth from Cretolamna appendiculata,reworked from the Campanian, Upper Cretaceous (SCB Dal-Sel-881), labial. (27) Lower Cretaceous (Upper Valanginian) orthicon ammonite fragment of (Bochianites sp. (SCB no. Sedim-18). (28) Lower Cretaceous (Upper Valanginian) orthicon ammonite fragment, a so-called "cat paw" (SCB no. Sedim-17). (29) Lower Cretaceous (Lower Hauteriveian) ammonite fragment of Endemoceras sp. (SCB no. Sedim-22). (30) Lower Cretaceous (Upper Hauteriveian) ammonite fragment of Aegocrioceras sp. (SCB no. Sedim-22). (31) Lower Cretaceous (Upper Valanginian) ammonite fragment of Oostrella sp. (SCB no. Sedim-18). (32) Lower Cretaceous elongated burrow preserved in iron-carbonate concretion (SCB no. Sedim-43).

Middle Eocene (Lutetian) marine deposits for the Dalum and Osteroden sites are presented herein, and a new formation name introduced. A compilation of the two sections from Dalum and Osteroden (Figure 3) has allowed the dating of these sediments, and also enabled sedimentological and paleoenvironmental interpretations to be made.

3.3. The New "Fürstenau Formation" (Ypresian to Lutetian, Early to Middle Eocene)

3.3.1. Type Stratum

Fürstenau Formation, Early to Middle Eocene, after the nearby city of Fürstenau.

3.3.2. Type Localities

The Dalum site near Fürstenau and the Osteroden sand pit near Bippen, north-west Germany.

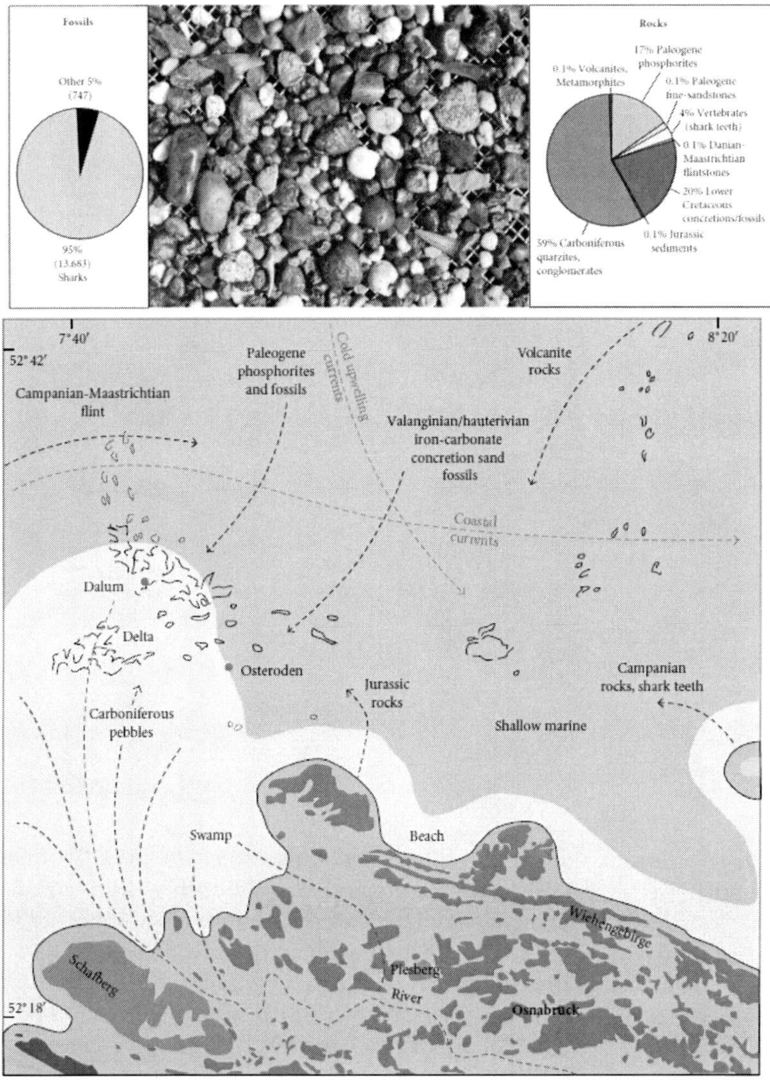

Figure 5: Basal Middle Eocene (Early Lutetian) gravel: quantitative analysis and transportation direction reconstruction. The main Paleozoic components (Carboniferous quartz) must have been transported by an ancient river running between today's Wiehengebirge and Teutoburger Wald mountain ranges, ending in a delta. Other components (Jurassic material) are from the Wiehengebirge range, whereas Lower Cretaceous pebbles and ammonite fossils are from north Germany. Flints were accumulated from different directions (Maastricht/Aachen area, and Dammer Berge; Material no. SCB Dal-Sedim-1–16).

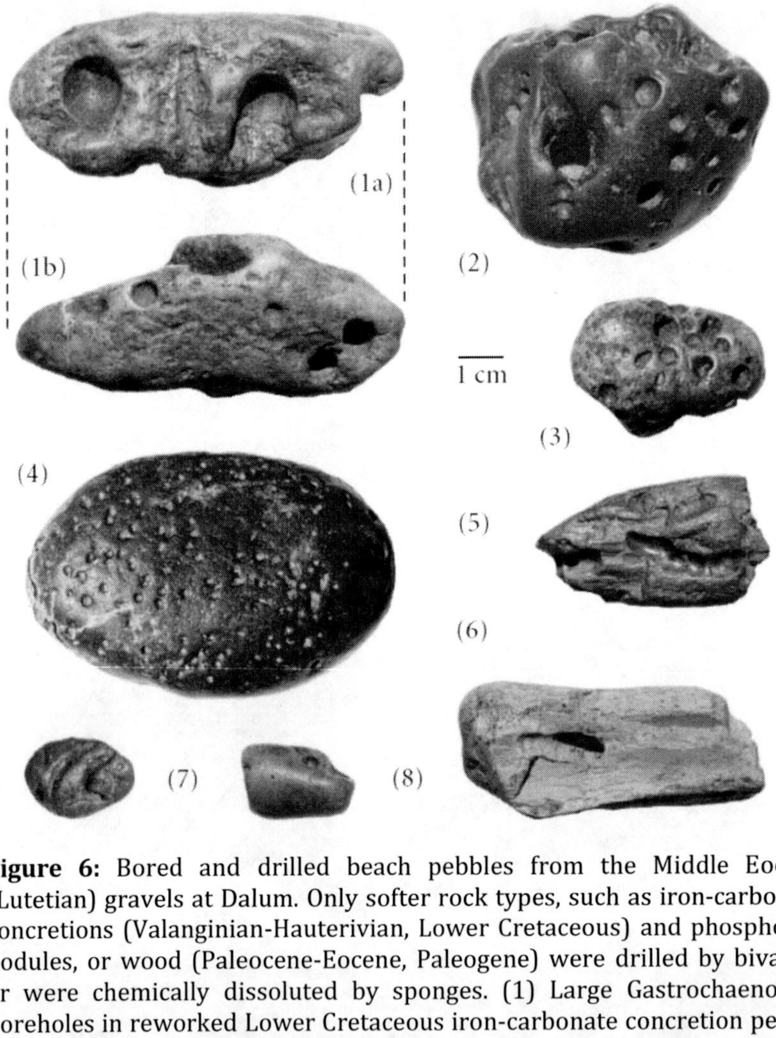

Figure 6: Bored and drilled beach pebbles from the Middle Eocene (Lutetian) gravels at Dalum. Only softer rock types, such as iron-carbonate concretions (Valanginian-Hauterivian, Lower Cretaceous) and phosphorite nodules, or wood (Paleocene-Eocene, Paleogene) were drilled by bivalves or were chemically dissoluted by sponges. (1) Large Gastrochaenolites boreholes in reworked Lower Cretaceous iron-carbonate concretion pebble (SCB no. Dal-Invert-2). (2) Small Gastrochaenolithes boreholes in reworked Eocene phosphorite (SCB no. Dal-Invert-1). (3) Small Gastrochaenolithes boreholes in reworked Lower Cretaceous iron-carbonate concretion pebble (SCB no. Dal-Invert-3). (4) Small boreholes made by Clione-type sponges in reworked Eocene phosphorite (SCB no. Dal-Invert-4). (5) Boreholes of the bivalve Teredo sp. in Paleogene (Eocene) silicified driftwood (SCB no. Dal-Invert-6). (6) Boreholes of the bivalve Teredo sp. in silicificated Eocene driftwood pieces (SCB no. Dal-Invert-7). (7) Boreholes in reworked Lower Cretaceous ammonite fragment, iron-carbonate concretion (SCB no. Dal-Invert-5a). (8) Small Gastrochaenolithes boreholes in reworked Lower Cretaceous orthicon ammonite fragment (Bochianites), iron-carbonate concretion (SCB no. Dal-Invert-5b).

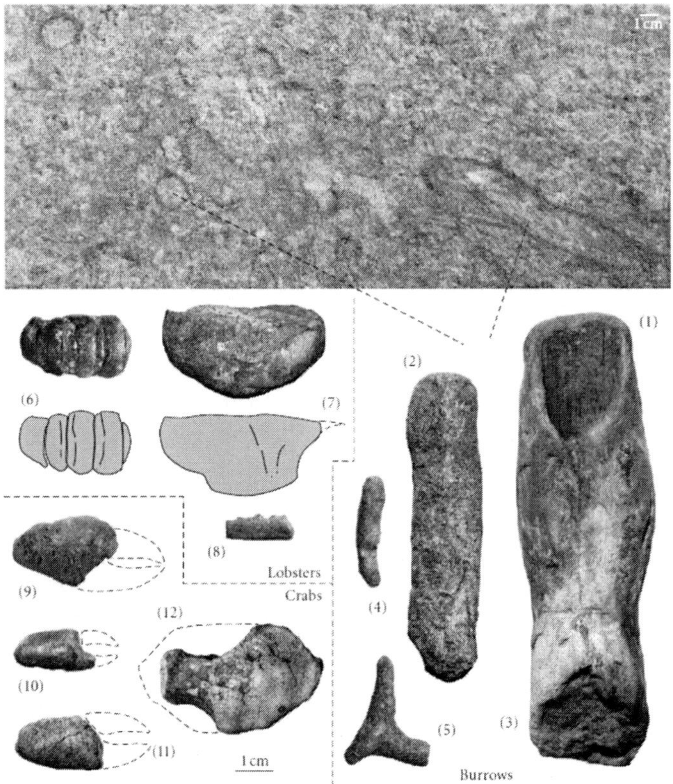

Figure 7: Crustacean remains and burrows from the Middle Eocene (Lutetian) gravels at Dalum. (1) Burrows by crustaceans, with large burrows of Ophiomorpha isp. and other smaller burrows in bioturbated glauconitic sands of the Fürstenau Fm. at the Osteroden sand pit (see also burrows in Figure. 3(d)). (2) Refilled Ophiomorpha isp. burrow fragment, phosphorite concretion (SCB no. Invert-41). (3) Encrusted Ophiomorpha isp. burrow fragment, phosphorite concretion (SCB no. Invert-38). (4) Refilled burrow fragment, phosphorite concretion (SCB no. Invert-40). (5) Triaxonic burrow fragment, phosphorite concretion (SCB no. Invert-39). (6) Abdomen fragment from the lobster Hoploparia sp., phosphorite concretion (SCB no. Invert-14), dorsal. (7) Carapace of the lobster Hoploparia sp., phosphorite concretion (SCB no. Invert-16), lateral. (8) Lower pincer of the lobster Hoploparia sp., phosphorite concretion (SCB no. Invert-13), lateral. (9) Crab Zanthopsis large right pincer, phosphorite concretion (SCB no. Invert-18), anterior. (10) Crab Litoricola/Portunites right pincer, phosphorite concretion (SCB no. Invert-17), anterior. (11) Crab Xanthilites sp. large right pincer, phosphorite concretion (SCB no. Invert-20), anterior. (12) Crab Xanthilites sp. carapace in phosphorite concretion (SCB no. Invert-19), dorsal.

Figure 8: Mollusc and anthozoa remains from the Middle Eocene (Lutetian) gravels at Dalum. (1) Gastropod Natica sp., phosphorite steinkern (SCB no. Invert-25), dorsal and lateral. (2) Gastropod Sconsia sp., phosphorite steinkern (SCB no. Invert-27), dorsal and lateral. (3) Gastropod Cyprea sp. phosphorite steinkern (SCB no. Invert-29), dorsal and lateral. (4) Gastropod Tectonatica sp., phosphorite steinkern (SCB no. Invert-26), lateral. (5) Gastropod Volutospina sp. phosphorite steinkern (SCB no. Invert-30), lateral. (6) Gastropod Littoriniscala sp., phosphorite steinkern (SCB no. Invert-32), lateral. (7) Gastropod Xenophora sp. phosphorite steinkern (SCB no. Invert-28), dorsal and lateral. (8) Gastropod Falsifusus sp. phosphorite steinkern-negative (SCB no. Invert-31), lateral. (9) Bivalve Striarca sp. phosphorite steinkern (SCB no. Invert-34), lateral. (10) Bivalve Pteriasp., phosphorite steinkern (SCB no. Invert-35), lateral. (11) Oyster Pycnodonte sp., shell fragment, metasomatically replaced, silicified (SCB no. Invert-36), lateral. (12) Anthozoa Paracyathus caryophyllus, metasomatically replaced, silicified (SCB no. Invert-37), lateral and dorsal.

3.3.3. Description

The Fürstenau Formation is exposed over a thickness of about 22 meters (Figure 3), the lower part of which (Early Eocene, Ypresian) consists of slightly glauconitic marine sands, with thinly bedded sandstone layers above (Dalum site). These are followed by 1 to 2 meters of thick gray-green clays with lenses of glauconitic sand (possibly still Early Eocene), the top of which have been eroded by a conglomerate bed 40 cm thick at Osteroden and 80 cm thick at Dalum (Figure 2(b)). These conglomerates can be dated herein into the basal Lutetian (basal Middle Eocene). They are fully decalcified at Dalum but are carbonatic at Osteroden where they contain calcareous shells of marine invertebrates, as well as aragonitic fish otoliths in places. The middle Middle Eocene marine glauconitic sands (NP 15 Zone) are composed of at least seven parasequence sets, starting with green, clay-rich, fine-grained sands and ending with green-gray, glauconitic, highly bioturbated (mainly Ophiomorpha), medium-grained sands (Figures 3, 7(1–3)). The gravels and the macrofauna from the Fürstenau Formation are described below.

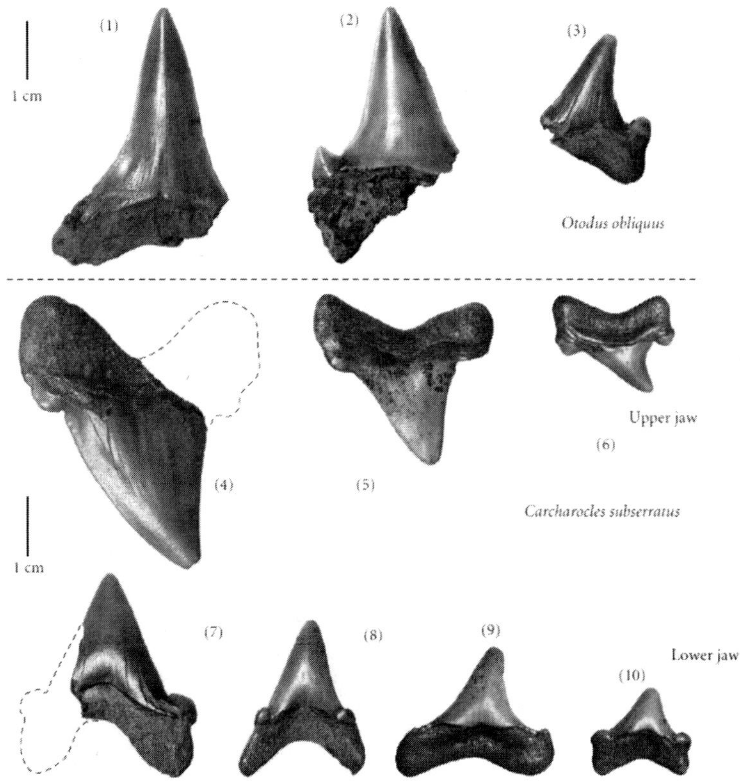

Figure 9: Shark teeth from lamniform (Odontidae) sharks (from different jaw positions) from the Middle Eocene (Lutetian) gravels at Dalum. (1–3). Otodus obliquus (Agassiz, 1843): (1) lower jaw anterior tooth (SCB no. Sel-856), (2) lower jaw anterolateral tooth (SCB no. Sel-857), and (3) lateral tooth (SCB no. Sel-858). (4–10). Carcharocles subserratus(Agassiz, 1843): (4) upper jaw anterior tooth (SCB no. Sel-869), (5) upper jaw lateral tooth (SCB no. Sel-870), (6) upper jaw posterolateral tooth (SCB no. Sel-872), (7) lower jaw anterior tooth (SCB no. Sel-873), and (8) Lower jaw anterolateral tooth (SCB no. Sel-874), (9) lower jaw lateral tooth (SCB no. Sel-876), (10) lower jaw posteriolateral tooth (SCB no. Sel-875; all in labial view).

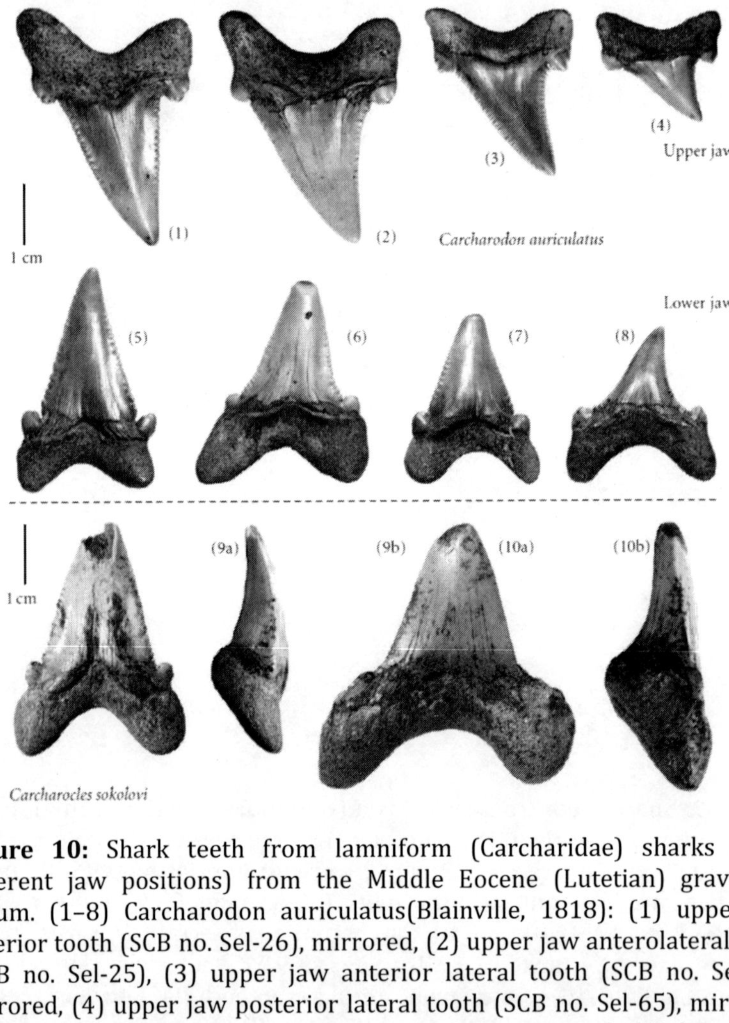

Figure 10: Shark teeth from lamniform (Carcharidae) sharks (from different jaw positions) from the Middle Eocene (Lutetian) gravels at Dalum. (1–8) Carcharodon auriculatus(Blainville, 1818): (1) upper jaw anterior tooth (SCB no. Sel-26), mirrored, (2) upper jaw anterolateral tooth (SCB no. Sel-25), (3) upper jaw anterior lateral tooth (SCB no. Sel-27), mirrored, (4) upper jaw posterior lateral tooth (SCB no. Sel-65), mirrored, (5) lower jaw anterior tooth (SCB no. Sel-4), mirrored, (6) lower jaw anterolateral tooth (SCB no. Sel-5), mirrored, (7) lower jaw anterior lateral tooth (SCB no. Sel-7), (8) lower jaw posterior lateral tooth (SCB no. Sel-66; (1-2(a)) labial and (b) lateral, (3–10), labial views). (9-10) Carcharocles sokolovi (Jaeckel, 1895): (1) lower anterior tooth (SCB no. Sel-85) and (2) lower anteriolateral tooth (SCB no. Sel-88).

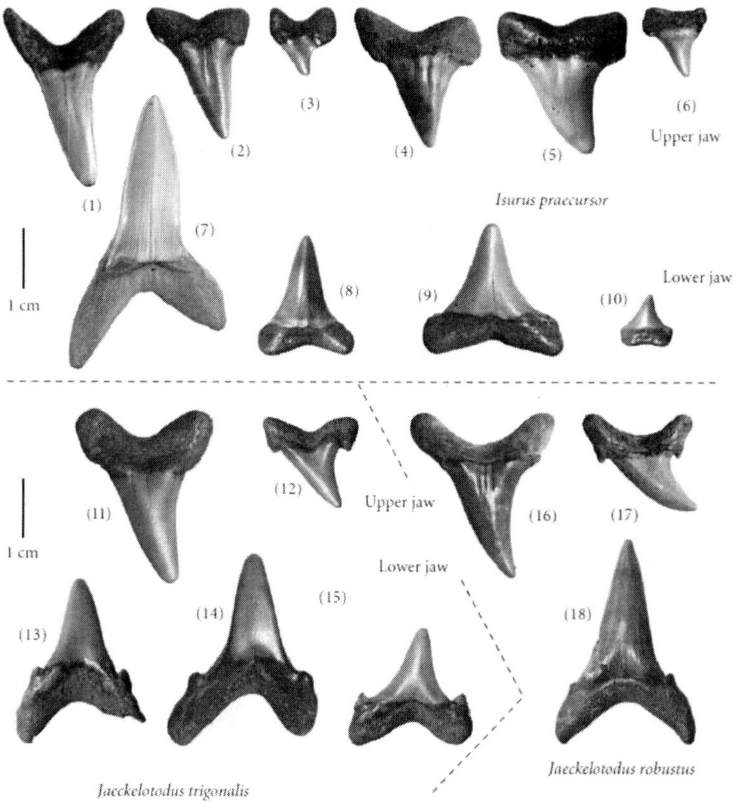

Figure 11: Shark teeth from lamniform (Lamnidae, Jaeckelotodontidae) sharks (from different jaw positions) from the Middle Eocene (Lutetian) gravels at Dalum. (1–10) Isurus praecursor (Lériche, 1905): (1) first upper anterior tooth (SCB no. Sel-101), 2. upper jaw anterior tooth (SCB no. Sel-117), (3) second upper jaw tooth (SCB no. Sel-112),(4) upper jaw anterolateral tooth (SCB no. Sel-124), (5) upper jaw anterior lateral tooth (SCB no. Sel-123), (6) upper jaw posterior lateral tooth (SCB no. Sel-125), (7) first lower jaw anterior tooth (SCB no. Sel-97), (8) second lower jaw anterolateral tooth (SCB no. Sel-109), (9) lower jaw middle lateral tooth (SCB no. Sel-104), (10) lower jaw distal lateral tooth (SCB no. Sel-108). (11–15) Jaeckelotodus trigonalis (Jaeckel, 1895): (11) upper jaw anterior tooth (SCB no. Sel-157), (12) upper jaw lateral tooth (SCB no. Sel-156), (13) lower jaw first anterior tooth (SCB no. Sel-160), (14) lower jaw anterolateral tooth (SCB no. Sel-158), (15) lower jaw lateral tooth (SCB no. Sel-159). (16–18) Jaeckelotodus robustus (Lériche, 1921): (16) upper jaw anterolateral tooth (SCB no. Sel-131), (17) upper jaw lateral tooth (SCB no. Sel-137), and (18) lower jaw anterior tooth (SCB no. Sel-129; all in labial view).

Figure 12: Shark teeth from lamnid (Odontaspididae) Striatolamia macrota (Agassiz, 1843) sharks (from different jaw positions) from the Middle Eocene (Lutetian) gravels at Dalum. (1) upper jaw first tooth (SCB no. Sel-195), (2) upper jaw second tooth (SCB no. Sel-179), (3) upper jaw third tooth (SCB no. Sel-212), (4) upper jaw fifth tooth (SCB no. Sel-195), (5) upper jaw middle lateral tooth (SCB no. Sel-172), (6) upper jaw middle lateral tooth (SCB no. Sel-171), (7) upper jaw posteriolateral tooth (SCB no. Sel-220), (8) lower jaw first (symphyseal) tooth (SCB no. Sel-167), (9) lower jaw second tooth (SCB no. Sel-168), (10) lower jaw third tooth (SCB no. Sel-169), (11) lower jaw fourth tooth (SCB no. Sel-170), (12) lower jaw anterior lateral tooth (SCB no. Sel-217), (13) lower jaw middle lateral tooth (SCB no. Sel-218), (14) lower jaw distal lateral tooth (SCB no. Sel-219), and (15) lower jaw posterolateral tooth (SCB no. Sel-214; and all in labial view).

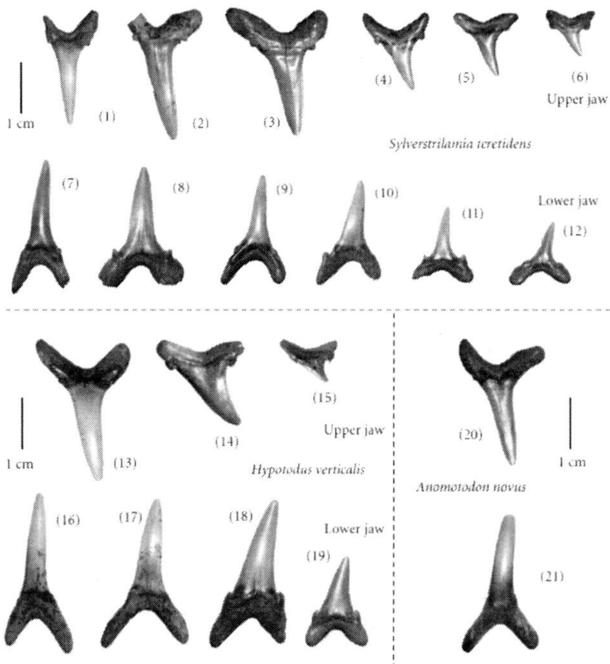

Figure 13: Shark teeth from lamniform (Odontaspididae, Mitsukurinidae) sharks (from different jaw positions) from the Middle Eocene (Lutetian) gravels at Dalum. (1–12). Sylvestrilamia teretidens (White, 1931): (1) upper jaw first anterior tooth (SCB no. Sel-832), (2) upper jaw second anterior tooth (SCB no. Sel-833), (3) upper jaw third anterior tooth (SCB no. Sel-834), (4) upper jaw middle lateral tooth (SCB no. Sel-835), mirrored, (5) upper jaw middle lateral tooth (SCB no. Sel-836), mirrored, (6) upper jaw posteriolateral tooth (SCB no. Sel-837), (7) lower jaw second anterior tooth (SCB no. Sel-838), (8) lower jaw third anterior tooth (SCB no. Sel-839), (9) lower jaw anterior tooth (SCB no. Sel-840), (10) lower jaw anterolateral tooth (SCB no. Sel-841), (11) lower jaw middle lateral tooth (SCB no. Sel-843),and (12) lower jaw posterior lateral tooth (SCB no. Sel-845), mirrored. (13–19). Hypotodus verticalis (Agassiz, 1843): (13) upper jaw anterior tooth (SCB no. Sel-823), (14) upper jaw middle lateral tooth (SCB no. Sel-824), (15) upper jaw posteriolateral tooth (SCB no. Sel-825), (16) lower jaw second anterior tooth (SCB no. Sel-827), (17) lower jaw third anterior tooth (SCB no. Sel-828), (18) lower jaw fourth anterior tooth (SCB no. Sel-829), (19) lower jaw middle lateral tooth (SCB no. Sel-826). (20–22) Anomotodon novus (Winkler, 1874): (20) upper jaw anterior tooth (SCB no. Sel-818), and (21) Lower jaw anterior tooth (SCB no. Sel-819), (22) lower jaw anterior tooth (SCB no. Sel-; 820) all in labial view).

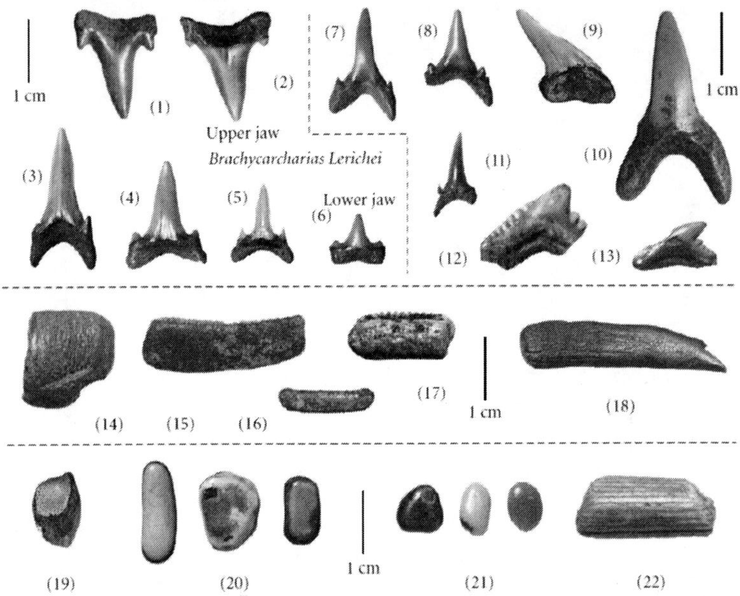

Figure 14: Shark teeth from rare shark species, rays, chimaera, and bony fish remaining from the Middle Eocene (Lutetian) gravels at Dalum. 1–6. Brachycarcharias lerichei(Casier, 1946): (1) upper jaw second anterior tooth (SCB no. Sel-849), (2) upper jaw third anterior tooth (SCB no. Sel-850), (3) lower jaw first anterior tooth (SCB no. Sel-851), (4) lower jaw third anterior tooth (SCB no. Sel-848), (5) lower jaw lateral tooth (SCB no. Sel-852), (6) lower jaw posterolateral tooth (SCB no. Sel-853). 7-8. Carcharias sp.: (7) lower jaw anterolateral tooth (SCB no. Sel-878), (8) lower jaw lateral tooth (SCB no. Sel-879). (9) Xiphodolamia ensis Leidy, 1877 upper jaw second tooth (SCB no. Sel-859). (10) Isurolamna affinis (Casier, 1946) lower jaw first anterior tooth (SCB no. Sel-865). (11) Otodus winkleri Lériche, 1905 upper jaw anterolateral tooth (SCB no. Sel-883). (12) Galeocerdo eaglesomei (White, 1955) tooth (SCB no. Sel-2). (13) Pachygaleus lefevrei(Dammeries, 1891) tooth (SCB no. Sel-1). (14) Tooth from a large unidentified ray (SCB no. Bat-7). (15) Aetobatis irregularis (Agassiz, 1843) tooth fragment (SCB no. Bat-4). (16) Leidybatis jugosus (Leidy, 1876) tooth (SCB no. Sel-902). (17) Myliobatis dixoni (Agassiz, 1843) tooth fragment (SCB no. Bat-6). (18) Pristis lathami Galeotti, 1837 rostral tooth (SCB no. Bat-9). (19) Edaphodon bucklandi Agassiz, 1843 (SCB no. Chim-1). (20) Phylodussp. teeth (SCB no. Tel-1). (21) Pycnodus sp. three lateral row teeth (SCB no. Tel-2). (22) Cylindracanthus sp. rostrum fragment (SCB no. Tel-8). (1–13 labial view).

Figure 15: Reptile and mammal remains form the Middle Eocene (Lutetian) gravels at Dalum. (1) Crocodile tooth from Diplocynodon (SCB no. Rept-1), labial and lateral. (2) Crocodile osteoderm fragment from Diplocynodon (SCB no. Rept-2), dorsal. (3) Turtle Trionyx osteoderm fragment (SCB no. Rept-3), dorsal. (4) Distal joint of a humerus from an unidentified medium-sized mammal (SCB no. Mam-1), (a) dorsal, (b) lateral, and (c) ventral. (5) Femur fragment from an unidentified large mammal (SCB no. Mam-2), dorsal and lateral.

4. ANALYSIS OF THE CONDENSED LUTETIAN PHOSPHORITE CONGLOMERATE BED

4.1. Pebble Types and Origin

The degree of rounding in the pebbles varies (Figure 4), but this property is not very useful for the Carboniferous quartzite pebbles as they are rederived Paleozoic fluvial clasts that were already rounded. Further analysis of the degree of rounding and estimates of transport distances are therefore unlikely to succeed, at least for that particular fraction of the pebbles.

Other pebble groups such as most of the flint pebbles, however, do indicate long-distance transport over a long period of time within a marine environment, while those comprised of local sediments, especially carbonates, have clearly only travelled relatively short distances.

The qualitative rock analysis reveals more then 30 different types of rock within the conglomerates from Dalum and Osteroden (Figure 4), for which provenances cannot always be determined. It has, however, been possible to identify the main rock types and their original ages (Table 1) which are important for an understanding of terrestrial or limnic influences, marine paleocurrents, and the erosional history of the region, and have allowed a new paleogeographical map to be drawn up (Figure 5).

Table 1: Dalum gravel analysis based on 3,927 pebbles (>4 mm).

Age	Rock type	Number of clasts
Paleogene (Early/Middle Eocene)	Phosphorite concretions	600
	Phosphorite burrows	1
	Phosphorite fossils	1
	Fine-grained white sandstone	10
	Vertebrate remains (mainly shark teeth)	150
Upper Cretaceous (Campanian-Maastrichtian)	Flint	10
Lower Cretaceous (Valanginian-Hauterivian)	Iron-carbonate concretion: rock fragments	980
	Iron-carbonate concretion: burrows	50
	Iron-carbonate concretion: ammonite fragments	10
Jurassic (Bathonian-Kimmeridgian)	Oolitic iron ores	1
	Oolitic carbonates	1
	Silica-iron-carbonate concretions	1
Carboniferous (Westphalian C-D)	Quartzite	2,100
	Quartzite conglomerates	10
Devonian	Black lydite	1
Volcanic rocks (age unclear)	Red volcanic rocks	1

4.1.1 Paleozoic Sedimentary Clasts

Clear to white and pale red or milky, well-rounded quartz pebbles (Figures 4(1) and 4(4)) are from the Westphalian C-D sandstone series of the Upper Carboniferous, which outcrops mainly within the Schafberg horst structure, but also at the lesser extent Piesberg and Hüggel horsts near Osnabrück [21]. These quartzite pebbles were already well rounded by previous transport during the Paleozoic and similar well-rounded quartzite pebbles can be seen in the conglomerates of the Piesberg and Schafberg horsts. These pebbles have therefore been redeposited after travelling only short distances, and do not provide any evidence of long-distance transport during the Paleogene, to their secondary deposition site.

4.1.2. Paleozoic Volcanic Clasts

Clasts of red volcanic rocks (Figures 4(11–12)) are only present as small pebbles (<3 cm) and their origin remains unclear. They were probably carried southwards from the Fennoscandian High by upwelling currents, because no such volcanic rock types are known to the south in the northern Rhenish Massif.

4.1.3. Lower Cretaceous Reworked Fossils and Clasts

The most abundant reworked fossils are ammonite fragments (Figures 4(27) and 4(31)), especially single chambered fragments (see the "cat paw" specimen in Figure 4(28)), which are all found in carbonate-iron concretions that originated from the Lower Cretaceous claystones of northern Germany. Most abundant are fragments of orthicon ammonites (Figures 4(27) and 4(28)).

The ammonite fragments, which are also found in the Lower Cretaceous claystones of north-west Germany [22], range in age from Valanginian to Hauterivian (Lower Cretaceous) and include Bochianites sp. and Oosterella sp. (Upper Valanginian), Endemoceras sp. and Leopoldia sp. (Lower Hauterivian), and Aegocriocerassp. (Upper Hauterivian). All ammonites must have already been previously reworked during the Lower Eocene [19], and again during the basal Middle Eocene transgression. Rock types with similar (medium-brown) preservation to these reworked fossils contain abundant fossilized burrow casts (Figure 4(32)) and must have originated from the Lower Cretaceous marine sediments of north-west Germany. These can be distinguished from Eocene burrow casts since the latter are different and occur in phosphatic concretions (cf. Figure 7(2–5)).

4.1.4. Upper Cretaceous Clasts

The Upper Cretaceous clasts are carbonates from the Teutoburger Wald (Cenomanian to Santonian) and Dammer Berg (Campanian) regions of

Germany (Figure 2(a)) that weather easily (except for the Campanian flint nodules, but these are absent from the Cenomanian to Santonian). The platy flints, which show little rounding, (Figure 4(23)) possibly originated from the Dammer Berge region. The Maastrichtian marine carbonate deposits in the studied region are not represented but typical black flint from further west in Belgium is common, including the "Maas-Eggs" (Figure 4(8)) that can sometimes be found in the Dalum gravels. Other red-to-brown flint variants resemble the "Rijckholt" flint (Figures 4(6) and 4(7)), typical of the Campanian limestones in northern Holland and on the Dutch-German border. All of these flint pebbles in the Dalum gravels (except for the platy flints) are well rounded and seem to have been transported to north-west Germany by marine paleocurrents from the west (Figure 5).

4.1.5. Jurassic Clasts

The black silica-ironstone concretions found in these gravels (Figure 4(16)), which resemble black Paleozoic lydite, are from the Middle Jurassic Soninnien beds and are well known in the Osnabrücker Bergland, as well as also being found in reworked Quaternary deposits of the region [23]. Rare, typically red, oolites (iron ores, Figure 4(17)) originate from the Middle Jurassic Macrocephaly oolith beds (also know as "Portha-Oolith or Wittekindsflöz") of the central part of the Wiehengebirge mountain range [24] (Figure 5) and, as with the carbonates, do not survive over long transport distances. In total, the sedimentary clasts from the Jurassic of the Osnabrücker Bergland, and in particular the Wiehengebirge mountain range, are poorly represented and comprise less then 1% of the total number of clasts (Figure 5).

4.1.6. Lower to Middle Eocene Clasts

The main hard rock types represented from the Lower and Middle Eocene are the phosphorite nodules (Figure 4(20)) that have been reworked from the soft Eocene marine sands (including most of the marine vertebrate remains that accumulated in the Dalum and Osteroden condensed phosphorite nodule gravels), and possibly some white fine-grained sandstones (Ypresian, Figure 4(13)). The more polished pebbles appear to have come from the older Lower Eocene (Paleocene) horizons that have been reported from other Paleogene sites in north-west Germany [13] and have later been reworked and bored by organisms during the Middle Eocene (Figures 6(2) and 6(4)), while the unpolished phosphorites (Figure 4(20)) may be younger (Middle Eocene) or else have not travelled as far. Also less-transported are sporadic phosphatic shark coprolites, some of which have been reworked from earlier Eocene horizons (Figures 4(21) and 4(22)).

The quantitative rock-type analysis (Figure 5) is based on 3,927 pebbles greater than 4 mm (Table 2, Figure 5). The main rock types identified were abundant Carboniferous (Westphalian C-D) quartzites and (less common) quartzite conglomerate clasts (59%), Lower Cretaceous (Valagin/ Hauterive) iron-carbonate concretions and fossils (20%), and Paleogene (Paleocene) Eocene phosphorite nodules (15%). These three main rock types derived from different directions and have been reworked from sediments of different ages (Figure 5). While the Carboniferous material must have been washed in from the south east by a river passing the Schafberg and reaching into a fluvial delta system, the Lower Cretaceous pebbles and fossils were washed opposite south and south east towards the coast, together with the Paleogene phosphorite and fossil material (Figure 5). All other rock types and components are less well represented and not important to the reconstruction of paleocurrents and erosional history.

Table 2: Fossil types at the Dalum site, based on 14,440 specimens in material >4 mm.

Fossil type	Amount
Plants	173
Burrows/boreholes in pebbles	184
Burrows (refilled, phosphatic)	43
Bivalves (fragments)	61
Gastropods	25
Crustaceans	29
Anthozoans	4
Chimaera	15
Rays	206
Sharks	13,690
Turtles	5
Crocodiles	3
Mammals	2

4.2. Fossil Content

In total, 14,437 fossils were recovered from the Dalum site excavation; these are listed by fossil type in Table 2. Nearly all of the Eocene invertebrate remains are only preserved in the reworked phosphorites; the only fossils preserved by silica replacement are the shells of larger oysters (with silica rings) and the solitary corals. The phosphatic coprolites, the silicified wood, all the bones, and in particular the teeth, were the most resistant specimens.

4.2.1. Invertebrates

The invertebrate fauna is very incomplete and largely represented by phosphorite steinkerns of shells and burrow casts that survived the enrichment processes in the Middle Eocene gravels of Dalum.

Drills and Boreholes

Drills and boreholes in pebbles [25] are quite common in pebble-rich beach Eocene paleoenvironments also found in the Oligocene pebble-rich beaches of north-west Germany [26]. The mechanical bivalve drills and chemical sponge boreholes in the Eocene pebbles from the Dalum site have only been made in soft to moderately hard rock types, and in particular in 85% of the reworked Lower Cretaceous iron-carbonate nodules (as in the Oligocene beach pebbles from the Osnabrück Bergland region) [26] and ammonite fragments; they are less common in Eocene phosphorites. The two typical ichnites of Gastrochaenolites isp. produced by bivalves (Figures 6(1–3), 6(7) and 6(8)), and Clione sponge burrows (Figure 6(4)), are common in Lower Cretaceous iron-carbonate concretion pebbles and lesser in Eocene phosphorite pebbles. Teredo bivalve boreholes are also present, but only abundant in driftwood (Figures 6(5) and 6(6)).

Anthozoans

Corals without reef structures are known throughout the southern North Sea basin, represented by the solitary coral Paracyathus caryophyllus. The few specimens of this coral that have been recovered have been preserved by metasomatic silica replacement (Figure 8(12)). This solitary, nonhermatypic coral is also found in Oligocene shallow marine, beach pebble paleoenvironments [26, 27].

Crustaceans

These are common in the Eocene North Sea basin [28, 29], but in the Dalum material they are only preserved in the reworked phosphorite concretions. The remains of Hoploparia sp. (Figures 7(6–8)) are significant as these lobsters could be largely responsible for the high degree of bioturbation in the glauconitic sands (Osteroden section, cf. Figures 3(d), 7(1)) and may have been responsible for the Ophiomorpha isp. burrows (Figure 7(1)), fragments of which are also found (as phosphorite concretion casts) within the gravels (Figures 7(2) and 7(3)). Smaller burrows, especially the y-shaped forms (Figures 7(4) and 7(5)), may also be related to smaller crustaceans. Crab remains are more common and at least three different species can be identified from their pincers and carapaces: Xanthilithes sp. (Figures 7(11) and 7(12)), Zanthopsis (Figure 7(9)), and Litoricola/ Portunites (Figure 7(10)).

Molluscs

Bivalves have been identified following the published works on the Belgian [30, 31] and English [7] sites. In the Dalum material almost all of these have been destroyed because of their aragonite shells, with only a few infaunistic forms such as Arca sp. and Striarca sp. (Figure 8(9)) preserved in phosphorite concretions, some epifaunistic forms such as Pteria sp. (Figure 8(10)), and the more abundant large oyster Pycnodonte sp. (only shell fragments preserved by metasomatic silica replacement), which seem to have settled on the gravels of the pebble beaches. The gastropods are more diverse with at least eight species represented (all reworked Eocene phosphorite steinkerns, Figures 8(1–8)), although most of the specimens are from the infaunistic Natica sp.; epifaunistic forms include Sconsia sp., Cyprea sp., Tectonatica sp., and Xenophora sp.

4.2.2. Vertebrates

The chimera, shark, and ray material was compared to the main important works on Paleocene-Eocene sharks of Europe [5, 7, 32–39]. The vertebrates remains are dominated by shark teeth (95%), followed by remains of rays (2%); the rest are chimera, bony fish, reptile, and (extremely rare) mammal remains (Table 2, Figure 5).

Chimera

Only one chimera species, Edaphodon bucklandi Agassiz, 1843, has been recorded to date; it is represented by jaw fragments (Figure 14(19)).

Rays

These are the second most common type of fish remains, represented mainly by broken teeth, but also by fin spines. At least four species can be recognised from their teeth including Aetobatus irregularis (Agassiz, 1843), Leidybatis jugosus (Leidy, 1876), Myliobatis dixoni (Agassiz, 1843), and one large unidentified species (Figures 14(14–17)), while Pristis lethami Galeotti, 1837, is represented by several rostral tooth fragments (Figure 14(18)).

Sharks

Out of the total of 13,690 Paleogene shark teeth recovered from the Dalum site, 6,946 (more then 50%) have been identified at a species level (Table 3). This large number of identifications is a result of the large quantity of material and many complete and comparable teeth recovered, from different jaw positions in different species. The preliminary qualitative analysis herein includes only 19 of the larger Paleogene shark species and five ray species (>4 mm material), which can be found listed in Table 3 together with their time ranges and abundances. Preliminary sampling of the material below 4 mm indicates twice as much fish biodiversity, including of course the smaller species whose teeth are, however, quite rare.

Large teeth from Otodus obliquus (Agassiz, 1843; Figure 9(1–3)) are all incomplete and mostly very rounded, but the sharp-edged teeth illustrated herein would appear to be more or less autochthonous (also rare sharp edged in abundant Striatolamia teeth), which is significant because this species now appears to extend from the Paleocene into the lower Middle Eocene in Europe, but not into the middle or upper parts of the Middle Eocene. All of the large teeth from Carcharocles subserratus (Agassiz, 1843; Figure 9(4-10)) are rounded, indicating most probably in all cases the reworking of horizons that appear to extend into the upper Paleocene or Lower Eocene. However, this material may partly belong to the most abundant type of white shark, Carcharodon auriculatus (Blainville, 1818), which is represented by both enrolled teeth and sharp serrated teeth, from various jaw positions (Figures 10(1–8)). The largest and most robust teeth recovered from the Dalum site are from the megatooth shark lineage—Carcharocles sokolovi (Jaeckel, 1895)—which is not well defined, and is rare in any material. All of the megatooth shark ancestor material will be discussed further in the future, when more teeth are available (only 13 have been recovered to date, and only two of these are in good condition: Figures 10(9) and 10(10)) and can be better distinguished from the other serrated teeth. Another, more abundant, species is Isurus praecursor (Lériche, 1905), with material from all jaw positions (Figure 11(1–10)). Jaeckelotodus is represented by two species, J. trigonalis (Jaeckel, 1895) from which all teeth are reworked (Figures 11(11–15)), and J. robustus (Lériche, 1921) which has both reworked and sharp-edged specimens (Figures 11(16–18)). The most abundant shark is the sand shark ancestor Striatolamia macrota(Agassiz, 1843), with teeth present from all the main jaw positions (Figure 12) and including teeth from juveniles right through to teeth from mature individuals. Only 0.1% of the teeth are sharp edged with little evidence of any transport. Smaller species are also abundant, such as Sylvestrilamia teretidens (White, 1931; Figures 13(1–12)), and Hypotodus verticalis (Agassiz, 1843; Figures 13(13–19)). Less abundant are teeth from Anomotodon novus (Winkler, 1874; Figures 13(20–22)) which are all reworked, as is also largely the case with teeth from Brachycarcharias lerichei (Casier, 1946; Figures 14(1–6)). This latter species sometimes also has sharp-edged teeth, such as Carcharias sp. (Figures 14(7) and 14(8)). Rare teeth from some other species are also present (Table 3, Figures 14(9) and 14(13)).

Table 3: Shark species from the Dalum site based on the number of identifiable teeth >4 mm.

Sharks	Quantity	Time range
Striatolamia macrota	5,901	Paleocene-Eocene
Carcharodon auriculatus	455	Middle/Upper Eocene
Isurus praecursor	179	Lower/Middle Eocene
Sylvestrilamia teretidens	95	Eocene
Hypotodus verticalis	64	Eocene
Carcharocles subserratus	48	Lower Eocene
Otodus obliquus	47	Paleocene-Lower Eocene
Jaeckelotodus robustus	46	Lower/Middle Eocene
Brachycarcharias lerichei	36	Eocene
Jaeckelotodus trigonalis	33	Lower/Middle Eocene
Carcharocles sokolovi	13	Middle Eocene-Lower Oligocene
Carcharias sp.	12	Eocene
Anomotodon novus	5	Eocene
Isurolamia affinis	4	Eocene
Galeocerdo latidens	3	Eocene
Pachygaleus lefevrei	2	Eocene
Xiphodolamia ensis	1	Eocene
Usakias wardi	1	Eocene
Odontaspis winkleri	1	Eocene
Total	6,946	
Selachii Indet.	6,737	
Total number of teeth	13,683	

Bone Fish Some vertebrae and teeth, mainly from Phylodus sp. (Figure 14(20)) and Pycnodus sp. (Figure 14(21)) and some rostrum fragments from Cylindracanthus sp. (Figure 14(22)) provide only a poor representation of the diverse fish fauna, which is normally estimated from the otoliths, but these are absent from Dalum.

Terrestrial Vertebrates

Only single vertebrate teeth have been described from Early to Middle Eocene marine deposits around the world [4, 7, 40], but large pachyostotic rib fragments from Dalum appear to be attributable to early marine Sirenian mammals when compared to Oligocene material [41], and these rare Middle Eocene marine mammal remains will be described in a future publication. The terrestrial vertebrate remains recovered from the Dalum material are of rare crocodile teeth, osteoderm fragments from Diplocynodon (Figures 15(1) and 15(2)), and freshwater turtle remains of Trionyx (Figure 15(3)), all of which are suggestive of freshwater lakes or swamps and therefore a possible deltaic environment. Some bone fragments are impossible to identify but appear to relate to mammals, judging by their spongious nature. A distal humerus fragment (Figure 15(4)) appears to belong to a medium-sized marine mammal, and a femur fragment (Figure 15(5)) may belong to a larger mammal.

5. DISCUSSION

5.1. Stratigraphy and Dating

The sediments from the 22-meter-thick sections at Dalum and Osteroden have a different lithology from the Lower to Middle Eocene Brussel and Lede Sand Formations of the Belgian Oosterzele and Balegem marine deposits [42]. They also contain less carbonate and no storm shell beds, but only massive clay and conglomerate layers, and hence a new formation name is proposed for these deposits: the Fürstenau Formation.

At the Dalum site only the older (Ypresian) Eocene layers are present, with initial marine sequences possibly overlain by a massive clay bed, although precise dating of this part of the section is not yet available. The conglomerate layer can be dated by the terrestrial mammal teeth (MP 11 zone), as at Balegem, in Belgium [4, 43], and by shark teeth indicating the presence or absence of certain species (e.g., O. obliquus). The more carbonate-influenced Middle Eocene conglomerates (two horizons) of the Lede Sand Formation at Balegem and Osterzele [44] have been described as being of Middle Lutetian age, and may therefore possibly be slightly younger than those from northern Germany [1, 5]. The condensed phosphorite gravel beds at Dalum and Osteroden may be isochronous with the two bivalve-enriched condensed beds in the middle Middle Eocene Lede Sand Formation of Belgium [42, 45]. The conglomerates at Dalum and Osteroden also appear to be isochronous with the Lower Geiseltalium, and would therefore be of a similar age to the Messel and older Geiseltal mammal faunas [9].

The marine clay-sand parasequences of the Fürstenau Formation have been dated by nannoplankton into the middle Middle Eocene (Lutetian, NP 14 zone [46, 47]) and appear to be nonisochronous with the shark-tooth-rich layers at Balegem (NP 15 zone) [43, 48]. The lithology of the newly introduced Fürstenau Formation differs from that of the marine localities in the southern North Sea and Paris Basins in having a much lower carbonate content, an almost complete absence of sandstone beds, and an absence of the bivalve (dominated by the bivalve Megacardita) and gastropod shell beds.

At least two shark species appear to be from reworked Upper Cretaceous deposits, represented by some teeth from Cretolamna appendiculata and a single tooth from Squalicorax kaupi (Figures 4(24) and 4(26)). Upper Paleocene and Lower Eocene tooth material appears to be particularly well represented by Carcharocles obliquus and Otodus subserratus. While O. obliquus is typically present in the Late Paleocene and Lower Eocene of Europe (cf. London Clay in: [7]) it is represented here by sharp-edged teeth that show little signs of transport, and so its time range may extend into the Early Lutetian. In contrast, O. obliquus and C. appendiculata, are both absent from the diverse selachian fauna of Middle Lutetian age in Belgium [5]. The Carcharodon auriculatus teeth from Dalum and Osteroden in Germany, and those from Balegem and Oosterzele in Belgium, differ in their serrations, especially on the lateral cusps which are more strongly developed in the younger material from Belgium [5]. The shark fauna from the Dalum and Osteroden gravels appears to be a little younger than the Middle Eocene Belgian faunas, which supports the dating of the Dalum and Osteroden gravels (but not all the teeth therein) into the Early Lutetian transgression (Early Middle Eocene [48], during which older Paleogene tooth material was reworked and concentrated.

5.2. Biodiversity and Facies

The mollusc population seems to be represented by forms from the sandy floor of a shallow sea, although only rudimentary taphonomic comparisons have been made to the Belgian and English sites [7, 30, 31]. There is anyway too little material available at this stage for a model to be made of the autochthonous benthic fauna on the sandy sea floor. The invertebrate fauna of the Middle Eocene beach gravel paleoenvironment is expected to be restricted, with a dominance of oysters [26], as has been demonstrated for the Oligocene pebble beaches of north-west Germany. Oysters such as Pycnodonte appear to have been attached to the gravels, but proof is yet to be found in the fragmented material from Dalum. The solitary corals are also in same state of preservation as the oysters and would not have survived any significant amount of transport, which means that they are also likely to have been attached to the gravels. A restriction of the marine invertebrate benthic fauna must also be expected due to the freshwater influence, but further analysis of the Osteroden site, with its well-preserved mollusc assemblages (Felker pers. com.) is still required.

The vertebrates are dominated by the ancestral sand shark Striatolamia macrota (Figure 12, Table 3), which appears to have also had nursery schools in those shallow waters, as indicated by the presence of teeth from all sizes of animals including many very small teeth from young individuals. Most interesting in the shark research are two different “white and megatooth shark” ancestors, one with massive and the other with slim tooth morphologies: the megatooth shark ancestor Carcharocles sokolovi (the massive-toothed form) [49], and the great white shark ancestor Carcharodon auriculatus (the smaller form) [49].

Two different “large serrated shark" forms can therefore be shown (from the well-dated and large quantities of teeth) to have already developed by the basal Middle Eocene, both of which will need to be further analysed in greater detail on the basis of the material from Dalum in order to further define the ancestry of the white/megatooth sharks and their evolutionary record, which remains as yet unclear, controversial, and a matter of some discussion [49, 50]. The Dalum material includes intermediate forms between Otodus obliquusand Carcharocles sokolovi, such as Carcharocles subserratus whose teeth seem to have all been reworked from Lower Eocene horizons, as have most of the O. obliquus teeth. The origin of the white/megatooth shark lineage will be investigated using about 1,000 serrated shark teeth from the two sites at Dalum and Osteroden, which is possibly the largest collection in general [51], and for the Eocene period in particular.

5.3. Paleogeographic Reconstruction and Paleoenvironment

Analysis of conglomerate pebbles and the reworked Mesozoic fossils has allowed a detailed paleogeographic map of the Osnabrück region, north of the Rhenish Massif, to be prepared, revising the former less refined models [13, 16, 20]. Previous pebble analysis work by Bartholomäus [2], in which the iron-carbonate clasts, ammonite fragments, and several pebble types were not recognised, has been considerably extended herein.

Large quantities of Paleozoic quartzite material (from Carboniferous sandstones and conglomerates) must have been transported, mainly from Schafberg area but also from the Piesberg area (see Carboniferous horst structures and Westphal C-D sandstones [21]), which appears to have been the result of major erosion under a tropical climate by an ancient (hypothetically reconstructed) river that flowed during the Middle Eocene between the mountain chains that today form the Wiehengebirge and

Teutoburger Wald ranges (Figure 5). This river also may also have transported Paleozoic gravels (including lydite) from the Rhenish Massif. Only the hard silica-rich rock types could have survived long-distance transport. With such a river and a tropical climate an estuarine or deltaic situation can be expected at its mouth [52], as has been reconstructed herein (on a preliminary basis) using only the two Dalum and Osteroden sections and their differences in gravel thickness. Dalum would have been more proximal (80 cm thickness) than Osteroden (40 cm thickness) in the deltaic fan (see [52]). More sections and drill cores from the Fürstenau area will need to be studied in the future in order to obtain an improved understanding of this paleoenvironment. The conglomerates can anyway not be simply treated as tempestites or as being of purely transgressive origin [1, 13], as they are demonstrated herein to be a combination of marine transgressive and fluvial deltaic deposits. The Lower Cretaceous (Valanginian/Hauterivian) iron-carbonate concretions, the smaller burrows, and the many ammonite fragments, with all softer rock types showing burrows and boreholes from Eocene times (Figure 5), were transported during the Early Eocene (Ypresian [19]) and subsequently reworked during the Middle Eocene from the Early Eocene marine deposits, or possibly even reworked twice, and finally accumulated north of the Wiehengebirge mountain range (to the north-west of Osnabrück) in the southern pre-North Sea basin coastal pebble facies that were mixed with deltaic deposits (Figure 5).

The region north-west of Fürstenau (i.e., the Dalum/Osteroden region) represents a deltaic, fresh-water-influenced, low-carbonate, siliciclastic glauconite sand to pebble facies in the shallow marine coastal environment of the southern pre-North Sea basin, to the north of the Rhenish Massif. In contrast, the Belgian (Balegem/Osterzele) region was more carbonate influenced and was located in shallow marine areas between the southwestern pre-North Sea and the Paris Basins [44]. The Osterzele and Balegem facies had similarities with the Middle Eocene facies from north-west Germany, but had less terrestrial influences whereas in the Lower Eocene London Clay Formation the terrestrial input was even higher into the coastal paleoenvironments [14].

6. CONCLUSION

The Middle Eocene (Lutetian) transgressive conglomerates from Dalum and Osteroden in north-west Germany are a mixture that includes reworked Lower Cretaceous (Valanginian-Hauterivian) marine claystones, with the derived ammonite fragments and pebbles making up 20% of the total number of clasts. About 80% of these reworked Lower Cretaceous pebbles contain burrows or boreholes (e.g., Gastrochaenolithes boreholes made by bivalves). Boreholes have also been left in driftwood by the bivalve Teredo.

In addition, Paleogene phosphorite nodules sometimes contain benthic marine invertebrate fauna (and also burrow casts). These were transported southward to the pebble beach coast, as were all the Lower to basal Middle Eocene vertebrate remains and, in particular, the shark teeth that are highly concentrated within these gravels, with a density of about 5,000 per cubic meter of gravel.

The other main components of the conglomerate are Paleozoic (Carboniferous) siliciclastic sandstones, and the (mostly well-rounded) quartzites that make up 59% of the pebbles, but these must have been transported by river and come from the opposite direction, deriving from the mainland to the south. An ancient river between the Wiehengebirge and Teutoburger Wald mountain ranges must have flowed to the west of Osnabrück and the Carboniferous Schafberg where a river passed to the west of the main quartz source. Even the resistant sandstone rocks were affected by weathering in the harsh tropical climate. A "Fürstenau-Delta" is postulated, with a conglomerate thickness of about 80 cm in Dalum (proximal) and 40 cm in Osteroden (distal), suggesting a deltaic fan close to the western end of the mountain chains on the Rhenish Massif mainland. The coastline during the Early Eocene (Ypresian) was a little further south, and later moved some tens of kilometres north of the Wiehengebirge mountain range, where it was during the Middle Eocene (Lutetian). The northern Rhenish Massif (the Westphalian Cretaceous Basin/Bay) was not flooded during a medium high stand of the Early to Middle Eocene pre-North Sea Basin, nor was the area between the Wiehengebirge and Teutoburger Wald mountain ranges subjected to marine flooding. The area between the Wiehengebirgs and Teutoburger Wald mountain ranges was, however, subjected to flooding later during the Oligocene (Latdorfian to Chattian), when the old river drainage system of Eocene times was flooded and destroyed.

The shark fauna is mainly dominated by the sand shark Striatolamia macrota, for which a full range of ages is well represented. Otodus obliquus is also another common large shark. A "great white" shark ancestor is represented by teeth from Carcharodon auriculatus and a megatooth shark ancestor by the larger and more massively rooted teeth of Carcharocles sokolovi. Other moderately common sharks include Isurus praecursor, Jaeckelotodus robustus, and J. trigonalis. The Otodus obliquus, and C. subserratus teeth, in particular, appear to indicate reworking of older Early Eocene marine sediments, which are known to exist in north-west Germany. All of the other large shark species (teeth >4 mm), out of a total of 19 that have been identified, are more rare. The larger sharks show a moderate degree of biodiversity and do not indicate tropical marine water temperatures, nor does the absence of coral reefs and the low carbonate production, which is in contrast to the terrestrial tropical forest mammal fauna biodiversity of the Rhenish Massif Mainland.

ACKNOWLEDGMENTS

The excavation field project was sponsored by the "Geo- and Naturpark TERRA.Vita" (H. Escher), and the Landkreis, Osnabrück. The private company "PaleoLogic" was responsible for the field work and sponsoring of the scientific research. Special thanks are due to A. Felde from the Reuter company for his excellent work with the backhoe, and to the company for its full support. The Kuhlhoff Education Center leaders Mr. A. Bruns and Mr. W. Hollermann, and the mayor of Bippen, Mr. H. Tolsdorf, are thanked for their managements to install the field sieving laboratory and for any support. Finally, the author would also like to thank H. Felker, a local hobby paleontologist, who provided support, helped with sieving the material, and allowed access to his large collection (obtained over 25 years from both localities (Dalum and Osteroden) for comparisons.

REFERENCES

1. F. Hocht, "Eine Lagerstätte kreidezeitlicher und paläogener Chondrichthys-Reste bei Fürstenau (Niedersachsen)," Osnabrücker Naturwissenschaftliche Mitteilungen, vol. 6, pp. 35–44, 1979.
2. W. A. Bartholomäus, "Ein kondensiertes Knochenlager im marinen Alttertiär von Fürstenau/SW-Niedersachsen," Fossilien, vol. 2, no. 93, pp. 79–82, 1993.
3. W. Schwarzhans, "The otoliths from the Middle Eocene of Osteroden near Bramsche, north-western Germany," Neues Jahrbuch fur Geologie und Palaontologie, vol. 244, no. 3, pp. 299–369, 2007.
4. J. L. Franzen and T. Mörs, "Das nördlichste Vorkommen paläogener Säugetiere in Europa," Palaontologische Zeitschrift, vol. 81, no. 4, pp. 447–456, 2007.
5. G. Van den Eeckhaut and P. De Schutter, "The Elasmobranch fauna of the Lede Sand Formation at Osterzele (Lutetian, Middle Eocene of Belgium)," Palaeofocus, vol. 1, pp. 1–57, 2009.
6. M. Lériche, "L'Eocene des Bassins Parisiennes et Belge, Livret-guide de la réunion extraordinaire de la Société Géologique de France," Bulletin de la Société Géologique de France, vol. 4, no. 12, pp. 692–807, 1912.
7. D. Rayner, T. Mitchel, M. Rayner, and F. Clouter, London Clay Fossils of Kent and Essex. Medway Fossil and Mineral Society, Caxton and Holmesdale Press Ltd, Kent, Ohio, USA, 2009.
8. J.L. Franzen, "The biostratigraphic and palaeoecologic significance of the Middle Eocene locality Geiseltal near Halle (German, Democratic Republic)," Münchner Geowissenchaftliche Abhandlungen A, vol. 10, pp. 93–100, 1987.
9. S. Schaal and W. Ziegler, Eds., Messel—ein Schaufenster in die Geschichte der Erde und des Lebens, Waldemar Kramer, Frankfurt, Germany, 1988.

10. F. W. Negendank, G. Irion, and J. Linden, "Ein eozänes Maar bei Eckfeld nordöstlich Manderscheid (SW-Eifel)," Mainzer Geowissenschaftliche Mitteilungen, vol. 11, pp. 157–172, 1982.
11. H. H. Schleich, "Neue Reptilfunde aus dem Tertiär Deutschlands. 13. Schildkröten- und Krokodilreste aus der eozänen Braunkohle des Untertargebaues Stolzenbach bei Borken (Hessen)," Courier des Forschungsinstitutes Senckenberg, vol. 173, pp. 79–101, 1994.
12. J.-L. Hartenberger, "Les Mammifères d'Egerkingen et l'histoire des faunes de l'Eocène d'Europe," Bulletin de la Société Geologique France, Serie VII, vol. 12, no. 5, pp. 886–893, 1970.
13. H. Hiltermann, "Tertiär," in Geologie des Osnabrücker Berglandes, H. Klassen, Ed., pp. 463–494, Osnabrück Germany, 1984.
14. C. King, "The stratigraphy of the London Clay Formation and Virginia Formation in the coastal sections of the Isle of Sheppy (Kent, England)," Tertiary Research, vol. 5, no. 3, pp. 121–160, 1984.
15. P. Mangelsdorf, Geologische Wanderkarte Landkreis Osnabrück, Landkreis Osnabrück, Hanover, Germany, 1984.
16. F. Bettenstaedt, "Paläogeographie des nordwestdeutschen Tertiärs," Erdöl und Tektonik in NW Deutschland, vol. 1941, pp. 143–172, 1949.
17. F. Bettenstaedt, "Tertiär Norddeutschlands," Leitfossilien der Mikropaläontologie, vol. 1962, pp. 339–379, 1962.
18. J. Rohling, "Beiträge zur Stratigraphie und Tektonik des Tertiärs in Südoldenburg," Decheniana A, vol. 100, pp. 1–103, 1941.
19. K. Staesche and H. Hilterman, "Mikrofaunen aus dem Tertiär NW Deutschlands," Abhandlungen der Reichsanstalt für Bodenforschung Neue Folge, vol. 201, pp. 1–26, 1940.
20. P. A. Ziegler, Geological Atlas of Western and Central Europe, Shell, Amsterdam, The Netherlands, 1990.
21. K.-H. Josten, K. Köwing, A. Rabitz et al., "Oberkarbon," in Geologie des Osnabrücker Berglandes, H. Klassen, Ed., pp. 7–77, Osnabrück, Germany, 1984.
22. E. Kempe, Die Tiefe Unterkreide im Vechte-Dinkel-Gebiet, Dobler-Druck, Alfeld,Germany, 1992.
23. C. Diedrich, "Kieselgeoden als wichtiger Rohstoff für die Beilproduktion im Neolithikum von Nordwestdeutschland—erste Nachweise von Produktionsplätzen der Trichterbecherkultur," pp. 1–36, 2004, www.jungsteinsite.de/.
24. W. Weitschat, "Lias und Dogger," in Geologie des Osnabrücker Berglandes, H. Klassen, Ed., pp. 335–385, Osnabrück, Germany, 1984.
25. M. Bertling and K. Hermanns, "Autochthone Muschelbohrungen im Neogen des Rheinischen Braunkohlenreviers und ihre sedimentologische Bedeutung," Zentralblatt für Geologie und Paläontologie I, vol. 1995, no. 1-2, pp. 33–44, 1996.
26. C. Diedrich, "Palaeoecology, facies and stratigraphy of the shallow marine macrofauna from the Upper Oligocene (Palaeogene) in the southern Pre- North-Sea Basin of Astrup near Osnabrück (NW Germany)," Cainozoic Research. In press.
27. M. Berndt and J. Welle, "Ahermatype Korallen aus dem Oberoligozän der Schächte Rheinberg und Sophia Jacoba 8 (Niederrheinische Bucht)," Neues Jahrbuch für Geologie und Paläontologie Abhandlungen, vol. 198, no. 1-2, pp. 47–68, 1995.
28. W. J. Quayle, "English Eocene Crustacea (Lobsters and Stomatopods)," Tertiary Research, vol. 30, no. 5, pp. 581–612, 1987.
29. J. S. H. Collins and J. Saward, "Three new genera and species of crabs from the Lower Eocene London Clay of Essex, England," Bulletin of the Mitzunami Fossil Museum, vol. 33, no. 2006, pp. 67–76, 2002.
30. M. Gilbert, "Les Bivalvia du Lédien (Eocène ,oyen supérieur) de la Belgique I, Paleotaxodonta, cryptodonta, Pteriomorpha," Bulletin d l`Institut royal des Sciences Naturelles de Belgique/Mededelingen van het Koninklijk Belgish Instituut voor Natuurwetenschappen, vol. 51, no. 3, p. 1, 1975.
31. M. Gilbert, "Les Bivalvia du Lédien (Eocène ,oyen supérieur) de la Belgique II. Heterodonta, et Anomalodesmata," Bulletin d l`Institut Royal des Sciences Naturelles de

Belgique/Mededelingen van het Koninklijk Belgish Instituut voor Natuurwetenschappen, vol. 52, no. 5, pp. 1–80, 1980.
32. O. Jaeckel, Die Eozänen Selachier aus Monte Bolca, 1894.
33. O. Jaeckel, "Unter-Tertiäre Selachier aus Südrussland," Mémoires du Comité Géologique de St.-Pétersbourg, vol. 9, pp. 19–35, 1895.
34. M. Lériche, "Contribution à lètude des poisons fossils du Nord de la France et des regions voisines," Mémoires de la Société Geologique du Nord, vol. 5, pp. 1–430, 1906.
35. E. Casier, "Contributions à l'étude des poisons fossiles de la Belgique VII—les Pristidés Eocènes," Bulletin de l'Institut Royal des Sciences Naturelles de Belgique/Mededelingen van het Koninklijk Belgisch Instituut voor Naturwetenshappen, vol. 25, no. 19, pp. 1–52, 1949.
36. E. Casier, "Fauna ichthyologique du London Clay," British Museum Natural History London, vol. 1966, pp. 1–496, 1966.
37. H. Cappetta, Chondrichthyes II. Mesozoic and Cenozoic Elasmobranchii. Handbook of Paleoichthyology, Gustav Fischer, Stuttgart, Germany, 1987.
38. H. Cappetta and D. Nolf, "Révision de quelques Odontaspididae (Neoselachii: Lamniformes) du Paléocene et de l'Eocène du Bassin de la mer du Nord," Bulletin de l'Institut Royal des Sciences Naturelles de Belgique/Mededelingen van het Koninklijk Belgisch Instituut voor Naturwetenshappen, Sciences de la Terre/Aardwetenshappen, vol. 75, pp. 237–266, 2005.
39. S. Adnet, "Nouvelles faunes de Sélachiens (Elasmobranchii, neoselachii) de l'Eocène moyen des Landes (Sud-Ouest France)," Implications Dans la Connaissance des Communautés de Sélachians Dèaux Profonds. Palaeo Ichthyologica, vol. 10, pp. 1–128, 2006.
40. S. G. Lucas, "Fossil mammals and the Paleocene/Eocene series boundary in Europe, North Amrica, and Asia," in Late Paleocene-Early Eocene Climatic and Biotic Evolution, M.-P. Aubry, S. Lucas, and W. A. Berggren, Eds., pp. 451–500, Columbia University Press, New York, NY, USA, 1998.
41. C. Diedrich, "The food of the miosiren anomotherium langenwieschei (Siegfried)—indirect proof of seaweed or seagrass by xenomorphic oyster fixation structures in the Upper Oligocene (Neogene) of the Doberg, Bünde (NW Germany) and comparisons to modern Dugong dugon (Müller) feeding strategies," Senckenbergiana Maritima, vol. 38, no. 1, pp. 59–73, 2008. View at Publisher ·
42. J. Hermann, E. Steurbaut, and N. Vandenberghe, "The boundary between the Middle Eocene Brussel Sand and the Lede Sand Formations on the Zaventem-Nederokkerzeel area (Northeast of Brussels, Belgium)," Geologica Belgica, vol. 3, no. 3-4, pp. 213–255, 2000.
43. T. Smith, B. De Wilde, and E. Steurbaut, "Primitive equoid and Tapiroid mammals: keys for interpreting the Ypresian/Lutatian transition in Belgium," Bulletin van het Koninklijk Belgisch Institut voor Natuurwetenshappen, Aardwetenschappen, Supplement, vol. 74, pp. 165–175, 2004.
44. B. Fobe and V. Spiers, "Sedimentology and facies distribution of the Lede Formation (Eocene) in Belgium and northern France," Contributions to Tertiary and Quaternary Geology, vol. 29, no. 1-2, pp. 9–20, 1992.
45. P. Jacobs and E. Sevens, "Middle Eocene sequence stratigraphy in the Balegem quarry (Western Belgium, Southern Bigth North Sea)," Bulletin de la Société belge de Géologie/Bulletin van de Belgische Vereniging voor Geologie, vol. 102, no. 1-2, pp. 203–213, 1994.
46. M.-P. Aubry, "Palaeocene/Eocene boundary events in space and time," Israel Bulletin of Earth Sciences, vol. 44, pp. 1–253, 1998.
47. H.-P. Lauterbacher, J. R. Ali, H. Brinkhuis et al., "The paleogene period," in A Geological Time Scale, G. M. Gradstein, J. G. Ogg, A. Smith et al., Eds., pp. 384–408, Cambridge University Press, Cambridge, UK, 2004.
48. W. A. Berggren and M.-P Aubry, "Late Paleocene-Early Eocene NW European and North sea magnetostratigraphic correlation network. A sequence stratigraphic approach," in

Correlation of the Early Palaeogene in Northwestern Europe, O. R. B. Knox, R. C. Corfiled, and R. E. Dunay, Eds., vol. 101, pp. 309–352, 1996.
49. Martin, "Systematics of the Lamnidae and the Origination Time of Carcharodon carcharias inferred from the comparative analysis of mitochondrial DNA sequences," in Great White Sharks: The Biology of Carcharodon Carcharias, A. P. Kimley and D. G. Ainley, Eds., pp. 67–78, Academic Press, San Diego, Calif, USA, 1996.
50. S. P. Applegate and L. Espinosa-Arrubarrena, "The fossil history of Carcharodon and its possible ancestor, Cretolamna: a study in tooth identification," in Great White Sharks—The Biology of Carcharodon Carcharias. Ainley, A. P. Klimley and G. David, Eds., pp. 19–36, Academic Press, San Diego, Calif, USA, 1996.
51. D. J. Long and B. M. Waggoner, "Evolutionary relationships of White Shark: a phylogeny of lamniform sharks based on dental morphology," in Great White Sharks—The Biology of Carcharodon Carcharias, A. P. Klimley and G. David, Eds., pp. 37–47, Academic Press, San Diego, Calif, USA, 1996.
52. H. Füchtbauer and G. Müller, Sedimente und Sedimentgesteine, Schweizbart' sche Verlagsbuchhandlung, Stuttgart, Germany, 1970.

CHAPTER 12

Epiphytic Microalgal Dynamics and Species Composition on Brown Seaweeds (Phaeophyceae) on the Northern Coast of Jeddah, Saudi Arabia

Salim Marzoog Al-Harbi*

Department of Marine Biology, King Abdulaziz University, Jeddah 21589, Saudi Arabia

ABSTRACT

Epiphytic microalgae grow on host seaweeds. Seaweeds are multicellular macro-algae, which are abundant in intertidal zones of coastal environments. The report on epiphytic microalgae is rare in the Red Sea. Thus, an investigation on epiphytic microalgae was carried on northern coast of Jeddah, the Red Sea, Saudi Arabia. The Red Sea is narrow oceanic basin which is lying between the African and the Asian continental shelves. During the study, 4 species of seaweeds were recorded in Phaeophyceae which were Laminaria sp., Padina fraseri, Sargassum muticum and Turbinaria ornata. A total of 83 epiphytic microalgae were identified, including 76 belong to Bacillariophyceae, 5 belong to Cyanophyceae, 1 belongs to Chlorophyceae and 1 belongs to Dinophyceae. On seaweeds of Laminaria sp., the dominant epiphytes were Leptocylindrus danicus,

Licmophora flabellata and Navicula ramosissma. On seaweeds Padina fraseri, the dominant epiphytes were Cylindrotheca closterium, Navicula sp., Nitzschia sp. and Prorocentrum lima and Cocconeis lineatus. On seaweeds of Sargassum muticum, the dominant epiphytes were Navicula distans, Nitzschia socialis and Navicula vanhoeffeni. On seaweeds Turbinaria ornata, the dominant epiphytes were Leptocylindrus minimus, Bacillaria paxillifer, Leptocylindrus danicus, Nitzschia hungarica and Thalassionema frauenfeldii. The host seaweeds influenced the associated diatoms mainly through its morphology and surface texture and roughness to provide a point of attachment and shelter for host-adapted species. Most of the dominant epiphytes were in smaller cell size diatoms, except Cylindrotheca closterium. The findings of the study focused on individual species of benthic marine diatoms associated with specific seaweeds species even in different regions of the Red Sea in different seasons of Saudi Arabia. These findings could be the important source for future explanation of marine eco-biogeographical phenomena in the Red Sea.

Keywords: Epiphytic microalgae; Species compostion; Benthic species; Brown seaweeds; Seaweeds host; Dynamic; Jeddah Coast; The Red Sea

1. INTRODUCTION

Epiphytic microalgae on seaweeds are sessile plants grown with attached or associated to seaweeds hosts. Seaweeds are multicellular and macroscopic macro-algae, which are abundant in intertidal zones of coastal environments. Primarily they grow in near-shore coastal waters with suitable substrates for attachment [1]. Seaweed is one of the best growing plants worldwide. Different species occupy different habitats based on depth, degree of exposure to wave action, shore beds and other environmental factors [2,3]. Seaweeds provide more space availability for settlement and colonization of epiphytes [4], and increase the probability of more individuals and species being present. They also facilitate additional habitats which sanction the coexistence of more species in a given area. Thus, increasing of the habitat complexity provides an intensifier of the density and diversity of organisms in marine systems [5,6]. The leaves of seaweeds provide substrates suitable for the attachment and growth of various numbers of microalgae [7]. Epiphytic flora attach on host tissue such as leaves of seaweeds with loosely attached and tightly attached or adnate component [8,9]. Epiphytic microalgal community on seaweeds is composed of variety species of diatoms, red, brown, green algae and Cyanophyta [10]. They are the common and important flora in terms of aquatic productivity [11], and a suitable food source for higher trophic levels [12,13]. Therefore, epiphytes and their host seaweeds assemblages serve as nursery habitats for many littoral fishes [14,15]. The dynamics, abundance and community structure of epiphytic microalgae are influenced by biotic factors such as leaf age, seasonal cycle of the host, and grazing pressure by herbivores, and also by the light, temperature, nutrients and water motion [7,16,17]. The abundant and diverse epiphytic microalgae include diatoms, dinoflagellates and cyanobacteria which make a significant contribution to food webs.

An investigation on epiphytic microalgae was carried out on northern coast of Jeddah, the Red Sea, Saudi Arabia. The Red Sea is narrow oceanic basin which is lying between the African and the Asian continental shelves. The northern end of the Red Sea connects with the Mediterranean Sea through the Suez Canal, and the southern region exchanges its waters with the open ocean (Gulf of Aden and the Arabian Sea) through Bab-el-Mandeb strait [18]. The Red Sea is a unique area in the tropics for several reasons which having the highest saline water body in the world's seas with the deep water salinity of 40.06 psu while the surface water salinity ranged from 40.10-40.65 psu [19,20]. The high salinity phenomenon occurs due to its location in an arid and hot climatic zone, isolation from open ocean, without riverine inputs, the high evaporation rate (>210 cm yr-1), and scant rainfall restricted from October to May, resulting in a negative hydrological budget. The Red Sea is also considered as an oligotrophic water body due to deficient of several major nutrients such as nitrate, ammonium, phosphate, and silicate [21]. The required nutrients for photosynthetic activity have to come through water intrusion (from the Gulf of Aden), the sub-surface (below the nutricline), or via aerial deposition. Higher nutrient concentrations occur in the southern areas, and gradually decline towards the north [22]. The reservoirs of nutrients in the Red Sea are trapped below the stratified zone due to the persistent pycnocline. The deep water renewal is generally prevented by high stratification levels and the resulting homogeneity of the northern Red Sea deep waters which limits nutrient availability and biological productivity in the euphotic zone. The nutrients supply occurred through the convergence zone formed by the Eastern and Western coastal currents collision. These two currents (one heading northwards and the other southwards) collision brings cold nutrients rich water from deeper parts of the water column to rise to the surface [21]. Moreover, Jeddah Coast is receiving huge waste water from fish farms, industries and domestic use as it is the second biggest city of Saudi Arabia, located on the Coast off the Red Sea and is extending rapidly.

There are few reports on the study of microalgae ecology, community and primary productivity on the Jeddah Coast of the Red Sea. Aleem [23] studied on community composition of microalgae at Obhur, Jeddah Coast of the Red Sea and he recorded 16 species of Cyanophyta and 27 species of Chlorophyta. Khomayis [24] studied on phytoplankton abundance with annual nutrients recycle was conducted in tourist resort area from South Corniche to North Obhur of Jeddah Coastal water. Diatom composition in marine fouling area of Obhur was studied by Khomayis and Haibi [25] in which they recorded 22 species of diatoms. Recently, Harbi and Affan [26] studied on seasonal dynamics of epiphytic microalgae and their host seaweeds Florideophyceae at Jeddah Coast, the Red Sea, and Saudi Arabia. In this study, brown seaweeds (Phaeophycease) and their epiphytic microalgal association was studied in relation with environmental factors at the northern part of Jeddah Coast. The aim of this study was to investigate the community composition and seasonal dynamics of leaf epiphytic microalgal abundance of host seaweeds species of Phaeophyceae at the northern coast of Jeddah, the Red Sea.

2. MATERIALS AND METHODS

2.1 Study area, sampling and sample analysis

The sampling was done seasonally at the peak of season which was in March (spring), June (summer), September (autumn) and December (winter) in 2015. The sampling was conducted at the northwest coast of Jeddah, Saudi Arabia, the Red Sea. The sampling sites were; (i) Yambu (23°59′45″N and 38°10′35″ E) am one of major sea port of the Red Sea in the Al Madinah province of western Saudi Arabia. It is approximately 300 kilometers northwest of Jeddah. There are oil refineries and petrochemical industry. (ii) Ar-Riyas (23°33′55″N and 38°36′17″E) is an industrial city of Madinah province, (iii) Mastura (23°08′80″N and 38º47′14″E) is famous for fishing, (iv) Rabigh (22°41′35″N and 39°00′34″E) has several industries such as cement factory, electricity station, large refinery, Petro Rabigh and (v) Thuwal (22°16′44″N and 39º05′21″E) had long been a fishing center. The town is experiencing an increase in industries due to its proximity to the King Abdullah Economic City (Figure 1).

Seaweeds samples were collected randomly from the above mentioned sites. The wet seaweeds leaves (except stalk and rhizoids) were kept in polyethylene bags and transferred to laboratory. Hundred grams wet sea weeds were put in plastic bottle with filtered seawater, and separation of epiphytic algae from their host was performed by manual shaking method. Then, the epiphytic microalgal samples were preserved with 2% of logul's solution for further study. For taxonomical study, the sample was observed under the phase-contrast microscope (Zeiss Axioplan, Germany) at 400X magnification. Taxonomic identifications were made with reference to Prescott [27], Tomas [28,29], Shim [30], Kobayasi [31] and Round et al. [32]. For quantitative study, a 1 mL sample was taken after mixing of the preserved sample and counted in a Sedgewick-Rafter (S-R) counter chamber with same microscope as mentioned above. Counted results were summarized as cells per 100 grams of seaweeds, and used for graphical presentation (Figure 2). The water temperature, salinity and pH were measured on spot with portable meter HACH (Model HQ14d).

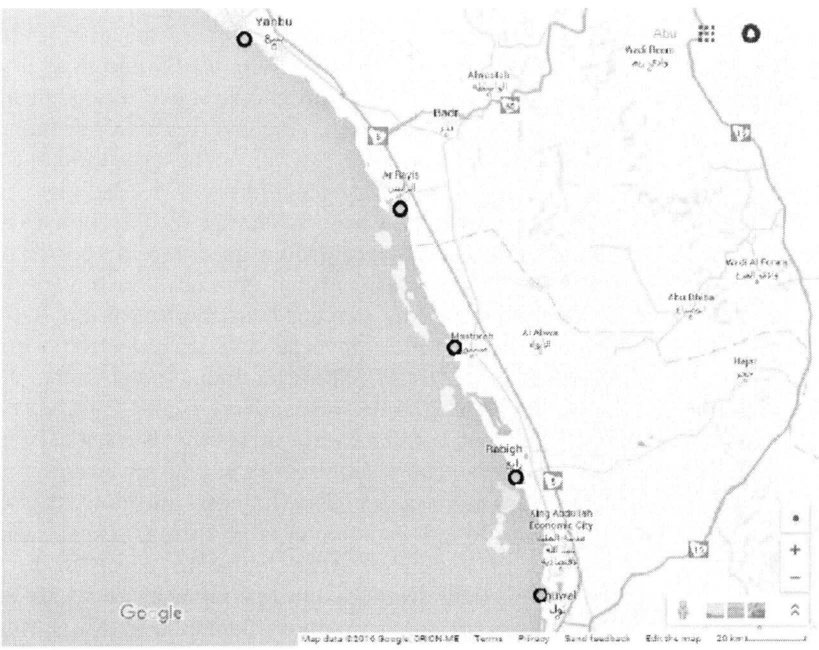

Figure 1: Location of the study sites; Northwest coast of Jeddah, Saudi Arabia, namely Yambu, Ar-Rayis, Mastorah, Rabigh and Thuwal, the Red Sea.

3. RESULTS

The water temperature was found to be highest in summer-autumn, followed by spring and winter. Temperature varied from 22.10-32.02°C with an average of 28.48°C, showing minimum and maximum in winter and summer at Yambu and Rabigh, respectively. Salinity fluctuated from 38.72-39.35 psu with an average of 38.92 psu. The salinity was lowest in spring at Ar-Rayis and the highest was in summer at Yambu (Figure 3). The pH varied from 8.40 to 8.63 throughout the study with an average of 8.52. The highest and the lowest pH was recorded at Rabigh and Mastorah in spring and summer, respectively. The high pH was in spring, followed by winter among all the stations (Figure 4).

3.1 Seaweeds and epiphytic microalgal species composition

Total 4 species seaweeds of Phaeophyceae were found throughout the year among different sites. Padina fraseri was found in four seasons at stations Thuwal, Masturah, Rabigh and Yambu. Laminaria sp. was found in spring and summer at Ar-Riyas and Rabigh. Sargassum muticum was found spring and autumn at Ar-Riyas, Masturah and Rabigh. Turbinaria ornata was found in winter, spring and autumn at Rabigh and Thuwal (Table 1). A total of 83 epiphytic microalgae were identified, including 76 belong to Bacillariophyceae, 5 belong to Cyanophyceae, 1 belongs to Chlorophyceae and 1 belongs to Dinophyceae which were grown on different species of seaweeds of Phaeophyceae. Among the identified epiphytic microalgae of the host seaweeds of Phaophyceae, the percent contribution of epiphytic Bacillariophyceae, Cyanophyceae, Chlorophyceae and Dinophyceae were 91.57, 6.02, 1.20 and 1.20%, respectively. Within Diatoms, the pennate and centric diatoms were 72.37% and 27.63%, respectively (Table 2). The cell abundance of epiphytes on host Laminaria sp. of seaweeds varied from 16.00×10^5 to 73.00×10^5 cells/100 g Laminaria sp. with an average of 45.00×10^5 cells/100 g Laminaria sp. The highest and lowest cell abundance of epiphytes was found in summer and spring, respectively (Figure 5). Three species of epiphytes were found to occur in high frequently during the study. Cylindrotheca closterium was found to be occurred in high frequently in spring, Licmophora flabellata was in spring and summer, and Navicula transitans was in winter (Table 2).

3.2 Seasonal dynamics of epiphytes on host seaweeds

On seaweeds of Laminaria sp., the dominant epiphytes were Leptocylindrus danicus, Licmophora flabellata and Navicula ramosissma in winter. The abundance of Leptocylindrus danicus, Licomophora flabellata and Navicula ramosissma were 16.01×10^5, 16.01×10^5 and 14.11×10^5 cells/100 g of Laminaria sp. and their percent contribution (21.93, 21.93, 19.33)% to the total epiphytic microalgae. Similarly, in spring, the dominant species of epiphytes were Cylindrotheca clostridium and Leptocylindrus danicus which contributed 12.18% and 51.11%, respectively to the total epiphytic microalgae (Table 3). On seaweeds of Padina fraseri, a total five dominant epiphytic microalgae were found and among them Cylindrotheca closterium, Navicula sp, Nitzschia sp. and Prorocentrum lima were dominant in spring, and in summer the dominant epiphytic microalgae was

Cocconeis lineatus. In winter, the highest percent contribution was 19% by Cylindrotheca closterium, followed by Navicula sp. (17%) and Nitzschia sp. (13%). Similarly, in summer the percent composition of Cocconeis lineatus and Trichodesmium sp., 10 and 10% to the total epiphytes (Table 3). On seaweeds of Sargassum muticum, four epiphytes were found to be dominated in spring, and another two epiphytes were found to be dominant autumn. In spring, Navicula distans and Nitzschia sociales contributed 23.00 and 23.00%, followed by Navicula vanhoeffeni (11.04%) and Leptocylindrus minimus (10.00%) in epiphytic microalgae. In autumn, Leptocylindrus minimus and Navicula transitans were the dominant species which contributed 37.00 and 10.00%, respectively to the total epiphytes (Table 3). On seaweeds of Turbinaria ornata, six species of epiphytes were found to be dominant from winter to autumn. In winter the highest percent contribution was 59.00% by Leptocylindrus minimus, followed by Bacillaria paxillifer (15.00%). Leptocylindrus danicus contributed 20.65 and 10.32% to the total epiphytes in spring and summer, respectively. In autumn, the highest contribution was 17.00% by Nitzschia hungarica, followed by Leptocylindrus danicus (14.00%) and Thalassionema frauenfeldii (12.00%) to the total epiphytes (Table 3).

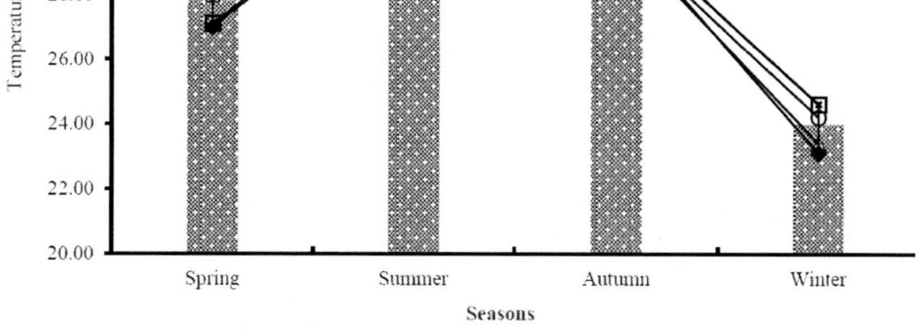

Figure 2: Seasonal variation of water temperature (°C) at the Northwest coast of Jeddah, namely Yambu, Ar-Riyas, Mastorah, Rabigh and Thuwal, the Red Sea, Saudi Arabia from spring to winter 2015.

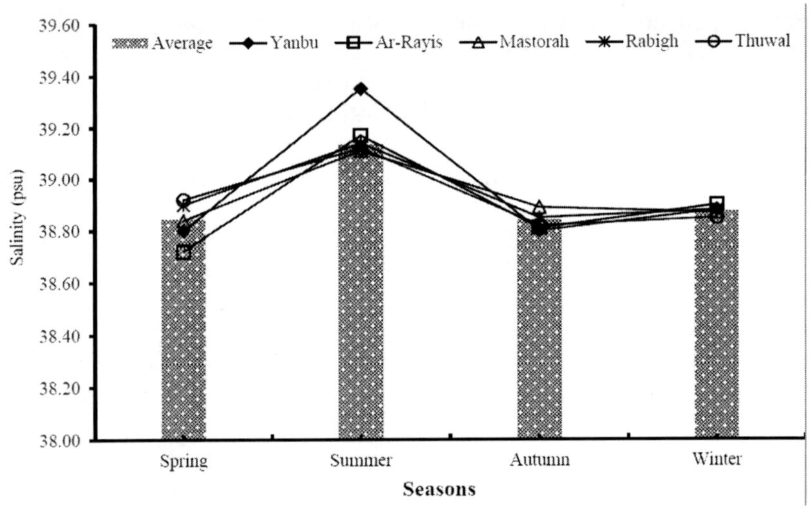

Figure 3: Seasonal variation of salinity (psu) at the Northwest coast of Jeddah, namely Yambu, Ar-Riyas, Mastorah, Rabigh and Thuwal, the Red Sea, Saudi Arabia from spring to winter 2015.

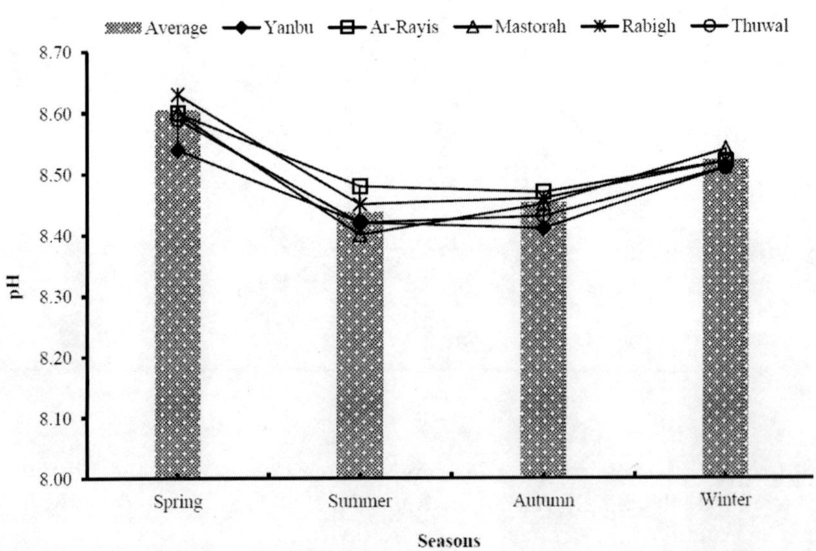

Figure 4: Seasonal variation of pH at the at the Northwest coast of Jeddah, namely Yambu, Ar-Riyas, Mastorah, Rabigh and Thuwal, the Red Sea, Saudi Arabia from spring to winter 2015.

Table 1: List of seaweeds found at different sites of Northwest coast of Jeddah, Saudi Arabia, namely Yanbu, Ar-Riyas, Mastorah, Rabigh and Thuwal, the Red Sea from spring to winter 2015.

Phaeophyceae				
Species name	**Winter**	**Spring**	**Summer**	**Autumn**
Laminaria sp.	-	R, A	R	-
Padina fraseri (Greville, 1830)	M, R, T	M, R, T	R, M	M, R, T, Y
Sargassum muticum (Fensholt, 1955)	-	R, M, A	M	M, R, A
Turbinaria ornata (Agardh, 1848)	R, T	R	-	R, T

*Legend of location; M, Masturah; R, Rabigh; T, Thuwal; Ar-Riyas; and Y, Yanbu.

Table 2: List of epiphytic microalgal taxa, frequency of occurrence and percent (%) contribution of each class on the seaweeds Phaeophyceae at the northern coast of Jeddah from spring to winter, 2015.

	Spring	Summer	Autumn	Winter	%
Bacillariophyceae					91.57
Pennate diatoms					72.37
Amphiprora alata Kützing				X	
A. angustata Hendey				X	
A. paludosa Smith	X	X		X	
A. paludosa Smith	R				
Amphora arcus Gregory 1855				X	
A. coffeaeformis Kützing	X	X			
A. commutata Grunow		X			
A. exigua Levkov			X		
Bacillaria paradoxa Gmelin	X		X		
B. paxillifer Müller		X	R	X	
Closterium jenneri Ralfs			R		
C. moniliferum Ehrenberg				R	
C. pronum Brébisson				X	
Cocconeis heteroidea Hantzsch		X	X		
C. heteroieda Hantzsch				X	
C. pediculus Ehrenberg	X				
C. pellucida var. minor Grunow	R				
Cylindrotheca closterium Ehrenberg	H	F	X	F	
Entomoneis paludosa var. subsalina Cleve			X		
Fragilaria crotonensis Kitton		X			

Gyrosigma acuminatum Kützing				R	
G. balticum Ehrenberg				X	
Licmophora paradoxa Lyngbye				X	
L. abbreviata Agardh		F	F	C	
L. flabellata Greville	H	R	H		
L. gracilis Ehrenberg		X		R	
L. paradoxa Agardh				R	
Navicula delicatula Cleve	F	X			
N. didyma Ehrenberg		X			
N. distans Smith	C	F	C	X	
N. granii Gran				X	
N. incerta Grunow	C		X		
N. jejunoides f. longissima Van	F	R			
N. peregrine Ehrenberg			R		
N. ramosissma Agardh	X			X	
N. transitans Cleve		F	F	H	
Navicula vanhoeffeni Smayda		X		C	
Nitzschia closterium Ehrenberg				X	
N. hungarica Grunow	X	R			
N. longissima Brébisson			F	C	
N. longissima Brébisson	X				
N. reversa Smith.			R	R	
N. seriata Cleve		X	R		
N. sigma Kützin				R	
N. socialis Gregory		R			
Pinnularia brebissonii Kützing				X	
P. viridis Nitzsch		X	R	R	

P. viridis Nitzsch	X				
Pleurosigma angulatum Smith	X	C		X	
P. directum Grunow				X	
P. normanii Ralfs	R	R	F	R	
Synedra ulna Nitzsch			X		
Tabellaria fenestrata Nitzsch	X	R		X	
Thalassionema frauenfeldii Grunow		X	R		
T. nitzschioides Grunow		X		X	
Centric diatoms					27.63
Biddulphia sp. Gray	X				
Chaetoceros affinis Lauder				X	
C. curvisetus Cleve	R				
C. socialis Lauder		R			
C. turgidus Kützing			X	X	
Coscinodiscus sp.	X				
C. lineatus Ehrenberg		X			
C. wailesii Gran & Angst		X			
Cyclotella meneghiniana Kützing			X		
Guinardia delicatula Cleve		X			
Hyalosynedra laevigata Grunow			X		
Leptocylindrus minimus Gran	X	R	F	C	
L. adriaticus Schroder	R				
L. danicus Cleve	F	F	C	C	

Melosira nummuloides Agardh	X				
Rhizosolenia delicatula Cleve	X	X		X	
R. lineola Cleve			X		
R. pungens Clev		X			
R. setigera Cleve		X	X	X	
R. stolterfothii Peragallo		X	X		
Striatella delicatula Kützing		X	R		
Chlorophyceae					1.20
Tetraselmis cordiformis Carter				X	
Cyanophyceae					6.02
Anabaena collarcito Thomas		X	R		
Merismopedia sp.			X		
Oscillatoria sp.	X	C			
O. formosa Gomont			X	C	
Trichodesmium sp.	R	F	F	R	
Dinophyceae					
Prorocentrum lima Ehrenberg	C	R	R	R	1.02

"Legend of occurrence frequency: Sporadically (X=1-20%), Rarely (R=21-40%), Commonly (C=41-60%), Frequently (F=61-80%) and High Frequently (H=81-100%).

4. DISCUSSION

Four species seaweed of Phaeophyceae were identified at the study area in different seasons in different stations. Different types of epiphytic microalgae were also found on same seaweeds of Phaeophyceae which might be occurred depending on local environmental conditions (nutrient, substratum, current, etc.). The epiphytes abundance and composition are the results of the interaction between the lifespan of the host and the reproductive lifespan of the epiphytes although environmental factors and predators control the abundance of epiphytic microalgae [33]. In this study, some of the species diatoms among 78 species of diatoms might be obligatory to be epiphytic in habitat. It is reported that the genera Cocconeis, Achnanthes, and Tabularia are obligate diatoms [32,34]. Among the epiphytic smaller diatoms, the genera of Leptocylindrus sp., Licmophora sp., Cocconeis sp., and some species Navicula ramossima, Navicula distans, Navicula transitans, Navicula vanhoeffeni and Nitzschia socialis were found to be obligate epiphytes at different seasons on different type of seaweed of Phaeophyceae. There is evidence that tube forming species of Navicula, Nitzschia and Amphora which create cells masses within a gelatinous matrix are epiphytes [32]. Thus, the species which showed 10% or more abundance among the microalgae found on the leaves of seaweeds of Phaeophyceae can be consider as real epiphytic microalgae though they occurred in different seasons on same or different genus of seaweeds of Phaeophyceae.

Diatoms were the most dominant epiphyte though some also from Dinophyta, Chlorophyta and Cyanophyta. Diatoms have advantage over other group of epiphytes because of their high fucoxanthin content. Fucoxanthin is known as the most efficient photosynthetic carotenoid absorbing light in the green waveband [35]. Thus, diatoms become dominant on the seaweeds in coastal waters with effect of available of particulate and dissolved organic matter and blue light. The blue light can rapidly attenuated with preferential transmission of the green-to- yellow wavelengths for photosynthesis [36]. Additionally, it is well know that vegetative ecosystems are ideal habitats for benthic diatoms and other epiphytes since seaweeds leaves provide greater surface area for the colonization and growth of diatoms [37,38] found that diatoms and bacteria are the first organisms which make colony on the submerged objects. Therefore, dominancy of diatoms on submerged seaweeds of Phaeophyceae from spring to autumn might be occurred due to suitable circumstance with favorable temperature and light as those seasons day length is longer in Saudi Arabia than that of winter.

Table 3: List of dominant (10% or above) epiphytes cell abundance (105 cell/100 g of seaweeds) and host seaweeds found from winter to autumn 2015 at northwest coast of Jeddah, the Red Sea.

	Winter		Spring		Summer		Autumn	
	10^5 cell	%	10^5 cell	%	10^5 cell	%	10^5 cell	%
Laminaria sp.								
Cylindrotheca closterium	-	-	1.95	12.18	-	-	-	-
Leptocylindrus danicus	16.01	21.93	8.18	51.11	-	-	-	-
Licomophora flabellata	16.01	21.93	-		-	-	-	-
Navicula ramosissma	14.11	19.33	-		-	-	-	-
Navicula sp.	-	-	3.22	20.11	-	-	-	-
Padina fraseri								
Cocconeis lineatus	-	-	-	-	3.90	10	-	-
Cylindrotheca closterium	-	-	1.93	19	-	-	-	-
Navicula sp.	-	-	1.73	17	-	-	-	-
Nitzschia sp.	-	-	1.33	13	-	-	-	-
Prorocentrum lima	-	-	1.00	10	-	-	-	-
Trichodesmium sp.	-	-	-	-	3.90	10	-	-
Sargassum muticum								
Leptocylindrus minimus	-	-	9.20	10	-	-	-	-
Leptocylindrus minimus	-	-	-	-	-	-	6.36	37
Navicula distans	-	-	23.00	25	-	-	-	-
Navicula transitans	-	-			-	-	1.69	10
Navicula vanhoeffeni	-	-	11.04	12	-	-	-	-
Nitzschia socialis	-	-	23.00	25	-	-	-	-
Turbinaria ornata								
Bacillaria paxillifer	0.74	15	-	-	-	-	-	-
Leptocylindrus minimus	2.94	59	-	-	-	-	-	-
Leptocylindrus danicus	-	-	39	20.65	10.32	10.32	-	-
Leptocylindrus minimus	-	-	-	-	-	-	4.53	14
Nitzschia hungarica	-	-	-	-	-	-	5.36	17
Thalassionema frauenfeldii	-	-	-	-	-	-	3.71	12

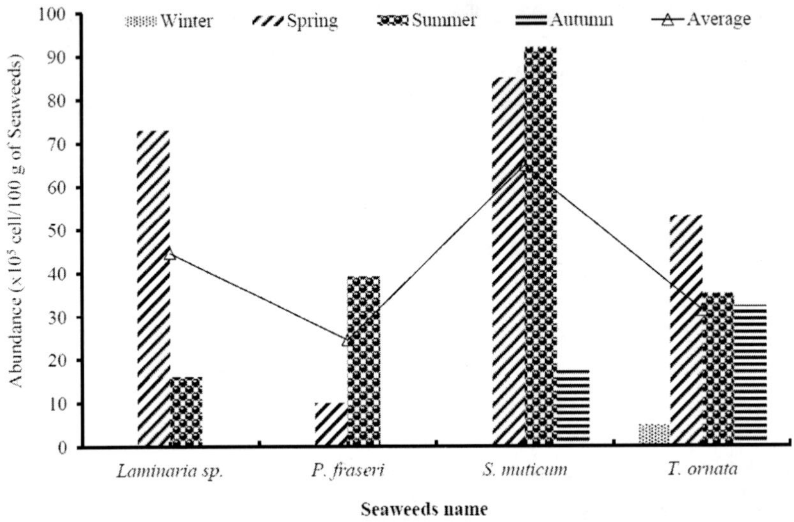

Figure 5: Variation of epiphytes cell abundance associated with seaweeds of Phaeophyceae available at northwest coast of Jeddah from winter to autumn 2015.

However, Ramm [39] reported that the reasons for a specific dependency of epiphytic microalgae on the substratum might be related with the shape of the thalli surface. The diatoms with small attaching area like Licmophora sp., is predominant on fine branched thalli while think branched thalli are preferred by diatoms with a large attaching area like Cocconeis sp., and tube living diatoms like Navicula sp., [39]. In this study, the seaweeds of Phaeophyceae having larger and branched thalli which provided suitable substratum for small and large attaching diatoms. In all seasons, almost same species (Leptocylindrus sp., Licmophora sp., Navicula sp. and Nitzschia sp.) of epiphytic microalgae were found to be dominant on the four species of seaweeds of Phaeophyceae. Thus, it can be said that these epiphytes are obligate on the seaweeds (Laminaria sp., P. fraseri, S. muticum and T. ornata) of Phaeophyceae. Except diatoms, a species of dinoflagellates (Prorocentrum lima) was found to be epiphytic on Padina fraseri seaweed and it was found to be dominant above 10% to the total epiphytic microalgae. It is well known that dinoflagellates are planktonic species, thus its presence as a dominant species might be related with the occurrence of its bloom and it attached with seaweeds temporarily during spring.

CONCLUSION

The host seaweeds influenced the associated diatoms mainly through its morphology and surface texture and roughness, providing a point of attachment and shelter for host-adapted species. Depth affected diatom community growth form structure. The seaweeds were collected from intertidal zone where the water depth was around half meter. Four species seaweeds of Phaeophyceae were found at the study area throughout the year. Most of the dominant epiphytes were in smaller cell size diatoms, except Cylindrotheca closterium. The findings of the study focused on the need for further, long-term and spatially extensive investigations to gather the necessary information about individual species of benthic marine diatoms associated with specific seaweeds species even in different regions of the Red Sea in different seasons of Saudi Arabia. These findings could be the important source for future explanation of marine eco-biogeographical phenomena in the Red Sea.

ACKNOWLEDGEMENT

This project was funded by the Deanship of Scientific Research (DSR), King Abdulaziz University, Jeddah, under grant no. 468/150/1434. The authors, therefore, acknowledge with thanks DSR technical and financial support.

REFERENCES

1. Murphy S, Eunice H, Pinn, Paul DJ (2013) The- short-beaked common dolphin (Delphinus delphis) in the -north--east Atlantic: distribution, ecology, management and conservation status. Oceanogra Mar Biol 51: 193-280.
2. Ballesteros E (1992) The coastal zoning and vegetables: That factors influencing species and communities in their distribution. Institute of Catalan Studies, Barcelona 101: 613
3. Sales M, Ballesteros E (2009) Shallow Cystoseira (Fucales: Ochro-phyta) assemblages thriving in sheltered areas from Menorca (NW Mediterranean): relationships with environmental fac-tors and anthropogenic pressures. Estuar Coast Shelf Sci 84: 476-482.
4. Jacobi CM, Langevin R (1996) Habitat geometry of benthic substrata: effects on arrival and settlement of mobile epifauna. J Exp Mar Biol Ecol 206: 39-54.
5. Heck KL, Orth RJ (1980) Seagrass habitats: the role of habitat complexity, competition and predation in structuring associated fish and motile macroinvertebrate assemblages. In: V. Kennedy (ed.), Estuarine perspectives. Academic Press, San Diego, pp: 449-464.
6. Stoner AW, Lewis FG (1985) The influence of quantitative aspects of habitat complexity in tropical seagrass meadows. J Exp Mar Biol Ecol 94: 19- 40
7. Mazzella L, Russo GF (1989) Grazing effect of two Gibbula species (Mollusca, Archaeogastropoda) on the epiphytic community of Posidonia oceanica leaves. Aquatic Botany 35: 357-373.

8. Burkholder JM, Wetzel RG, Komparens KL (1990) Direct comparison of phosphate uptake by adnate and loosely attached microalgae within an intact biofilm matrix. Appl Environ Microbiol 56: 2882-2890.
9. Cattaneo A, Kalff J (1978) Seasonal changes in the epiphyte community of natural and artificial macrophytes in Lake Memphremagog (Que. and Vt.). Hydrobiol 60: 135-144.
10. Frankovich TA, Zieman JC (1994) Total epiphyte and epiphytic carbonate production on Thalassia testudinum across Florida Bay. Bull Mar Sci 54: 679-695.
11. Burkholder JM, Wetzel RG (1990) Epiphytic alkaline phosphatase on natural and artificial plants in an oligotrophic lake: re-evaluation of the role of macrophytes as a phosphorus source for epiphytes. Limnol Oceanogr 35: 736-747.
12. Kitting CL, Fry B, Morgan MD (1984) Detection of inconspicuous epiphytic algae supporting food webs in seagrass meadows. Oecology 62: 145-149
13. Sullivan MJ, Moncreiff CA (1990) Edaphic algae are an important component of salt marsh food-webs: evidence from multiple stable isotope analyses. Mar Ecol Prog Set 62: 149-159
14. Aburto-Oropeza O, Sala E, Paredes G, Mendoza A, Ballesteros E (2007) Predictability of reef fish recruitment in a highly variable nursery habitat. Ecology 88: 2220-2228
15. Cheminee A, Sala E, Pastor J, Mangialajo L, Cottalorda JM, et al. (2010) Algal forests and the replenishment of Mediterranean rocky fishes. Consequences of the decline of Cystoseira forests on Mediterra-nean juvenile fish populations. ICES (ASC edn.), p: Q17
16. Prado P, Alcoverro T, Mart´ınez-Crego B, Verg´es MA, P´erez, et al. (2007) Macrograzers strongly influence patterns of epiphytic assemblages in seagrass meadows. J Expt Mar Biol Ecol 350: 130-143
17. Mabrouk L, Hamza A, Brahim MB, Bradai MN (2011) Temporal and depth distribution of microepiphytes on Posidonia oceanica (L.) Delile leaves in a meadow off Tunisia. Mar Ecol 32: 148-161
18. Sofianos SS, Johns WE (2002) An Oceanic General Circulation Model (OGCM) investigation of the Red Sea circulation, 1. Exchange between the Red Sea and the Indian Ocean. J Geophys Res 107: 3196
19. Marcos SA (1970) Physical and chemical oceanography of the Red Sea. Oceanogr Mar Biol Ann Rev 8: 73-202
20. Manasrah R, Badran M, Lass HU, Fennel WG (2004) Circulation and winter deep-water formation in the northern Red Sea. Oceanology 46: 5-23
21. Acker J, Leptoukh G, Shen S, Zhu T, Kempler S (2008) Remotely-sensed chlorophyll a observations of the northern Red Sea indicates seasonal variability and influence of coastal reefs. J Mar Sys 69: 191-204
22. Weikert H (1987) Plankton and the pelagic environment. In: Edwards AJ, Head SM (eds.) Key Environments: Red Sea. Pergamon Press, Oxford, pp: 90-111
23. Aleem AA (1978) Contributions to the study of the marine algae in the Red Sea. III. Marine algae from Obhur in the Vicinity of Jeddah, Saudi Arabia. Bull Facul Sci KAU Jeddah 2: 99-118
24. Khomayis HS (2002) The annual cycle of nutrient salts and chlorophyll a in the coastal waters of Jeddah, Red Sea. J KAU: Mar Sci 13: 131-145
25. Khomayis HS, Harbi SMA (2003) Pcriphyton flora in the coastal water of Jeddah, Saudi Arabia. J KAU: Mar Sci 14: 3-18
26. Harbi SMA, Affan MA (2016) Seasonal dynamics of epiphytic microalgae and their host seaweeds florideophyceae at jeddah coast, the Red sea, Saudi Arabia. Pak J Bot 8: 1289-1298
27. Prescott GW (1973) Algae of the western Great Lakes area. Dubuque, W.C. Brown
28. Tomas CR (1993) Marine Microalgae, a guide to naked flagellates and coccolithophorids. Academic Press, Harcourt Brace, San Diego, California, USA.
29. Tomas CR (1997) Identifying marine microalgae. Academic Press, Harcourt Brace, San Diego, California, USA.
30. Shim JH (1994) Illustrated encyclopedia of fauna and flora of Korea, vol. 34 Marine microalgae. Ministry of Educa-tion, Seoul, Korea, p: 487

31. Kobayasi H, Idei M, Mayama S, Nagumo T, Osada K, et al. (2006) Atlas of Japanese Diatoms based on electron microscopy. Uchida Rokakuho Publishing Co. Ltd., Tokyo.
32. Round FE, Crawford RM, Mann DG (1990) The diatoms: biology & morphology of the genera. Cambridge University press, Cambridge, p: 747
33. Borowitzka MA, Lavery P, Keulen M (2006) Epiphytes of seagrasses, in Seagrasses: Biology, Ecology and Conservation. In: Larkum AWD, Orth RJ, Duarte CM (eds.) Springer, Amsterdam, The Netherlands.
34. McIntire CD, Moore WW (1977) Marine littoral diatoms: ecological considerations. In: Werner D (ed.) The Biology of Diatoms. Blackwell, Oxford, pp: 333-371.
35. Ondrusek ME, Bidigare RR, Sweet ST, Defreitas DA, Brooks JM (1991) Distribution of phytoplankton pigments in the North Pacific Ocean in relation to physical and optical variability. Deep-sea Res 38: 243-266
36. Gin KYH, Zhang S, Lee YK (2003) Phytoplankton community structure in Singapore's coastal waters using HPLC pigment analysis and flow cytometry. J Plankton Res 25: 1507-1519
37. Zieman JC, Fourqurean JW, Iverson RL (1989) Distribution, abundance and productivity of seagrasses and macroalgae in Florida Bay. Bull Mar Sci 44: 292-311
38. Novak R (1984) A study in ultra-ecology: microorganisms on the seagrass Posidonia oceanica (L.) Delile. Mar Ecol 5: 143-190
39. Ramm G (1976) Structure of epiphytic diatom populations of the phytal of the Kiel Bight (Western Baltic). In: Simonsen R. (ed.) Proceedings of Fourth Symposium on Recent and Fossil Marine Diatoms. Ganter Verlag KG, Germany, pp: 379-387.

CHAPTER 13

Automated measurements of whitecaps on the ocean surface from a buoy-mounted camera

M. Bakhoday-Paskyabi[a,*], J. Reuder[a], M. Flügge[a,b]

[a] *Geophysical Institute, University of Bergen and Bjerknes Center for Climate Research, Bergen, Norway*

[b] *Christian Michelsen Research AS, Fantoftvegen 38, 5072 Bergen, Norway*

ABSTRACT

We quantify the percentage of sea surface covered by whitecaps from images taken by a non-stationary camera mounted on a moored buoy using an Adaptive Thresholding Segmentation (ATS) method and an Iterative Between Class Variance (IBCV) approach. In the ATS algorithm, the optimal value for the threshold is determined as the last inflection point of the smoothed cumulative histogram of the scene. This makes the method more effective in finding the optimal value of the threshold and reduces the computational efforts compared to the conventional Automated Whitecap Extraction (AWE) technique. In the IBCV method, the optimum criterion for determining the value of the threshold corresponds to the measure of separability between the segmented water and whitecap pixels. In our experiments, the fraction of each image covered by the whitecap is determined using the aforementioned dynamical thresholding techniques for images taken under complex forcing and lighting conditions. Comparisons between different techniques suggest the effectiveness of the proposed methodologies, in particular the ATS algorithm to separate the whitecap features from the darker water pixels.

Keywords: Image processing; Whitecap; Wave breaking; Remote sensing; Non-stationary camera

1. INTRODUCTION

For wind speeds higher than 3 m s^{-1}, breaking of surface gravity waves ejects sea spray droplets into the marine atmospheric boundary layer and entrains air into the water column in a broad ranges of scales, from dissolving bubbles (with radii less than 100 μm) to the large buoyant and rising bubbles near the ocean surface (Thorpe, 1982 ; Deane and Stokes, 2002). The surface whitecaps, which constitute 2%–5% of the global coverage of the ocean surface (Blanchard, 1963) are the most important manifestation of such enhanced interactions between atmosphere and ocean during the sea surface breaking events (Melville, 1996 ; Scanlon and Ward, 2013). The ejection of droplets from breaking waves, the bursting of bubbles floating on the sea surface, and several air–sea exchange processes are closely linked to the fractional areas of the whitecap coverage at the sea surface. Within the marine atmospheric boundary layer during high wind events, turbulent mixing carries the ejected sea spray droplets up into the air where they are effectively transported over the sea surface and possibly collide with fixed and floating offshore structures (Jones and Andreas, 2012). Formation and evolution of surface whitecaps also influence many air–sea exchange processes, such as gas- and heat-transfer (Andreas, 1992), and wave energy dissipation into the water column (Hanson and Phillips, 1999). Therefore, quantifying the properties of whitecaps, in particular active whitecaps as strong manifestations of wave breaking crests, substantially contributes to the detailed understanding of several air–sea processes.

After the pioneering study of Monahan (1969) for the extraction of the areas of whitecaps from film, and subsequently video images, several investigations have then focused on the determination of the fraction of the sea surface covered by whitecaps in an attempt to link the whitecap fraction to other measurable physical quantities, such as wind speed (Monahan, 1971). Various empirical parameterisations of the whitecap fraction at the sea surface may be specified as a function of wind speed at 10 m height, U_{10}, i.e.

$$W_c(\%) = W_c^A(\%) + W_c^B(\%) = \alpha \times (U_{10} - \beta)^n. \tag{1}$$

where W_c is the total whitecap fraction in %, and α, β, and n are tuneable coefficients which can be determined by a curve-fitting technique (Stramska and Petelski, 2003). Several studies have confirmed that the coefficients α, β, and n in Eq. (1) depend on the location of the measurements, the specific experimental conditions, the source of wind speed data (e.g. sonic anemometer), the temperature difference between air and sea, the sea-state conditions, and the atmospheric stability in the lower layer of the marine atmospheric boundary layer (Monahan and O'Muircheartaigh, 1980; Monahan and O'Muircheartaigh, 2012; Salisbury et al., 2014 ; Paget et al., 2015).W_c^A andw_c^Bindicate the active (highly reflective and bright surface elements) and residual (less bright and thinner vertically than the active whitecaps) whitecaps, respectively. While the active whitecaps are of importance for the parameterisations of dynamical processes at the sea surface, such as momentum transfer, dissipation, and spume drop production (Melville and Matusov, 2002), the residual whitecaps mainly influence heat exchange and some bubble-mediated processes such as gas transfer (Andreas et al., 2008) and jet droplet production (Woolf et al., 1987).

A key tool in camera-based whitecap measurement is the automatic identification of the whitecap covered areas from a sequence of images monitoring the sea surface. Callaghan et al. (2008) and Callaghan and White (2009) proposed an Automated Whitecap Extraction (AWE) approach to determine the total whitecap fraction value by the means of intensity thresholding. They showed that the uncertainty in determining the whitecap fraction is directly related to the number of images used in the processing and the sampling frequency. Particularly, they suggested a minimum measurement frequency between 0.25 and 0.33 Hz to achieve a - statistical uncertainty of 5% for each value of W_C. Although the AWE technique is typically applied for calculation of the whitecap fraction, the assumptions behind this method are violated in the presence of strong uneven illumination and other atmospheric interferences. Therefore, we present two alternative fully automated techniques to identify the fraction of whitecaps from the background non-whitecap areas based on the intensity segmentation of each individual pixel. The optimal value of the threshold for each scene is obtained based on the statistical information of the greyscale histogram, and the separability distance between the two segmented classes, i.e. the water and the whitecap regions.

This paper is organised as follows. Section 2 presents the image processing background, and Section 3 describes the experiment and the measurement site. The results are presented and discussed in Section 4. The last section will provide a brief summary and outlook.

2. DIGITAL IMAGE PROCESSING

The common digital image processing technique to calculate the whitecap coverage consists of several steps, e.g. Callaghan and White (2009):

1) The coloured images are converted into greyscale.
2) A threshold level is determined for each scene based on the sea surface brightness characteristics. This allows for the identification of whitecap regions as pixels with bright intensities above the determined optimal threshold.
3) The whitecap information are averaged over a certain number of images to determine statistically certain estimates for the values of whitecap coverage.

Due to the environmental complexity, occlusion effects, sun-light reflection, non-stationarity of whitecaps and change in the illumination at each scene, the principle assumption behind the thresholding methods that there should exist a distinct distribution difference between the water and the whitecap pixels may be violated. Therefore, it is essential to reduce the effects of such disturbances by applying appropriate pre-processing and quality control procedures before starting the feature extraction (Sugihara et al., 2007 ; Callaghan and White, 2009).

In order to validate the performance of methods, we discuss the application of the ATS and the IBCV algorithms to the ideal composite histograms generated by the sum of the Gaussian probability density functions as

$$h_s(x) = \sum_{i=1}^{n} \frac{a_i}{\sqrt{2\pi}\sigma_i} \exp\left[-\frac{(x-\mu_i)^2}{\sigma_i^2}\right] \tag{2}$$

where $h_s(x)$ is the simulated histogram as a function of the bin value x, n denotes the number of modes, a_i (here=1 for $i=1\cdots n$) are the respective proportion, and $\mu_i$and σ_i represent the mean and standard deviation of the ith mode. Table 1 lists the parameters used for checking the performance of the proposed threshold-detection algorithms.

Table 1. The used parameters to generate the idealistic histograms.

n	μ_1	μ_2	μ_3	σ_1	σ_2	σ_3
3	−10	0.0	10	2.5	2.5	2.5
2	−10	0.0	–	2.5	2.5	–
1	−10	–	–	2.5	–	–

2.1. Extraction of whitecap fraction

The population of pixels with intensity k is defined as

$$J_k = \sum_{(u,v)\in R} [I(u,v) == k], \tag{3}$$

where I(u,v) is the greyscale value of a pixel located at the point (u,v) in the image frame, and R is typically a local image region including the defined Region-of-Interest (ROI). The percentage of whitecap coverage is then calculated from

$$W_c(\%) = 100 \times \frac{\sum_{k\geq T} J_k}{\sum_{k\geq 0} J_k}, \tag{4}$$

where T denotes the value of threshold which is applied to divide the pixel space into the white objects and dark background. After determining the threshold, each pixel in the image is labelled as 1 if it corresponds to the object (whitecap) and 0 if it corresponds to the background (water). In the next section, we present different methods to determine the optimum threshold value.

2.2. Automated image segmentation

Although there are a wide range of sophisticated classification approaches, the pixel-based thresholding schemes are still common due to the simplicity of their algorithms (Sugihara et al., 2007 ; Callaghan and White, 2009). However, in images of the ocean surface in which scenes exhibit a strong change of illumination, the thresholding-based techniques are no longer appropriate for mapping of the pixel values due to particular surface events, i.e. the bursts of breaking crest and residual whitecaps. In such situations, the distinction between bright areas and the background water still remains a computational challenge for the vision-based techniques.

The major issue in the thresholding method is the accurate choice of the threshold for the separation between two groups of pixels, i.e. object (whitecap) and non-object pixels. A widespread approach is based on the determination of the local minimum of the intensity frequency histogram by assuming a bimodal histogram distribution, such as Otsu's method (Otsu, 1979). However, in a bimodal histogram where the valley between two peaks is wide and in the absence of strong illumination effect, the local minimum strategy will results in a range of choices for the value of the threshold instead of an optimal value. Furthermore, the histogram is not always bimodal and the optimal schemes consequently predict poor values for threshold in such cases.

2.2.1. Segmentation using AWE method

Callaghan and White (2009) developed the AWE algorithm to determine a suitable threshold intensity value to separate whitecaps from the background water pixels. This technique contains two major components: an image structure as the fraction of pixels with intensities greater than a given threshold; and an optimum threshold-selection system based on the first, the second, and the third-order derivatives of the image structure. According to our analysis for the images taken from a non-stationary camera, the image structure is controlled by the sunglint, sky reflection, and the strength of uneven illumination, see also Callaghan and White (2009). These features add brightness and texture to the vast majority of images used in our analysis and confine the identification of transition regions between whitecaps and background water due to the erratic nature of the image structure. The resulting disturbances in the image structure, especially when the threshold approaches the higher intensity values, cannot be sufficiently counteracted by the conventional smoothing procedures such as the iterative cropping methods or using contrast filtering techniques (Callaghan and White, 2009). We found the variations of $W_c(\%)$ as a function of the threshold value less sensitive to the atmospheric and platform motion-induced contaminations than the application of the image structure for our images. Hence, to avoid further quality control protocols to omit the noisy portions of the image structure, we replace hereafter the image structure in the AWE method by the percentage of the whitecap coverage (Eq. (4)) as a function of the threshold, see also Sugihara et al. (2007). The optimum threshold is then determined based on the first- or the second-order derivatives of the $W_c(\%)$ as a function of threshold. This modification leads to a very robust detection algorithm with minimum requirement to further image pre-processing operations.

In the AWE method used in this study, the percentage of whitecap coverage, $W_c(\%)$, is first determined over different normalised threshold values ranging between 0.01 and 1 (Callaghan and White, 2009). A decision graph for $W_c(\%)$, relative to the threshold values, is then constructed to extract the optimal threshold as a point at which the graph approaches a shallow tail (i.e. the last inflection point). Although the technique is effective, it misidentifies the whitecap area under certain conditions, such as strong pixel intensity and wave-induced camera motions, contamination from sunglint and sky reflection, raindrops, and uneven changes of illumination in the scene. These disturbances may lead to a substantial reduction in the frequency of images used for the extraction of $W_c(\%)$.

2.2.2. Segmentation using ATS

Motivated by Callaghan and White (2009), we propose an optimal threshold determination technique that utilises the information from a cumulative histogram of the greyscale images, see also Kleiss and Melville (2011) and Schwendeman and Thomson (2015). In this method, i.e. ATS, the threshold value is accomplished as the intensity value at which the curvature of the histogram graph is maximal or the change in the slope of the cumulative sum approaches asymptotically to zero. It is expected that using curvature information will result in a better estimation of the threshold compared to the histogram-based threshold detection techniques, particularly in the situations where the scene is multimodal or unimodal with weak well-separated peaks.

For the normalised (discrete) histogram h, the Cumulative Distribution Function (CDF) at point k is defined by

$$H(k) = \sum_{i=0}^{k} h(i). \tag{5}$$

Let H be a continuous function of variable x, the curvature of H is then given by

$$C(x) = H''(x)\,[1+H'(x)]^{-3/2} \tag{6}$$

where H' and H'' are the first- and the second-order derivatives of H, respectively. If the curvature approaches to zero at point x, an inflection point at x is achieved at which the cumulative sum finds a change of slope and sign. In real maritime images without a clear gap between background and foreground textures, C is very noisy and oscillatory in the vast majority of cases at the tail area. This makes the detection of the correct inflection point very complicated. Therefore, to effectively determine the inflection point of H (zero of C at the tail area), the following inequality needs to be checked

$$\min\{x : |H''(x)| \, ? \, \gamma\} \tag{7}$$

where $|\cdot|$ denotes the absolute value function, and $\gamma > 0$ is selected as a constant small number (here $\gamma = 10^{-4}$) or is determined adaptively to avoid the effects of noise at the tail of $C(x)$ for each scene. It is obvious that the efficiency of this method is directly controlled by the quality of the CDF graph to avoid spurious threshold estimation. Consequently, we apply a four-point running mean to the raw histograms prior to constructing the threshold-detection algorithm.

In this study, images mostly exhibit a long tail with whitecaps forming a rather weak peak compared to the corresponding peak of the background water pixels. We can then determine the optimum value of T_{opt} using a triangle algorithm for $H''(x)$ or $H'(x)$ (Zack et al., 1977). Since $C(x)$, $H'(x)$, and $H''(x)$ share similar behaviour at the tail area, we apply the triangle algorithm to the graph of $H'(x)$, which seems less noisier than other two functions, as follows:

1) Fit a line between the peak of $H'(x)$ at binT_{max} and the end of the longer tail of the $H'(x)$ graph at binT_{min}.
2) Determine the perpendicular distance between the fitted line and every point in the graph of $H'(x)$ between T_{min} and T_{max}.
3) Set the point corresponding to the maximum distance as the optimum threshold, T_{opt}.

An example of a non-modal histogram segmentation using the ATS threshold detection algorithm is given in Fig. 1. Since the values of $H''(x)$ and $H'(x)$ approach asymptotically to zero, the ATS threshold voting-system adjusts the threshold through a few iterations from the triangle method so that the value of $|H''(T_{opt})-H'(T_{opt})|$ for the optimum threshold T_{opt} is within a small offset, Fig. 1(d). It is worth to mention that the results in this study are less sensitive to this final adjustment.

2.2.3. Iterative Between-Class Variance (IBCV) segmentation

This technique is similar to Otsu's method for the selection of the optimal threshold. According to Otsu's method, the normalised grey level histogram is regarded as a discrete probability distribution function $p(i)=n_i/M$, where n_i is the number of pixels at grey level i and M is the total number of pixels in the ROI. By dividing the histogram into 2 classes using the threshold level T, the BCV is defined as follows (Otsu, 1979):

$$\mathrm{BCV}(T) = p_1(T)p_2(T)\,[m_2(T)-m_1(T)]^2 \tag{8}$$

where p_1 and p_2 are the probabilities of the respective classes (water and whitecap regions), and m_1 and m_2 denote the corresponding mean intensity values of the two classes. The optimal threshold can then be determined iteratively by maximising the BCV. It should be noted that the maximising (minimising) problem can be extended for a multi-threshold methodology (Kittler and Illingworth, 1986). For example, if we divide the histogram into three classes separated by two thresholds T_1 and T_2, the following Iterative BCV (IBCV) algorithm will approach to the optimal threshold value:

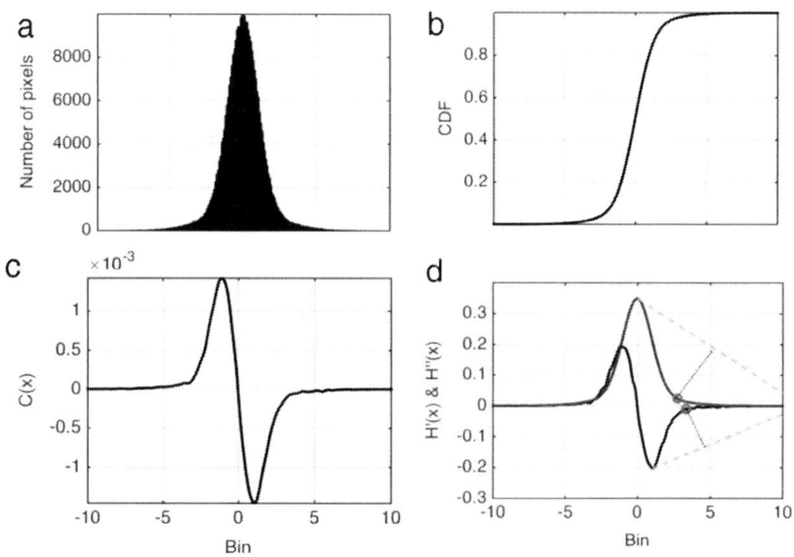

Figure. 1. Example of threshold detection of a unimodal histogram using ATS method. (a) Non-modal histogram as a function of bin; (b) the CDF graph; (c) the graph of curvature $C(x)$; and (d) graphs for the $H'(x)$, blue curve, and $H''(x)$, black curve, together with the triangle threshold detection phase of ATS. Green lines are the fitted lines and parameters used to generate the histogram in Fig. 1(a) are given in Table 1 for $n=1$. (For interpretation of the references to colour in this figure legend, the reader is referred to the web version of this article.)

1) initialise T_1 and T_2, e.g. the smallest and the largest possible values, respectively.
2) 2 Divide an image into three regions using the values of T_1 and T_2: R_1 (with all pixel value $\leq T_1$), R_2 (with all pixels values greater than T_1 and less than T_2), and R_3 (with all pixel values $\geq T_2$).
3) Calculate the average intensity values m_1, m_2 and m_3 for each region.
4) Compute a new value of T_1 and T_2 as

$$T_1 \longleftarrow \frac{m_1 + m_2}{2}; \quad \text{and} \quad T_2 \longleftarrow \frac{m_2 + m_3}{2}.$$

5) Repeat steps above from 2 to 4 until the differences between two successive values of T_1 and T_2 become minimal.

6) Set $T = \alpha_1 T_1 + \alpha_2 T_2$

where α_1 and α_2 are weighted average coefficients (here set to 0.5). When the density distribution of the histogram shifts towards the water background end, the variations of grey-level tend to alter the modal shape of the histogram (i.e. unimodal shape) in a manner that the IBCV method fails to detect a good threshold. In order to increase the practicality of the IBCV algorithm, we shall incorporate an automatic search threshold phase after step 6. The IBCV algorithm is then proceeded by the following illumination-correction phase applied for the real images in this study:

1) Search from the histogram value at bin T (estimated from step 6) towards the end of the longer (foreground) tail of the histogram.

2) Estimate the value for the whitecap fraction at each search bin.

3) Set as the optimal threshold, T_{opt}, the bin at which the absolute error between two successive values of the estimated $W_c(\%)$ at the sequence of the whitecap fraction is within a small fixed offset.

Fig. 2(a) shows an ideal histogram with a trimodal distribution at which two adjacent distributions with the same variance overlap heavily in the middle. It is observed that the IBCV technique results in two equally good threshold values (vertical red lines) with less tendency to bias the true locations of the thresholds. Fig. 2(b) shows a bimodal histogram for which the classical IBCV method makes a decent estimate of the threshold value. Fig. 2(c) illustrates an application of the method when there is an unclear modality pattern in the shape of the composite histogram. While the weighted average of two determined thresholds provides the optimum threshold, the optimal value of the threshold is highly sensitive to the disturbances induced by the uneven change of illumination at each real maritime scene. In such cases, the method may effectively be converged to the optimum value by utilising the illumination-correction stage, see also Section 4.4.

3. EXPERIMENTAL APPROACH

In November 2013, a field cruise was conducted with the research vessel R/V Håkon Mosby to Marstein Fyr, approximately 30 km southwest of Bergen, Norway (Fig. 3(a)). The objective of the cruise was the study of turbulent air–sea interaction processes in a coastal area. Several oceanographic instruments were deployed on both moving (submerged buoys) and fixed (bottom frame) sea going platforms. Moreover, a Furgo

Oceanor Wavescan buoy was moored at the measurement site to record gravity wave statistics. In addition to the oceanographic measurements, a direct covariance flux (sDCF) system was mounted at both the front bow of the research vessel and on top of the Wavescan buoy's meteorological mast (Fig. 3(b)). The DCF system, consisting of a sonic Gill R3 anemometer and an Inertial Measurement Unit (IMU), recorded atmospheric heat and momentum fluxes at approximately 8 m (vessel) and 3.5 m (buoy) above the sea surface. A Campbell scientific CC5MPX digital network camera was mounted next to the buoy's DCF system for visual inspection of the sea surface gravity waves and whitecapping events. The camera with the standard lens of 4–12 mm and a field of view from 27° to 80° was mounted with a downward tilting angle of approximately 10° relative to the horizon and programmed to collect snapshot images, 1 frame per second (fps), during daylight hours. Each full colour image with the size of 992×1280×3 is cropped on all sides (i.e. 793×1081×3) to reduce the atmospheric interferences depending on displacement from the vertical. The cropped colour image is then converted to the greyscale and the colour depth is reduced to 8 bits resulting in a normal greyscale image with 256 levels.

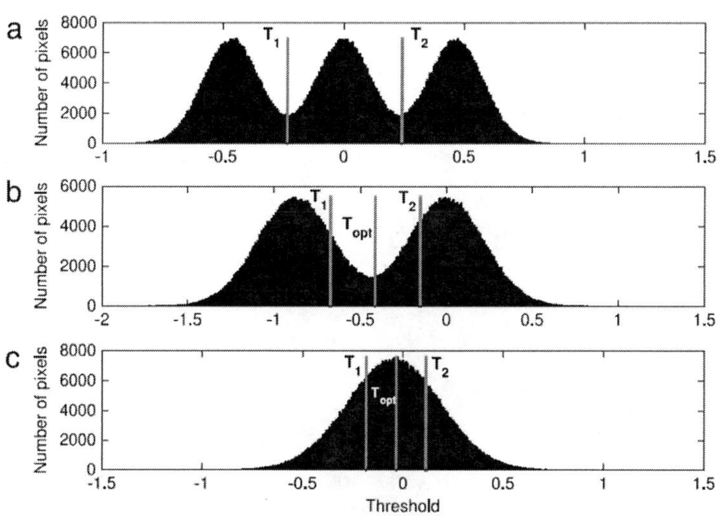

Figure. 2. Examples of threshold detection using the IBCV algorithm in the absence of illumination-correction phase for: (a) Trimodal histogram; (b) bimodal histogram; and (c) unimodal histogram using the classical IBCV thresholding search. Here, T_{opt} is the optimal threshold estimated from the classical IBCV method, i.e. from the step 6 of the algorithm. Parameters used to generate each histogram are given in Table 1. (For interpretation of the references to colour in this figure legend, the reader is referred to the web version of this article.)

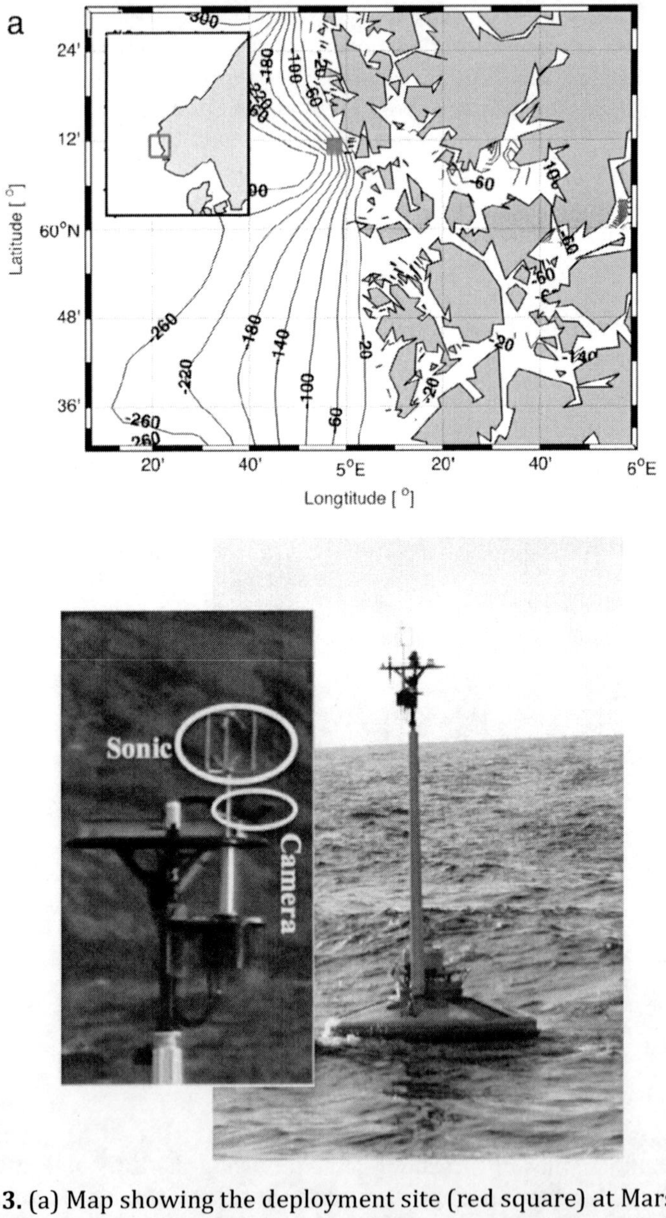

Figure. 3. (a) Map showing the deployment site (red square) at Marstein on the western coast of Norway. The inset shows the experiment site location in Norway; and (b) pictures of the Wavescan buoy and its mast with the installed sensors and the camera. (For interpretation of the references to colour in this figure legend, the reader is referred to the web version of this article.)

The weather and sea conditions encountered during the cruise are summarised in Fig. 4. Wind speed and wave height varied considerably during the campaign, with maximum wind speeds reaching 20 m s^{-1} and a significant wave height up to 6 m during a storm, which passed over the deployment site on November 28. With the approach of the low pressure system, the wind direction turned from northerly winds at the beginning of the campaign towards southeasterly and finally westerly directions during the passage of the storm. Until November 26, and after the passage of the storm, the sea-state was characterised by swell waves (Fig. 4(b)) and varying cloud cover which provided acceptable conditions for the video camera to capture the wave field. During the storm, the tilt angles of the buoy were too large and the recorded images could not be further processed.

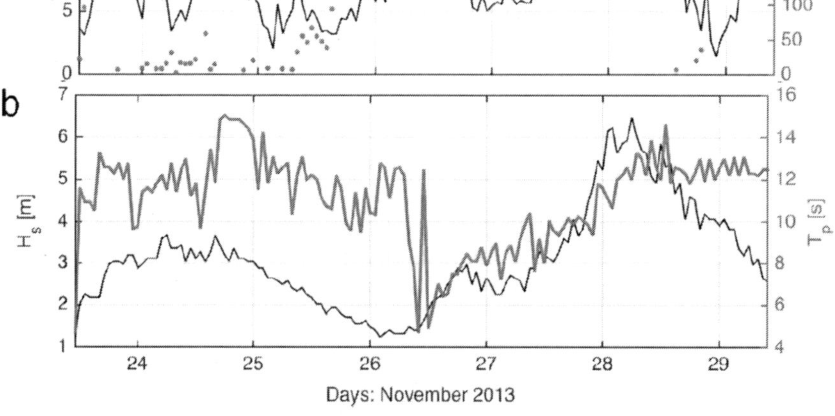

Figure. 4. Environmental forcing during the experiment. Time series of (a) wind speed and direction at 10 m height from the ship's meteorological mast; and (b) the significant wave height, H_s, and the wave peak period, T_p, recorded from the Furgo-Oceanor Wavescan buoy deployed nearby the ship station.

4. RESULTS AND COMPARISONS

This section presents the results for the determination of the whitecap fraction using video images. Here, we use the images taken from the camera mounted on the meteorological mast of the moored buoy. While the algorithms have been rigorously tested for different sequences of images, we use a six-minute series of images in this section which is sufficient for a proof of concept demonstration.

4.1. Image structure

Fig. 5 shows an example of the smoothed image structure (using a four-point running average) of an image and its first- and second-order derivatives. The image structure shows the variation of the Percentage Increase in number of Pixels (PIP) versus threshold at level j as

$$\mathrm{PIP}(j) = 100 \times \frac{\sum_{k=j}^{j+\Delta t} J_k}{\sum_{k \geq j} J_k}, \tag{9}$$

where J_k has been defined in Eq. (3), and Δt=0.01 denotes the threshold increment (Callaghan and White, 2009). It is observed that the shape of the image structure relies on the grey-level variations which substantially alter the illumination pattern of the image, Fig. 5. Therefore, prior to applying the detection algorithm, it is necessary to compensate the effects of the non-uniform illumination and sunglint or discard those contaminated images from the image dataset (Callaghan and White, 2009). However, a typical feature for almost all images used in this study is the strong uneven illumination of the background water with intensity values comparable to the intensity values of whitecaps, Fig. 5(a). This effect makes it very difficult to efficiently distinguish between the whitecap areas and water pixels using the PIP and its derivatives, Fig. 5(b), (c), and (d). Our analysis based on the cropped images suggests that the replacement of the image structure by W_c(%) in the AWE method may result in a more robust procedure for images taken from a non-stationary camera in the presence of contaminations induced by the sky reflection and uneven illumination. Hereafter, we use the AWE method based on W_c(%) instead of employing the image structure, and postpone further investigations based on the image structure using this dataset elsewhere.

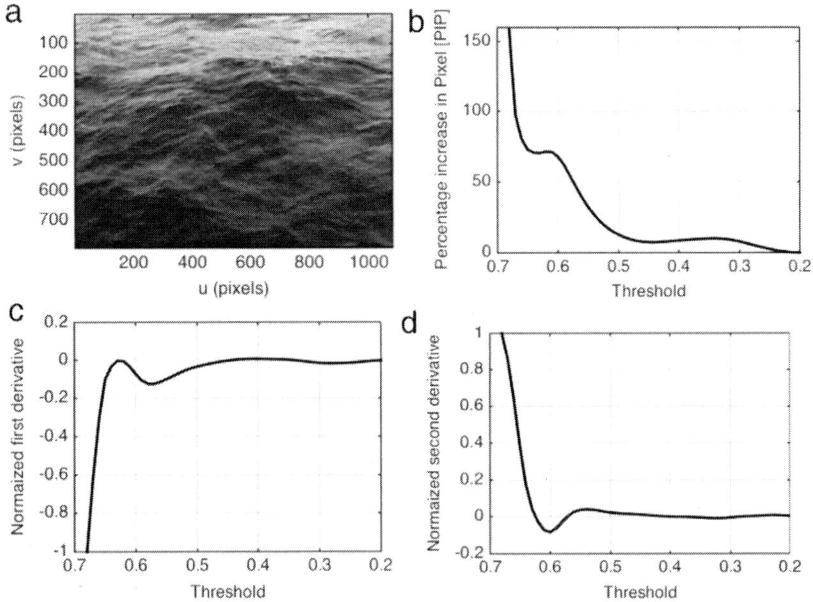

Figure. 5. The image structure of an image with strong non-uniform illumination pattern. (a) The original image; (b) the image structure estimated from Eq. (9); and (c and d) the first and the second-order derivatives of the image structure.

4.2. The AWE method

Fig. 6 illustrates a snapshot greyscale image including visible active whitecap events with a relatively long lifetime. This plunging plume can be captured in a digital image-processing algorithm by defining a global threshold to separate the background pixels (non-whitecapping) from the whitecapping pixels. In order to check the sensitivity of whitecap coverage to the threshold magnitude, three different threshold values in the range of $[0,1]$ are applied to the original greyscale image. The area of foam and the whitecap coverage are estimated by Eq. (4). In Fig. 6(b), the brightness of the binary image is relatively high due to the sunlight, and the non-whitecapping areas are identified as part of the whitecapping region with a coverage of approximately 80% for $T=0.3$. By using a threshold of $T=0.53$ (Fig. 6(c)), the contributions due to the sunglint substantially decline in the upper part of the image. Changing the value of threshold to 0.6 (Fig. 6(d)) results in further discrimination between whitecapping and non-whitecapping regions.

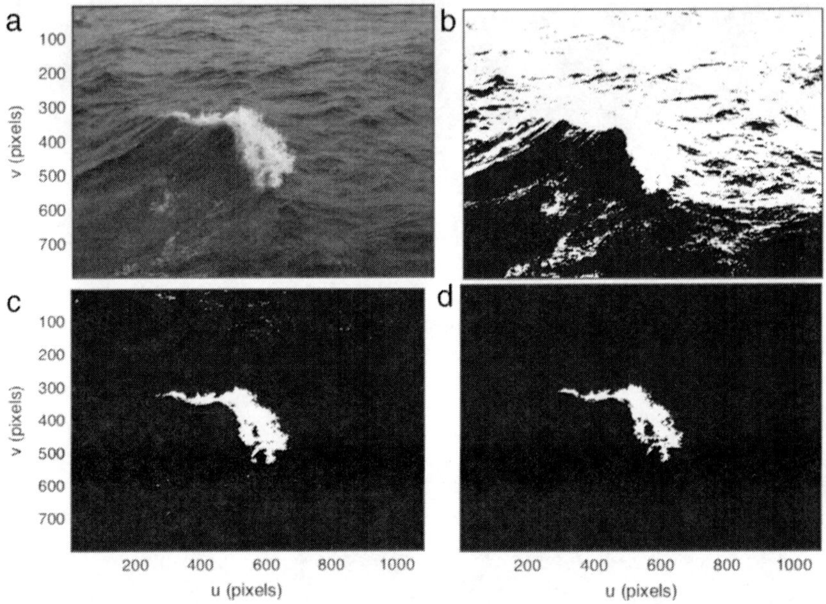

Figure. 6. Picture of whitecaps variations as a function of threshold. (a) Original greyscale image; (b) the binary image with $T=0.3$ and $W_c=80\%$; (c) the binary image with $T=0.53$ and $W_c=19\%$; and (d) the binary image with $T=0.6$ and $W_c=15\%$.

Fig. 6 highlights the importance of choosing an appropriate threshold level to distinguish between different illumination conditions. In Fig. 7, we show details of the optimal threshold-selection system based on the AWE approach.

The histogram of the greyscale image illustrates a clear separation between bright and dark pixels. The higher peak at $T=0.21$ represents the pixel corresponding to the darker colour with the larger intensity variation, and the information of the histogram at the valley between two peaks may be captured by the peak-selection algorithm as the optimum value of the threshold (Fig. 7(c), and Section 2.1). Fig. 7(c) illustrates the relationship between the threshold level and the fraction of whitecap coverage for the greyscale image, with a threshold increment of 0.01. The whitecapping coverage curve (red) remains flat until a threshold of approximately 0.15, before declining rapidly in the mid-section. Beyond the threshold value of $T=0.5$, the curve asymptotically approaches towards zero. Finally, the optimum value of the normalised threshold, T_{opt}, is chosen as the last inflection point of W_c which is calculated using the smoothed second-order derivative of $W_c(\%)$ (black line). While in this example, the AWE method perfectly determines the separation between whitecaps and water pixels, the histogram-based methods generally fail to successfully distinguish between the whitecapping and the non-whitecapping segments due to the valley caused by the illumination-induced contamination (Fig. 7(b)).

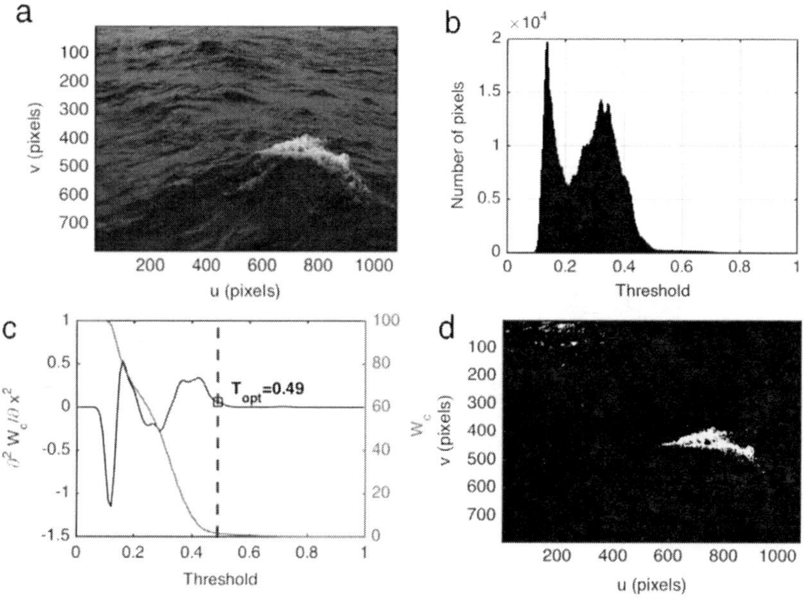

Figure. 7. (a) Original grey scale image; (b) image histogram; (c) variation of W_c based on the AWE method and its smoothed second-order derivative as function of the threshold, $\partial^2 W_c/\partial x^2$, where x denotes the dummy variable related to the histogram bin; and (d) the binary image with optimum threshold of 0.53. The threshold increments set to 0.01. (For interpretation of the references to colour in this figure legend, the reader is referred to the web version of this article.)

4.3. The ATS scheme

An example of using the ATS scheme for the extraction of bright pixels is shown in Fig. 8. This adaptive technique attempts to pick up the largest value for the threshold by assuring the appropriate detection of whitecap areas. The image greyscale histogram exhibits a valley region between two clear peaks as a result of contaminations from the uneven illumination. In such situations, there are several pixels in the whitecap patch which share the similar intensity distribution with the background water, and there is no distinct difference between the whitecap and water areas (see also Fig. 9(c)). Therefore, applying the Otsu's method will result in a non-optimal value for the threshold (Fig. 8(b)). An outline of the threshold selection procedure in the ATS method is shown in Fig. 8(c). After smoothing a greyscale image (using four-point running mean) to remove the noise and very high frequency content, the cumulative histogram, H, and its second-order derivative, $\partial^2 H/\partial x^2$, where x is the histogram bin variable, are calculated from the smoothed image. The strong gradient of both whitecap and water regions can then be separated using a proper selection of the threshold value as a point in which the change in the $\partial^2 H/\partial x^2$ slope approaches zero. While the optimal point can be chosen using small value of γ in Eq. (7), the optimum value for the threshold is objectively selected using the fully automated triangle method presented in Section 2.2.2.

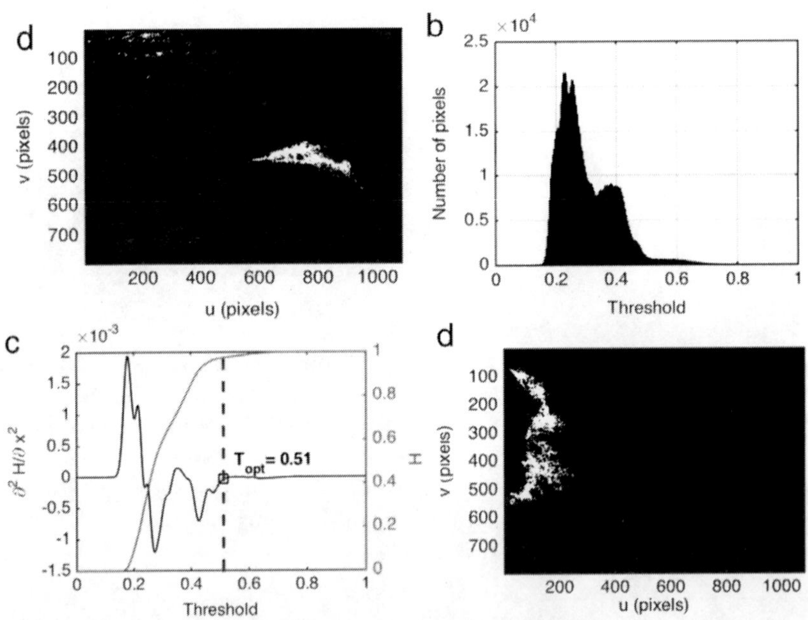

Figure. 8. (a) Original greyscale image; (b) image histogram; (c) variation of cumulative histogram, H, and its smoothed second-order derivative as a function of grey level, $\partial^2 H/\partial x^2$, where x denotes the dummy variable related to the histogram bin; and (d) the binary image with the optimal value of threshold, normalised in the range of $[0,1]$, calculated from the ATS algorithm.

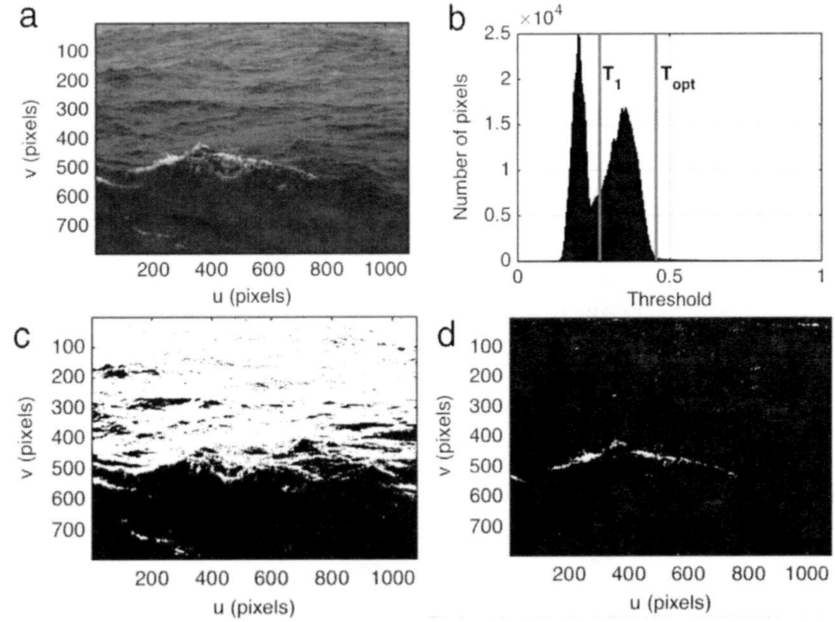

Figure. 9. An example of applying the IBCV threshold algorithm for the images with strong inhomogeneous background intensity distribution. (a) Original greyscale image; (b) image histogram and the threshold values from the standard IBCV algorithm (T_1) and the method including the illumination-correction phase, T_{opt}; (c) the binary image obtained by the standard IBCV algorithm; and (d) the binary image obtained by the illumination-correction version of the IBCV algorithm.

4.4. The IBCV method

We perform an analysis as shown in Fig. 9 to demonstrate the ability of the IBCV algorithm for an image with a strong background illumination pattern. It is observed that in the presence of strong inhomogeneous background conditions, the standard IBCV algorithm exhibits poor performance to capture the optimal value of the threshold. This may be attributed to the modulation of the image intensity histogram by the sudden and uneven changes of the image illumination, and nonuniform reflected sky conditions (Fig. 9(b) and (c)). While there are several approaches to handle the uneven illumination effects, we have found the IBCV method less vulnerable to the uneven illumination changes for each scene (Fig. 9(d)). It should be noted that we use the IBCV method together with the illumination-correction phase to obtain the optimal value of the threshold for all of maritime images in this study.

4.5. Comparisons and performance analysis

A comparison of the different approaches used in the whitecap detection process is shown in Fig. 10. Here we show the results from the AWE (Fig. 10(b)), the ATS (Fig. 10(c)), and the IBCV (Fig. 10(d)) techniques. All techniques determine suitable thresholds to separate between whitecaps and unbroken background water under uneven light conditions. While most of the whitecap pixels have been detected, none of the techniques could identify several sparse residual whitecaps in the scene. This is because these pixels have intensity values between the estimated threshold value and the intensity of the dark water pixels which may not be easily recognised by using the conventional thresholding-based techniques. Application of clustering algorithms, such as the k-mean algorithm, prior to the threshold-based segmentation may mitigate this drawback.

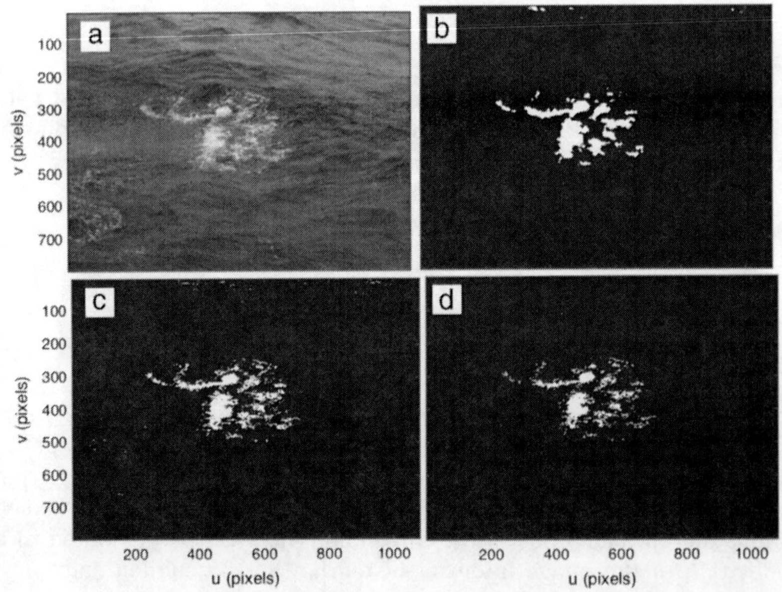

Figure. 10. The comparison between different W_c extraction techniques. (a) Original greyscale image; (b) the binary image estimated by the AWE method; (c) the binary image obtained from AST approach; and (d) the binary image estimated by the IBCV method.

From Fig. 11, it can be seen that there is high positive correlation between the optimum thresholds estimated from two automated techniques and one from the ground truth analysis, which is determined manually using the thresholding-based technique. Generally, there might be some situations in which the ATS and the AWE schemes may result in erroneous values for the threshold when getting shifted towards the lower values of intensity. This can happen when the scene contains low noise levels and the cumulative spectrum does not exhibit any gradient response. However, such situations were not observed in the images used in this study.

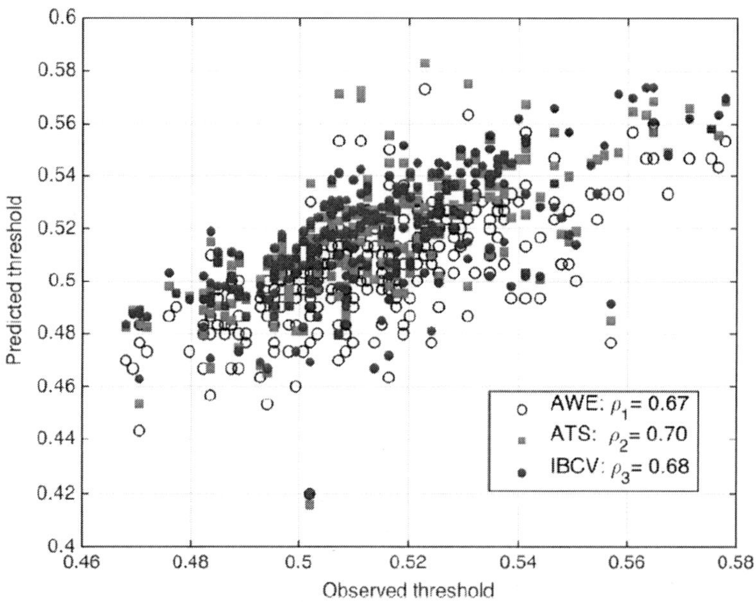

Figure. 11. The scatter plot between the normalised predicted thresholds, from the AWE method, the ATS scheme, the IBCV method, and the manually segmented ground truth data for 261 images. ρ_1, ρ_2, and ρ_3 indicate the values for the correlation between the observed and the predicted normalised thresholds.

In order to provide more insights into the accuracy of the linear regression and correlation coefficients (ρ), we conduct statistical testing of the significance such as the goodness-of-fit of the linear model (e.g. $y = a_0 x + b_0$) by calculating the error bound, and the significance of correlation coefficient by calculating the p-value and t-test. By assuming that both the ground truth and the estimated thresholds are Gaussian distributed, the t-value (i.e. Student's t-distribution) for the threshold values is determined using the following t-statistic as $t_{calc} = p\sqrt{N-2)/(1-p2)}$ where N is the number of images used for the inferential analysis (Emery and Thomson, 2001). The correlation coefficient is then significant if t_{calc} is greater than the critical t-value, t_{crit}(for the significance level of 0.05 and $N-2$ degrees of freedom). The p-value for all threshold-detection methods is approximately zero, suggesting that the correlation coefficient is significant. Accordingly, the t-test confirms further the statistical significance of all methods for the significance level of 0.05, i.e. $t_{calc} >> t_{crit}$. Table 2 includes estimates for the standard deviation of the error in predicting the observed threshold at 5% level of significance. While all methods suggest nearly identical information on the accuracy of the threshold-detection procedure, the ATS method gives overall better improvement in predicting of the true threshold compared to the other techniques.

Table 2. The results of statistical testing based on linear regression and correlation coefficient. a_0 and b_0 denote the slope and intercept of the linear model, respectively, ρ_i is the correlation coefficient for the ith method, the confidence interval error for the significant level of 0.05 is presented by Error (95%), the significance of the correlation analysis is given by p-value, and t_{calc} and t_{crit} are the t-test statistics.

Method	a_0	b_0	Error (95%)	p-value	ρ_i	t_{calc}	t_{crit}
AWE	0.72	0.14	0.0331	0.0	0.67	14.49	1.65
ATS	0.77	0.12	0.0323	0.0	0.70	15.83	1.65
IBCV	0.75	0.13	0.0330	0.0	0.68	15.09	1.65

The superiority of the ATS and IBCV techniques relative to the AWE method used for this study is that they do not need any prior information about the values of the threshold and W_c(%) to generate the optimal threshold. Therefore, these methods are able to predict the optimum value of the threshold with less computational complexity than the AWE method, while preserving the same level of accuracy. Fig. 12 illustrates the computational performance of three presented threshold-detection methods implemented on a MAC OS X 296 10.9.5 with Intel Core i7 3 GHz CPU, and 16 GB 1600 DDR3 RAM. According to Eqs. (4) ; (9), we conclude that the conventional AWE method (i.e. Callaghan and White, 2009) and the one used in the present study have approximately identical computational complexity. Consequently, using the ATS and the IBCV methods cause considerably less CPU usage than the time spent by the conventional AWE method to detect the optimum value of the threshold for each image.

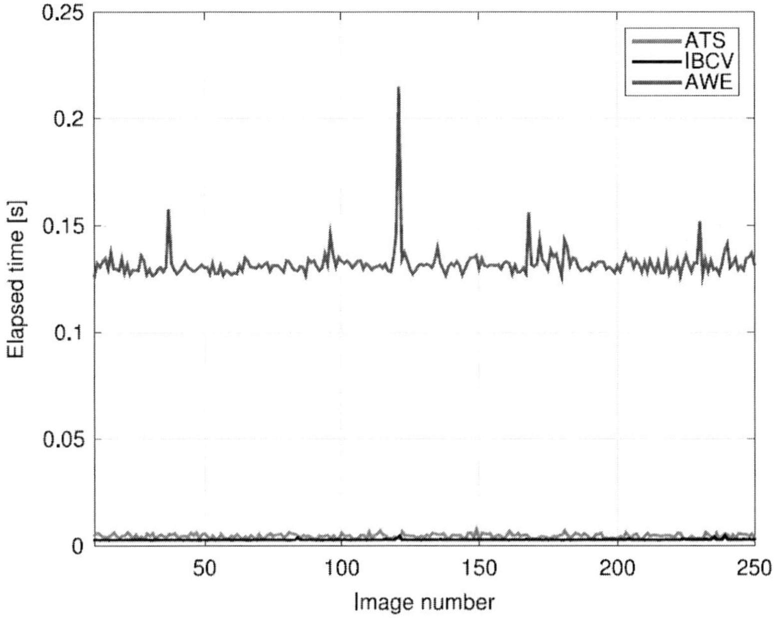

Figure. 12. Elapsed time of three threshold-detection methods for a sequence of images.

To further assess the accuracy and performance of the automated segmentation techniques, we calculate the evaluation metrics for each image by comparing its segmented binary image with the manually segmented ground truth, which is, of course, subject to human error and the inter-observer variability. The information of a Confusion Matrix (CM) for each scene is used to estimate the number of True Positive (TP), False Positive (FP), True Negative (TN), and False Negative (FN) pixels (Fig. 13(b), (c), and (d)). The calculated metrics are then used to estimate the precision and recall given by:

$$\text{Precision} = \frac{\text{TP}}{\text{TP} + \text{FP}}; \qquad \text{Recall} = \frac{\text{TP}}{\text{TP} + \text{FN}}, \tag{10}$$

where TP=CM(1,1), FP=CM(2,1), FN=CM(1,2), and TN=CM(2,2). Fig. 13 represents an example of the CM information estimated from the manually segmented image and the corresponding binary image obtained by applying the AWE, ATS, and IBCV techniques, respectively. The results show that the optimal threshold from the AWE method results in a high value for the true positive rate with the false negative rate of approximately 1.48% (Fig. 13(b)). The value of the FN rate is completely reduced by using the ATS and IBCV methods which preserve the higher values of TP rates than that of from the AWE method for this sample. Generally, the ATS classifier with 99% true positive rate and 0.43% false negative rate achieves better performance than the other automated segmentation techniques (see also Fig. 11). Furthermore, the implementation of the performance test for three segmentation schemes over 261 images shows that all automated techniques achieve a high level of performance with an approximate value of 98% for both precision and recall, with respect to the manually segmented ground truth. It should be noted that although promising, the performance analysis is constrained to the choice of the ground truth datasets, the general limitations of the thresholding methods, and the uncertainty of using a non-stationary camera in the evaluation-scheme. Investigation of such uncertainty resources is beyond the scope of the present study and will be addressed in more details elsewhere, i.e. (Bakhoday-Paskyabi et al., submitted for publication).

5. SUMMARY

In this study, we analysed a sequence of images acquired from a non-stationary camera mounted on a moored Wavescan buoy to measure the coverage of whitecaps at the sea surface. We proposed two threshold-based algorithms to discriminate between the whitecap and non-whitecap populations at each maritime scene with no need to know the camera's intrinsic and extrinsic parameters. The first method is an Adaptive Thresholding Segmentation (ATS) technique which uses the cumulative histogram information to extract the value of the optimal threshold as the last inflection point of the smoothed cumulative sum. From the implementation of this method, we observed that this technique not only makes it possible to efficiently divide each image into whitecap and non-whitecap classes, but also demonstrated that it is computationally very fast and robust relative to the conventional Automated Whitecap Extraction (AWE). Furthermore, we used another automated multi-thresholds algorithm to determine the optimal threshold by maximising the between-class variance (BCV) of the whitecap and the water pixels. An iterative strategy used in this algorithm allowed an efficient and robust bi-level segmentation of images with less computational demand and higher accuracy than the conventional Otsu's search method. A correction phase was further calculated in the standard Iterative BCV (IBCV) method to avoid the influence of image distortion by the uneven illumination. Comparisons between different techniques showed that the ATS method gives an overall better performance than the IBCV and the AWE methods.

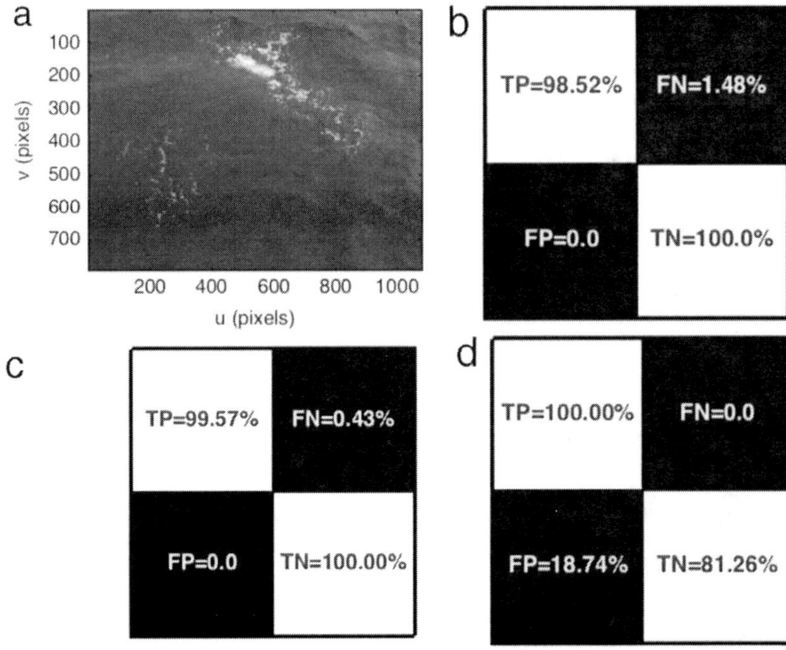

Figure. 13. The comparison between the estimated confusion matrix for different segmentation techniques compared with the manually (threshold-based) segmented ground truth: (a) Original greyscale image; (b) the AWE method; (c) the ATS method; and (d) the IBCV scheme. The performance metrics are defined as: T-Positive (TP): the whitecap pixel (foreground) is correctly identified as whitecap pixel; F-Positive (FP): the water pixel (background) is incorrectly diagnosed as the whitecap pixel; T-Negative (TN): the water pixel is correctly diagnosed as the water pixel; and F-Negative (FN): the whitecap pixel is incorrectly identified as the water pixel. The diagonal elements in the CM matrix correspond to the correct classifications and the off-diagonal elements denote the incorrect classifications.

Although the accuracy of the algorithms were precisely checked by using the confusion matrix entries, the analysis needs yet to be proven based on more accurate ground truth images less affected by the human errors and limitations of the conventional segmentation techniques. Since the investigation on the accurate choice of an empirical relation between the wind and $W_c(\%)$ and the best approach to the elimination of the contaminations induced by the non-stationary camera are beyond the scope of the present work, we will address them elsewhere.

ACKNOWLEDGEMENTS

This work has been performed as part of the Norwegian Center for Offshore Wind Energy (NORCOWE) funded by the Research Council of Norway (RCN 1938211560) and the OBLO (Offshore Boundary Layer Observatory) project (RCN: 227777). We thank Helge Bryhni (UoB/GFI) and Steinar Myking (UoB/GFI) for preparing the instruments and for their deployments. We also express our gratitude to Ilker Fer (UoB/GFI) and the crew of the RV Håkon Mosby for their support during the cruise.

REFERENCES

1. Andreas, E.L., 1992. Sea spray and the turbulent air-sea heat fluxe. J. Geophys. Res. 97, 11429–11441.
2. Andreas, E.L., Persson, P.P.G., Hare, J.E., 2008. A bulk turbulent air-sea flux algorithm for high-wind, spray conditions. J. Phys.Oceanogr. 38, 1581–1596. http://dx.doi.org/10. 1175/2007JPO3813.1.
3. Bakhoday-Paskyabi, M., Flügge, M., Schwendeman, M., Reuder, J., 2016. Visual attitude estimation from a shipboard movingcamera: Applications in turbulent flux measurements. J. Atmos. Ocean. Technol. submitted for publication.
4. Blanchard, D., 1963. The electrification of the atmosphere by particles from bubbles in the sea. J. Prog. Oceanogr. 1, 71–202.
5. Callaghan, A.H., de Leeuw, G., O'Dowd, C.D., 2008. Relationship of oceanic whitecap coverage to wind speed and wind history. Geophys. Res. Lett. 35, http://dx.doi.org/10.1029/2008GL036165.
6. Callaghan, A.H., White, S., 2009. Automated processing of sea surface images for the determination of whitecap coverage. J. Atmos. Ocean. Technol. 26, 383–394.
7. Deane, G.B., Stokes, M.D., 2002. Scale dependence of bubble creation mechanisms in breaking waves. Nature 418, 839–844. http://dx.doi.org/10.1038/nature00967.
8. Emery, W.J., Thomson, R.E., 2001. Data Analysis Methods in Physical Oceanography, second ed. Elsevier Science, Netherlands, p. 638.
9. Hanson, J.L., Phillips, O.M., 1999. Wind sea growth and dissipation in the open ocean. J. Phys. Oceanogr. 29, 1633–1648.
10. Jones, K.F., Andreas, E.L., 2012. Sea spray concentrations and the icing of fixed offshore structures. Q. J. R. Meteorol. Soc. 138, 131–144.
11. Kittler, J., Illingworth, J., 1986. Minimum error thresholding. Pattern Recognit. 19 (1), 41–47.
12. Kleiss, J.M., Melville, W.K., 2011. The analysis of sea surface imagery for whitecap kinematics. J. Atmos. Ocean. Technol. 28, 219–243.
13. Melville, W.K., 1996. The role of surface-wave breaking in air-sea interaction annual review of fluid mechanics. Annu. Rev. Fluid Mech. 28, 279–321. http://dx.doi.org/10.1146/annurev.fl.28.010196.001431.
14. Melville, W.K., Matusov, P., 2002. Distribution of breaking waves at the ocean surface. Nature 417, 58–63. http://dx.doi.org/10. 1038/417058a.
15. Monahan, E.C., 1969. Fresh water whitecaps. J. Atmospheric Sci. 26, 1026–1029.
16. Monahan, E.C., 1971. Oceanic whitecaps. J. Phys. Oceanogr. 1, 139–144.
17. Monahan, E., O'Muircheartaigh, I.G., 1980. Optimal power-law description of oceanic whitecap coverage dependence on wind speed. J. Phys. Oceanogr. 10, 2094–2099.
18. Monahan, E., O'Muircheartaigh, I.G., 2012. Oceanic whitecaps and the 10m-elevation wind speed: Toward improved powerlaw descriptions for use in climate modeling. In:

Presented at 18th Air–Sea Interaction Conference. Amer. Meteorol. Soc., Boston, Mass, 19–13. Jul.
19. Otsu, N., 1979. A threshold selection method from gray-level histograms. IEEE Trans. Mach. Learn. Appl. 17–21.
20. Paget, A.C., Bourassa, M.A., Anguelova, M.D., 2015. Comparing in situ and satellite-based parameterizations of oceanic whitecaps. J. Geophys. Res. Oceans 120, 2826–2843. http://dx.doi.org/10.1002/ 2014JC010328.
21. Salisbury, D.J., Anguelova, M.D., Brooks, I.M., 2014. Optimal power-law description of oceanic whitecap coverage dependence on wind speed. Geophys. Res. Lett. 41, 1616–1623. http://dx.doi.org/10.1002/2014GL059246.
22. Scanlon, B., Ward, B., 2013. Oceanic wave breaking coverage separation techniques for active and maturing whitecaps. Methods Oceanogr. http://dx.doi.org/10.1016/j.mio.2014.03.001.
23. Schwendeman, M., Thomson, J., 2015. Observations of whitecap coverage and the relation to wind stress, wave slope, and turbulent dissipation. Geophys. Res. Oceans 12, http://dx.doi.org/10.1002/2015JC011196.
24. Stramska, M., Petelski, T., 2003. Observations of oceanic whitecaps in the north polar waters of the Atlantic. J. Geophys. Res. 108 (C3), http://dx.doi.org/10.1029/2002JC001321.
25. Sugihara, Y., Tsumori, H., Ohga, T., Yoshioka, H., Serizawa, S., 2007. Variation of whitecap coverage with wave-field conditions. J. Mar. Syst. 66, 47–60.
26. Thorpe, S.A., 1982. On the clouds of bubbles formed by breaking wind-waves in deep water, and their role in air-sea gas transfer. J. Philos. Trans. R. Soc. Lond. Ser. A 304, 155–216.
27. Woolf, D.K., Bowyer, E.K., Monahan, E.C., 1987. Discriminating between the film-drops and jet-drops produced by a simulated whitecap. J. Geophys. Res. 92, 5142–5150.
28. Zack, G.W., Rogers, W.E., Latt, S.A., 1977. Automatic measurement of sister chromatid exchange frequency. J. Histochem. Cytochem. 25 (7), 741–753.

CHAPTER 14

Stable Isotope Methods for the Study of the Nitrogen Cycle

Evgenia Ryabenko[1, 2]

[1] GEOMAR/Helmholtz Centre for Ocean Research Kiel, Kiel, Germany

[2] Institute of Groundwater Ecology, Helmholtz Zentrum München – Germany Research Center for Environmental Health, München, Germany

1. INTRODUCTION

Nitrogen, a limiting element for biological productivity, plays a key role in regulating the biogeochemical processes in the ocean. In today's ocean, all of the major reactions in the N cycle are mediated by assimilatory or dissimilatory functions of marine organisms. Because marine organisms preferentially incorporate lighter stable isotope of nitrogen ^{14}N instead of ^{15}N, each major metabolic reaction in the N cycle involves irreversible kinetic fractionation of nitrogen.

The isotopic composition of a pool of nitrogen can be used to identify the relative importance of sources that are isotopically distinct, or processes that add or remove nitrogen with a characteristic pattern of isotopic discrimination. The strongest isotopic fractionations are associated with dissimilatory processes that mediate the transfer of nitrogen from one inorganic pool to another [1]. In contrast, processes such as primary production, which move nitrogen directly into and through the food web are associated with comparatively weak isotopic fractionations [1]. The extent of nitrogen isotope fractionation also depends upon the kinetics of individual metabolic reactions, concentration of products and reactants, environmental conditions (e.g., oxygen concentrations) and the microbial species involved.

The crucial challenge in using nitrogen isotope methods is the complexity of the marine nitrogen cycle and the potential influence of multiple processes on the isotopic Composition of several biologically active pools of nitrogen. Successful use of nitrogen isotopes in resolving N cycle fluxes and processes requires an understanding of the general distribution of nitrogen isotopes in marine systems, the nature of isotopic fractionation, and a careful consideration of the processes at work in the system of interest. In this chapter I will introduce the reader to the two different approaches of nitrogen isotope analysis: natural abundance or tracer methods (^{15}N-labeling). The natural abundance of the stable isotope of nitrogen can provide critical insights into processes acting on a variety of scales with minimal alteration and manipulation of the system being studied. The isotopic signal collected in this type of approach is an integrated value that incorporates large spatial and temporal variation of the relevant processes. Conversely, isotope tracer methods involve short-term incubations of small, isolated samples of water under in situ or simulated in situ conditions. Tracer methods are primarily used for rate experiments that complement larger-scale natural abundance studies.

I begin the chapter with basic definitions of isotopic fractionation and analytical considerations of the sample measurements. I follow with an overview of the processes of nitrogen sources and sinks. I conclude with a discussion of how isotopic data can contribute to the current debate on the balance of N-inputs and losses in the nitrogen cycle.

2. FUNDAMENTALS

2.1. Isotopes and calculation of their ratio

Isotopes are atoms of an element that share the same number of protons but a different number of neutrons. In the scientific nomenclature, isotopes are specified in the form $^{m}_{n}E$, where “m” indicates the mass number (the sum of protons and neutrons in the nucleus) and “n” refers to the atomic number of an element “E”. There are more than 10 nitrogen isotopes known. Most of these isotopes are radioactive and highly unstable with longest half-life time for N713N713 of 10 minutes. The only two stable nitrogen isotopes are N714N714 and N715N715, which have seven protons each and seven or eight neutrons in their nucleus, respectively. ^{15}N is the less frequent stable isotope, constituting of 0.365% of the global nitrogen pool [2]. Consequently, it is more practical to measure the difference or ratio of two isotopes instead of the absolute quantity of each.

Isotopic compositions are expressed in terms of "delta" (δ) values which are given in parts per thousand or per mil (‰). Nitrogen isotope ratios, for example, are expressed as the ‰-difference to atmospheric N_2, which has a constant $^{14}N/^{15}N$ of 272 ± 0.3 [3]. The $\delta^{15}N$-value in the sample is then calculated by the following equation (1).

$$\delta^{15}N(vs.air) = \left(\frac{(^{15}N/^{14}N)_{sample}}{(^{15}N/^{14}N)_{air}} - 1 \right) \times 1000 \quad (1)$$

The δ-values do not represent absolute isotope abundances but rather the ‰-difference to a widely used reference standard, such as VSMOW (Vienna Standard Mean Ocean Water). The δ-value is then calculated from equation (2), by measuring the isotope ratios (R) for the sample and the reference standard:

$$\delta^{15}N = \left(\frac{R_{sample}}{R_{standard}} - 1 \right) \times 1000 \quad (2)$$

where R_{sample} and $R_{standard}$ represent the isotope ratio $\delta^{15}N_{(vs.air)}$ in the sample and in the standard respectively, calculated using equation (1). By convention, R is the ratio of the less abundant isotope over the most abundant isotope (i.e. $^{15}N/^{14}N$ for nitrogen).

As previously discussed, the $\delta^{15}N$ value changes under the influence of chemical and physical processes. If the process is complete the resulting $\delta^{15}N$ value in the product is equal to the value of the reagent. Only if the reaction is incomplete the fractionation of isotopes happens, which means that the $\delta^{15}N$ values in the product and substrate differ.

2.2. Isotope fractionation effect

There are two different fractionation processes, both of which will be discussed here: equilibrium and kinetic fractionation processes.

Equilibrium fractionation processes are reversible processes. They are mainly driven by changes in the internal energy of a molecule, i.e. vibrations of the atoms within a molecule with respect to each other and rotations around the molecular axes. The equilibrium fractionation factor $\alpha_{eq.}$ is related to the equilibrium constant K as shown in equation (3), where n is the number of exchanged atoms.

$$\alpha_{eq.} = k^{1/n} \tag{3}$$

During equilibrium reactions, the heavier isotope preferentially accumulates in the compounds with a higher number of bonds. During phase changes, the ratio of heavy to light isotopes in the molecules also changes. For example, as water vapor condenses in rain clouds (a process typically viewed as an equilibrium process), the heavier water isotopes (^{18}O and ^{2}H) become enriched in the liquid phase while the lighter isotopes (^{16}O and ^{1}H) remain in the vapor phase. In addition, the equilibrium isotopic effect decreases as the system temperature increases.

Kinetic fractionation processes are also associated with incomplete processes like evaporation, dissociation reactions, biologically mediated reactions and diffusion [4]. Kinetic isotope fractionation reflects the difference in the bond strengths or motilities of the isotopic species. The degree of isotopic fractionation associated with a reaction is commonly expressed with α, which is the ratio of rate constants for molecules containing the different isotopes:

$$\alpha = {}^{14}k/{}^{15}k \tag{4}$$

where ^{14}k and ^{15}k are the rate constants for molecules containing the light and heavy isotopes, respectively. Most biological reactions discriminate against the heavier isotope, yielding $1.00<\alpha<1.03$. This makes it convenient to define an isotopic enrichment factor (ε) that highlights more clearly the range of variation:

$$\varepsilon = (\alpha - 1) \times 1000 \tag{5}$$

An alternative method for describing isotopic enrichment is the Rayleigh relationship. This is an exponential relationship that describes the partitioning of isotopes between two reservoirs as one reservoir decreases in size. The equation can be used if the following conditions are met: 1) material is continuously removed from a mixed system; 2) the fractionation accompanying the removal process at any instance is described by the fractionation factor α, and 3) α does not change during the process. Under these conditions, the evolution of the isotopic composition in the residual (reactant) material is described by:

$$R = R_0 f(\alpha - 1) \tag{6}$$

where R is the isotopic ratio of the product, R_0 is the initial ratio of the reactant, f is the fraction of the substrate pool remaining and α is the kinetic fractionation factor.

2.3. Measurements of δ^{15}n

Stable isotope measurements are becoming a routine tool in studies of marine ecosystems due to the increasing availability of mass spectrometers. Biological oceanographers usually using continuous flow systems for ^{15}N measurements that integrate a preparatory system (e.g., an elemental analyzer) with a mass spectrometer [5]. The mass spectrometer typically measures the isotopic composition of N_2 generated by combustion and carried in a stream of He gas to an open split interface that introduces a small fraction of the gas stream into the ion source for measurement. This analytical approach requires little or no manipulation of organic samples, though care must be taken to avoid contamination with exogenous nitrogen, particularly when tracer-level ^{15}N experiments are being carried out in the vicinity. Since N_2 is typically the analyte, minimizing atmospheric contamination in the preparatory system is also quite important. Alternatively, N_2O gas may be used for isotope measurements, which practically eliminates the potential contamination from the atmosphere and gives additional information of oxygen isotope composition of nitrogen compound (e.g nitrate or nitrite). In this case, nitrogen compounds must be chemically converted into N_2O prior the isotope analysis [6-11]. Special care has to be taken due to the potential exchange of oxygen isotopes between nitrate and water leading to deviation of initial value (e.g. [10]).

Tracer-level ^{15}N experiments used for rate measurements typically add about 10% of the ambient source concentration of ^{15}N-compound. The initial planning of tracer studies requires consideration of several practical issues. First of all, one has to determine the length of incubation, which is connected to turnover rate of the relevant pools and the detection limits for the analytical methods. The turnover rate (pool size^{-1}×uptake rate) has to be estimated for both source and product pools to ensure that tracer can be detected in the product pool, while dilution of the source pool is either insignificant or small [12, 13]. Increasing incubation time will increase the amount of tracer in the product but it will reduce the potential of isotope dilution affecting the source pool. Given a best-guess turnover rate for the product pool, one can estimate the relative trade-offs between incubation time and changes in isotopic labeling of the source and product pools at a given enrichment level (7).

$$\text{Turnover rate} \times \text{incubation time} \times \text{source enrichment} = \text{change in enrichment, e.g., } 0.1\ \text{d-1} \times 0.2\ \text{d-1} \times 10\% = 0.2\%. \tag{7}$$

Generally the shortest incubation possible would resolve this dichotomy for biogeochemical processes such as nitrification or denitrification. However, complications may arise for uptake studies since short incubations primarily represent the transport (movement across the cell membrane) of N-compounds, whereas longer incubations primarily represent assimilation into organic compounds such as amino acids or proteins.

Finally, one has to consider the degree of replication and treatment of the initial, or time zero, concentrations and enrichment of both the product and source pools. To account for small, natural (%) level changes in these pools, studies either assume that particles have constant natural abundance values, or they assess the natural abundance periodically during the study. Alternatively, tracer can be added to the incubation bottle, which is then deployed and then immediately filtered at an initial time point for comparison with the final values. One then needs to decide how to interpret the initial time point, as there is often a substantial time between tracer addition and filtration (often 30 min). During this interval, the population may experience abnormal conditions (e.g., high light exposure or rush uptake due to nutrient perturbation) prior to filtration. It was also shown recently by Mohr *et al.* that measurements of N_2-fixation rates by introducing a tracer in bubble form can lead to significant underestimation of the process [14]. The equilibration, i.e. the isotopic exchange between the $^{15}N_2$ gas bubble and the surrounding water is controlled primarily by diffusive processes. The major variables that influence the rate of isotopic exchange include the surface area to volume ratio of the bubble, the characteristics of the organic coating on the bubble surface [15], and temperature and the rate of renewal of the water-bubble interface [16]. To avoid this problem, the authors recommend adding an aliquot of $^{15}N_2$-enriched water to incubators used in rate measurements.

Samples of oceanographic interest often pose additional analytical challenges simply because they are difficult to obtain in large quantities and most mass spectrometry systems require fine-tuning and careful attention to leaks in order to process samples containing less than a few micromoles of nitrogen. In practice, open ocean samples including suspended particles, small zooplankton, and sinking organic matter are often available only in μmole or sub-μmole quantities. These low concentrations create a strong incentive to minimize the mass requirements of the analytical systems used for marine samples, and to develop methods for correcting for the influence of any analytical blank, which will disproportionately affect small samples. Even with substantial care to reduce this source of contamination, a nitrogen blank on the order of 0.05–0.15 μmol N is typical for systems in current use [17]. For analysis of the smallest samples, the influence of this blank can be reduced provided that an appropriate range of mass and isotopic standards is run, allowing estimation of the size and $\delta^{15}N$ of the blank.

Operationally, isotope ratio mass spectrometers measure the difference in ^{15}N abundance between a sample and a reference gas calibrated to atmospheric N_2. This reference gas calibration may be carried out directly by comparison to atmospheric values [18], or indirectly using any of a number of organic and inorganic secondary standards (e.g., nitrate and ammonium salts, acetanilide, glutamic acid) available from NIST or IAEA. For precise system calibration and correction of blank effects, it is also important to analyze a size series of a working standard that is chemically similar to the samples of interest.

Correcting for an analytical blank is a multi-step process involving separate regression analyses to evaluate the size of the blank and the dependence of $\delta^{15}N$ values on size. This information can then be used to remove the isotopic influence of the blank. Any real sample can be treated as a mixture of sample material and contaminants of various origins. If the blank contribution is constant in composition and magnitude across analyses, then a simple mass balance can be constructed for the mixture analyzed:

$$M_{mix}\delta^{15}N_{mix} = M_{sample}\delta_{15}N_{sample} + M_{blank}\delta^{15}N_{blank}, \quad (8)$$

where M is the size of an individual pool (sample, blank, or the mixture). In general, the nitrogen analytical blank is small enough that blank corrections are minor for samples containing more than 1.5–2 μmol of nitrogen.

3. SAMPLING FOR $\Delta^{15}N$ MEASUREMENTS

Modern water column nitrogen isotope analysis is used on broad variety of samples as particles and dissolved nitrogen species, e.g. nitrate, ammonia or N_2 and N_2O. For each of these methods, special care must be taken during sample collection and preparation.

3.1. Particles

The simplest and most common method to remove particles from incubation water is by filtration. During the filtration special care should be taken to minimize the pressure difference to avoid strong cell rupture [18]. Several studies showed that Teflon and aluminium oxide filters capture up to 60% higher amount of particulate organic nitrogen (PON) than the GF/F filters [19, 20]. GF/F filters have been also found to retain only ~50% of bacteria [21]. In addition to incomplete retention of particles, glass-fiber filters adsorb dissolved organics on the high surface area of the filter [22]. For ^{15}N studies, although mass retention of DON is not a major issue, retention of the highly labeled incubation water on the filter has the potential to affect the isotopic ratio of the retained particulates so that wetting the filter prior to filtration and rinsing afterwards with filtered seawater is a common practice. Filtered seawater is used to prevent osmotic shock of the living particles. Regardless of the type of filter, the particle load needs to be preserved and converted into the gaseous form prior to mass spectroscopic analysis. $\delta^{15}N$ of particles can be analysed directly via combustion (conversion into N_2) or via persulfate oxidation (conversion into NO_3^-) [23].

3.2. Nitrate and nitrite

The most abundant dissolved nitrogen specie in the ocean is nitrate (NO_3^-) followed by nitrite (NO_2^-). In order to perform the isotopic analysis both have to be converted into forms amenable to enter the mass spectrometer. The methods differ by level of reduction, with conversion from NO_3^- to NO_2^-, NO, N_2 or complete reduction to ammonium.

Although NO_2^- concentrations are usually very small compared to NO_3^-, there are cases when both should be analyzed without an interference of $\delta^{15}N$ signals of one another. To do this one has to separate NO_3^- and NO_2^- in the sample by excluding one of them from the sample. Olson [24] used sulfamic acid, while Yakushiji and Kanda [25] took advantage of the reaction of sulfanilic acid with NO_2^-. Rather than forming the final azo dye, the diazonium ion can be destroyed by heating which effectively destroys the initial NO_2^- present in the sample. Two new techniques remove nitrite by reaction with sodium azide in acetic acid buffer to produce N_2O [7] or via reaction with ascorbic acid to produce NO which is then removed bubbling [11].

Nitrate conversion to ammonia commonly uses DeVarda's alloy approach [26]combined with moderate heating. However, this method can add a significant blank that varies with the batch, and combustion of the alloy failed to reduce the blank while also reducing the reduction efficiency [27]. Tanaka and Saino [28] replaced Devarda's alloy with an aluminum reagent that reduced the blank and permitted analysis of nitrate at low concentrations.

Biological reduction of NO_3^- to NO_2^- or gaseous forms has been used by several investigators. Risgaard-Petersen et al. [29] and [30] used cultures of denitrifying bacteria to respire NO_3^- and NO_2^-toN_2 which was then stored for analysis by mass spectrometry. Sigman et al [8] changed this method by using mutant strain of a denitrifying bacteria that only produced N_2O to avoid the problem with contamination of N_2 from the air. The chemical alternative using sodium azide buffer to produce N_2O from NO_2^- was developed by McIlvin and Altabet [7] for natural abundance measurements in seawater samples. It was further improved for samples with low salinities [6] and can be used to analyze nitrate as well as nitrite in the samples.

3.3. Ammonium

Ammonium is not necessarily dominant by mass, but it is often predominant with respect to flux. NH_4^+ is produced primarily by the breakdown and remineralization of organic forms of nitrogen and consumed by autotrophic assimilation and re-incorporation into organic molecules, chemosynthetic nitrification to NO_2^-, and anammox oxidation by NO_2^- to N_2. For isotopic analysis ammonia has to be extracted quantitatively from the solution and converted to suitable gaseous species (e.g. N_2).

Initial approaches to isolating ammonium for isotopic analysis by steam distillation were derived from the soil literature. In all methods involving NH_4^+, great care is required to minimize ambient contamination (e.g., cleaning compounds such as Windex), from sample/reaction containers and from reagents.

Distillation of NH_4^+ under alkaline conditions with subsequent conversion to dinitrogen gas has been used by several groups but is time consuming and has problems with cross-contamination and fractionation [31-34]. In the diffusion method, ammonia passes through a membrane into an acidic medium and is trapped as an NH_4^+ salt [34-37]. This approach, however, is also time-intensive and not very reliable at low concentrations [34]. For freshwater samples Hg precipitation and cation exchange are reported to show very good results for isotope analysis [32, 38]. An organic reaction with NH_3 producing indophenol permits organic or solid-phase extraction works well for ^{15}N-labeled samples [32] but cannot be used for samples with $[NH_4^+] < 5\mu M$ [38]. Zhang and Altabet [39] developed a robust method for sea water samples at low concentrations (0.5 μM or lower), which involves oxidation of ammonia into NO_2^- and subsequent reduction to N_2O.

3.4. N2o and n2

In addition to the previously discussed methods where production of N_2 or N_2O from other forms of nitrogen is an analytical step, there are related methods for analysis of the gases where the gas itself is the form of interest. N_2O can be formed during nitrification and denitrification processes, while N_2 is the end product of denitrification and anammox. Isotope paring technique (IPT) for analysis of N_2 requires analysis of all three isotopic variants of N_2 ($^{28}N_2$, $^{29}N_2$, $^{30}N_2$) [40]. Either a quadrupole or magnetic sector mass spectrometer can be used for the isotope pairing analysis, but the quadrupole can also simultaneously determine the N_2/Ar ratio. Extraction of N_2 from water samples, generally stored in a gas-tight container sealed with a septum (e.g., Exetainer), involves partitioning of the gas into a suitable headspace such as helium. The gas is allowed to equilibrate and the headspace is sampled for introduction of the gas phase into the mass spectrometer [41].

Rather than extraction into a headspace, several investigators have employed membrane inlet mass spectrometers (MIMS) to directly sample N_2 for denitrification and nitrogen fixation studies [30, 42]. MIMS can also be used directly to measure denitrification by monitoring changes in the N_2/Ar ratio over time since the Ar composition reflects changes in temperature and other physical factors while change in the N_2 concentration reflects the balance between nitrogen fixation and denitrification [43]. An advantage of combining MIMS with the isotopes is the ability to separate the two contributions to the net flux. MIMS also obviates the need for the generally slow, separate extraction step. The membrane inlet configuration varies but always features a gas-permeable membrane that separates the mass spectrometer from the solution [44, 45]. A probe inlet [44]permits analysis of depth zonation of rate processes in sediments but has only been employed for N_2/Ar measurements.

Analysis of $^{15}N_2O$ during nitrification and denitrification rate studies has become increasingly popular. Punshon and Moore [46] further developed the method to extract $^{15}N_2O$ produced from either $^{15}NH_4^+$ or $^{15}NO_3^-$ by natural samples. Using purge-trap gas chromatography with a quadrupole mass spectrometer, they were able to determine production rates for N_2O from nitrification. The origin of N_2O (denitrification or nitrification) can be determined using isotopic signature of oxygen isotope. $\delta^{18}O$ isotope signature in N_2O molecule is a product of mixing from dissolved O_2 and H_2O and any fractionation process associated with its production [47]. For instance, production of N_2O via hydroxylamine oxidation (nitrification) will only include oxygen atoms from dissolved O_2, while via denitrification it will have $\delta^{18}O$ isotope of NO_2^- or NO_3^-, originating from both dissolved O_2 and H_2O [47, 48].

The site preference isotopic signature (isotopomers) is also used to identify pathways of N_2O production. In the molecule of N_2O the ^{15}N can be at a central (α) or outer (β) position. The intermolecular distribution of ^{15}N is expressed as site preference (SP) (eq.9)[49].

$$SP=\delta^{15}N^{\alpha}-\delta^{15}N^{\beta} \quad (9)$$

In contrast to $\delta^{18}O$ signature, SP is not dependent on the substrate's isotopic composition but reflects the microbial production mechanism [48]. For example, production of N_2O via hydroxylamine oxidation results in an SP of

33‰, while via denitrification the SP is 0‰ [50]. Good summery on isotopomers for soil studies is presented in the book of Environmental Isotope Geochemistry [51] showing the variable range of SP values for nitrification. In general, isotopomers are valuable tool for distinguishing between nitrification and denitrification processes of N_2O production.

4. MAIN PROCESSES OF NITROGEN CYCLE AND THEIR INFLUENCE ON ISOTOPE BALANCE

Most nitrogen in marine environments is present in five forms: N_2, a quite stable molecule that requires specialized enzymatic systems to break and use it; NO_3^-, the most oxidized form of nitrogen and the dominant form of N within oxic environments; NH_4^+, the most reduced natural form of N and the dominant biologically available form found in anoxic environments; particulate nitrogen, predominant within sediments and primarily in the form of organic N, and dissolved organic N (DON), which is a complex mixture of compounds with a wide range of compositions [52-55]. Nitrate, nitrite, ammonium, and organic nitrogen are typically grouped together as "fixed N" in discussions about the availability of nitrogen. Although each form has a different level of reactivity, a complex web of reactions links these different compounds. The most important processes discussed here are: N_2 fixation, nitrification, assimilation and N-loss (conversion of fixed N to N_2). As it was mentioned already earlier, all biologically mediated processes lead to nitrogen isotope fractionation. For example, denitrification converts NO_3^- to N_2, which has the net effect of depleting the ocean's supply of combined nitrogen and doing so with strong discrimination against ^{15}N. This results in an increase in the $\delta^{15}N$ of the major pool of oceanic combined nitrogen (NO_3^-) concomitant with a reduction in its overall size. In contrast, N_2-fixation adds combined nitrogen with a low $\delta^{15}N$ to the ocean, counteracting both the mass and isotopic effects of denitrification (figure 1). In the subsequent parts of this chapter I will discuss all these processes in detail and their influence on the nitrogen isotope budget in the ocean.

4.1. Nitrogen fixation in the ocean

Microbial N_2 fixation in the ocean inputs "new" N into the surface waters by reducing N_2 gas into two molecules of ammonia. The reaction is logically consumes energy and is catalyzed by nitrogenase enzyme complex, which is extremely oxygen sensitive (req. 10).

$$N_2 + 8H^+ + 16\,ATP + 8\,e^- \rightarrow 2\,NH_3 + H_2 + 16\,ADP + 16\,P_i \qquad (10)$$

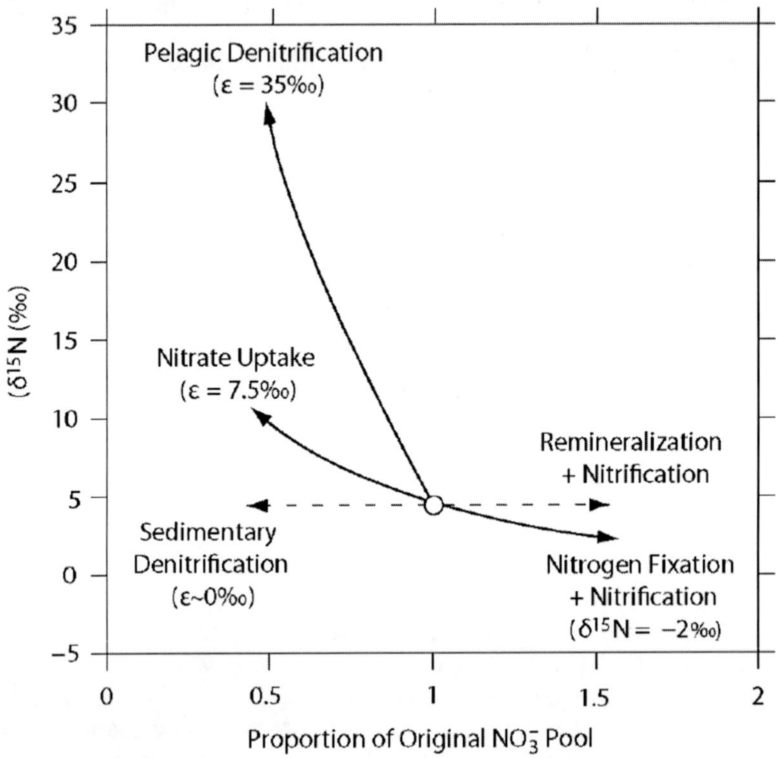

Figure 1. The impact of different processes on the $\delta^{15}N$ of oceanic NO_3^- [56]. Axes show deviation of $\delta^{15}N$ signal from oceanic average and loss/input of nitrogen due to different processes.

The nitrogenase proteins are highly similar among diazotrophs, and the well-conserved *nif*H gene is commonly used for phylogenetic and ecological studies [57]. In the open ocean N_2 fixation research has focused most intensively on the cyanobacterium, *Trichodesmium* [58, 59], which often occur as aggregates (often referred to as colonies) visible to the naked eye ('sea sawdust'). It can also occur, however, as individual filaments. In early reports N_2 fixation was observed in the pelagic zone and associated with the cyanobacterial epiphytes (*Dichothrix fucicola*) of the brown macroalga, *Sargassum* [60] and cyanobacterial endosymbionts of certain oceanic diatoms [61, 62]. *Trichodesmium* has a unique physiology that employs spatial and temporal segregation and increased oxygen consumption to allow it to simultaneously fix N_2 and CO_2 (and thus evolve O_2)[63].

Evidence of N_2 fixation began to first accumulate in the late 1970s with several pioneering N isotope studies examining the natural abundance of the ratio of $\delta^{15}N$ in the western Pacific Ocean [59, 64]. The average isotopic signature of nitrate in the deep ocean lay about 5‰ [27, 65, 66]. Nitrogen fixation is the process by which N_2 is incorporated from the air into the cell. As the $\delta^{15}N_{air}$ is set to 0‰, the kinetic isotope effect of the process is small (-2‰ to +2‰). Surface particulate and dissolved inorganic N pools in the Kuroshio often had a ^{15}N-depleted isotopic signature, indicative of a N_2fixation source [67]. Using *nif*H gene studies, N_2-fixing was found also in the picoplankton, as well as in heterotrophic bacteria from the guts of copepods [68]. Heterotrophic N_2 fixation does not consume CO_2 during the process as diazotrophs do but instead produces it. However, the mechanism of heterotrophic N_2 fixation is not yet understood and it is still unclear and the potential contribution of these organisms to marine N_2 fixation is uncertain.

4.1.1. Controlling factors for n_2 fixation and its distribution

Numerous factors, physical, chemical and biotic, can affect the extent of N_2 fixation in an ecosystem [69-72]. Many factors that affect nitrogenase activity co-vary, such as light, temperature, oxygen and turbulence, making it difficult to determine the relative impact of each. Species-level responses to these variables also differ. The heterocystous diazotrophs, for instance, are typically restricted to special marine environments. Metabolically active populations of *Trichodesmium* have been observed at 18.3 °C [54], but activity was low, and substantial growth is typically not seen until water temperature exceeds 20 °C [73]. In the cooler Baltic waters *Nodularia* spp. and *Aphanizomenon* spp. Are observed, which are restricted to 16–22 °C and 25–28 °C, respectively [74]. In warmer waters of the tropics and subtropics, heterocystous diazotrophs (such as *Richelia intracellularis*) are only encountered as symbionts of diatoms like *Hemiaulus* spp., and *Rhizosolenia* spp. [75].

For photoautrophic diazotrophs, nitrogenase activity is intimately linked to photosynthesis [76]. Thus, light is an obvious and important factor potentially regulating or constraining this process. Carpenter [73] has summarized much of the early marine work largely relating to *Trichodesmium* spp. with respect to their relationship to light. Whereas many non-heterocystous cyanobacteria fix nitrogen during the night, and thereby uncouple N_2 fixation from photosynthesis, *Trichodesmium* fixes nitrogen exclusively during the light period and shows a strong diel pattern of activity with maxima during midday [59, 63, 69, 77]. Natural populations of *Trichodesmium*, often found in the upper layers of the euphotic zone, appear to be adapted to high light with a relatively shallow compensation depth (typically 100–200 mmol quanta m^{-2} s^{-1}) for photosynthesis [78].

To calculate how much nitrate is produced via N_2-fixation, Redfield stoichiometry can be used [79]. With the given $NO_3^-{}_{exp}$ and the measured concentration of the actual dissolved inorganic nitrogen (DIN = NO_3^-+ NO_2^-+ NH_4^+), the amount of NO_3^-, which is removed by denitrification ($NO_3^-{}_{def}$), can be calculated as follows:

$$NO_3^-{}_{def} = NO_3^-{}_{exp} - DIN \quad (11)$$

An alternative stoichiometric method involves the quasi-conservative tracer N* [80, 81]. Gruber and Sarmiento used high quality data from JGOFS and GEOSECS databases to develop a general relationship between fixed inorganic nitrogen (DIN) and phosphate (PO_4^{3-}) for the world's oceans:

$$N^* = DIN - 16\,[PO_4^{3-}] + 2.90 \quad (12)$$

The value of 2.90 is the deviation from the amount of DIN predicted by the Redfield stoichiometry (N:P =16) and the world-ocean N:P regression relationship. Negative values of N* are interpreted to show the net *denitrification* whereas positive values show the net *nitrogen fixation*.

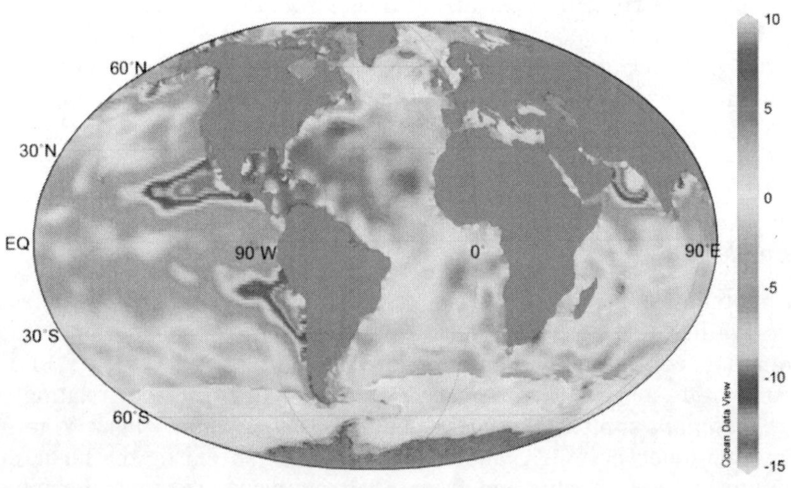

Figure 2. $N^* = [NO_4^{3-}] - 16^*[PO_4^{3-}] + 2.9$ on the 26.5 sigma theta surface (World Ocean Atlas 2005). Note the large negative N* values in the Tropical Pacific and Arabian Sea and positive N* values in the Tropical Atlantic Ocean.

Strong positive N* values have been observed in the tropical and subtropical North Atlantic (fig.2) and are proposed to be a result of the input into these areas of diazotrophic (i.e. nitrogen fixer) biomass with a higher N:P content than that typical of eukaryotic phytoplankton of the upper ocean [80-82]. The observation of relatively high concentrations of DON in surface waters of regions of the tropical oceans have also been attributed to nitrogen fixation [83-86]. At the Hawaiian Ocean Time (HOT) series station, pools of DON increased during a period in which microbial nitrogen fixation also became more prominent [87, 88].

Iron limitation may be important for cyanobacterial N_2 fixation as Fe is a cofactor in photosynthesis and nitrogenase, leading to is ~5 times higher Fe requirement for diazotrophy than that for ammonia assimilation [77]. Even though the concentration of dissolved Fe (dFe) is low in many of the environments where N_2 fixation occurs [89-92], a number of studies have speculated on diazotroph response to mineral dust Fe fertilization [93, 94]. A 1999 Saharan dust event coincided with increases in dissolved Fe concentrations on the west Florida shelf and a 100-fold increase in *Trichodesmium* biomass [95]. N_2 fixation rates were not measured, but DON concentrations doubled, presumably due to exudation by N_2 fixers [95]. A Saharan dust addition experiment to surface water samples collected along a west African cruise transect (35°W–17°W), found a minimal increase in CO_2 fixation and a large stimulation in N_2 fixation, suggesting that diazotrophs were co-limited by both P and Fe [96]. It should be noted however, the analysis of aerosol dust shows that while providing Fe, the dust also supplies P and combined N [96-98]. Work on *Trichodesmium* spp. culture has identified mechanisms to deal with P limitation. For instance, under P stress conditions it uptakes a greater amount of DIP and dissolved organic P (DOP), decreasing the requirement for cellular P.

N_2 fixation is an energetically costly process in comparison to ammonia assimilation, thus it is logical to assume that presence of high concentrations of fixed N play a role of an inhibitor. Indeed, under mM concentrations of ammonia (concentrations essentially never observed in the real surface ocean), N_2 fixation is inhibited in most diazotrophs [99]. However, mM concentrations of NO_3^- stopped N_2 fixation in some strains [100] but have little to no effect in others [101]. The higher energy requirement for the nitrate reduction compared to 'direct' assimilation of ammonia might explain these differences. Addition of ammonia to natural samples of *Trichodesmium* spp. can shut down nitrogenase activity [69]. However, pure culture studies have shown mixed results, with some cultures indicating no reduction of their activity [102] and others showing a reduction in activity under short-term additions of fixed N sources [103]. Interestingly, indication for N_2 fixation has been found also in the nitrate-rich upwelling areas off Chile and Peru [104, 105], indicating that N_2fixation may not be as strongly regulated by fixed N as previously presumed.

4.1.2. Nitrogen fixation in sediments

Within the marine realm, N_2 fixation in sediments has been mostly recognized in pelagic environments or in benthic microbial mats, sea grass sediments, and coral reef systems [106]. A recent study by Bertics et al. [107] revealed that N_2-fixation, coupled to sulfate reduction, is stimulated by bio-irrigation in organic-rich coastal sediments. The authors suggested that N_2-fixation has been largely overlooked in sediments, and that its occurrence might be significant, contradicting the general belief that sediments are areas of net nitrogen loss through denitrification and anaerobic ammonium oxidation [108, 109].

4.2. Nitrification

Nitrification process oxidizes NH_4^+ into NO_2^- and further to NO_3^-. Ammonium rarely occurs at high concentrations in the surface ocean as it is rapidly recycled between heterotrophic and N_2 fixing organisms and many heterotrophic and photosynthetic plankton species. N_2 fixing organisms play a role of ammonia source as they excrete NH_4^+ directly or release organic N that is microbialy degraded to NH_4^+, while plankton utilize ammonia as a N source for building up its biomass. Similarly, accumulation of NO_2^- is rarely observed in oxygenated habitats, although NO_2^- is an essential intermediate in several oxidation and reduction processes in the N cycle.

Nitrification is carried out through a combination of two microbial processes: ammonia oxidation to NO_2^- and nitrite oxidation to NO_3^-. The first step of the process is enabled by nitrifying organisms, ammonium-oxidizing bacteria and archaea (AOB and AOA), while the second step is carried out by nitrite-oxidizing bacteria (NOB). No known organism can carry out both reactions. The overall reaction of NH_4^+ oxidation (eq. 13) for AOB shows that the process consumes molecular oxygen and produces NO_2^- and hydrogen ions. Molecular oxygen is required in the first step of the reaction, which is catalyzed by a monooxygenase (NH_3 monooxygenase, AMO). Oxygen is also consumed by the terminal oxidase as a result of electron transport generating adenosine-5' -triphophate (ATP) for cellular metabolism. The immediate product of AMO is hydroxylamine, which is further oxidized by hydroxylamine oxidoreductase (HAO) to NO_2^-. In contrast, AOA apparently do not possess the hydroxylamine reductase gene, so the pathway of ammonia oxidation in these organisms must be quite different.

$$NH_4^+ + O_2 + H^+ + 2\,e^- \xrightarrow{AMO} NH_2OH + H_2O \xrightarrow{HAO} NO_2^- + 5\,H^+ + 4\,e^- \quad (13)$$

Depending on the conditions, NO, N_2O and even N_2 have been reported as secondary products in autotrophic NH_4^+ oxidation by both marine and terrestrial strains [110, 111]. Although N_2O and NO can be produced in vitro by HAO from hydroxylamine [112], the reduction of NO_2^- appears to be the dominant pathway in all cells [113-115]. One of the first studies of kinetic isotopic effect of nitrification in pure cultures of *Nitrosomonas europaea* was made by the Mariotti group [116], where a fractionation factor of 34.7 ± 2.5‰ was determined for conversion of NH_4^+ into NO_2^-. The later work of Casciotti [117] identified that the fractionation factor for different *Nitrosomonas* cultures of ammonia-oxidizing bacteria was linked to amino acid sequences. The isotopic effect in these cultures ranged from 14.2‰ to 38.2‰.

N_2O is produced by AOB through two separate pathways: hydroxylamine decomposition and nitrite reduction, so-called "nitrifier denitrification" [114]. The isotopic compositions ($\delta^{15}N_{bulk}$, $\delta^{18}O$, $\delta^{15}N_\alpha$, $\delta^{15}N_\beta$, and site preference = $\delta^{15}N_\alpha$ - $\delta^{15}N_\beta$) of the produced N_2O can provide insight into the mechanisms of N_2O production under different conditions [48, 50, 118]. For example, N_2O production through nitrifier denitrification has low $\delta^{15}N_{bulk}$ and low site preferences relative to that produced by hydroxylamine decomposition [119]. As discussed earlier, the N_2O site preference (SP) is determined by the enzymatic mechanism [49, 120]. The site preference of N_2O produced during nitrification is +30‰ to +38‰, while N_2O production via denitrification and nitrifier-denitrification has a SP of -10‰ to +5‰ [48, 118, 121, 122]. The large difference between the SP values of these two mechanisms of N_2O production provides a good tool with which to distinguish their contributions.

The biochemistry of the NO_2^- oxidation is simpler than that of NH_4^+ oxidation because it is only involves a transfer of two electrons without intermediates (eq. 14). The additional oxygen atom in NO_3^- is derived from water, and the molecular oxygen that is involved in the net reaction results from electron transport involving cytochrome oxidase.

$$NO_2^- + H_2O \xrightarrow[\text{oxydoreductase}]{\text{nitrite}} NO_3^- + 2H^+ + 2e^- + 0.5\,O_2 \xrightarrow[\text{oxidase}]{\text{cytochrome}} NO_3^- + H_2O \qquad (14)$$

A recent study of isotopic fractionation during nitrite oxidation to nitrate [1] showed a unique negative isotopic effect of -12.8‰. Because of the inverse isotope effect, when nitrite oxidation is taking place, the $\delta^{15}O_{NO2}$ (and $\delta^{18}O_{NO2}$) values are expected to be lower than the NO_2^- initially produced by ammonia oxidation or nitrate reduction. In most parts of the ocean, however, NO_2^-does not accumulate and the isotope effects of different processes complicate the interpretation. In the euphotic zone NO_2^- is either oxidized to NO_3^- or assimilated into particulate nitrogen (PN) and the heavy isotope can be preferentially shunted in one direction vs. the other [119].

4.2.1. Controlling factors of nitrification and its distribution

Nitrifying bacteria are traditionally considered to be obligate aerobes; they require molecular oxygen for reactions in the N oxidation pathways and for respiration. A significant positive correlation between apparent oxygen utilization (AOU) and N_2O accumulation is often observed in marine systems [123-125]. The relationship implies that nitrification is responsible for N_2O accumulation in oxic waters where it is released as a byproduct. The relationship breaks down at very low oxygen concentrations (~6 mM, [126]), where N_2O is consumed by denitrification.

There is abundant evidence from culture studies that both AOB and NOB are photosensitive. Several studies of nitrification rates in surface seawaters from various geographical regions show profiles that are consistent with light inhibition of both NH_4^+ and NO_2^- oxidation [127-129]. The highest nitrification rates occur in a region near the bottom of the euphotic zone. It is in this interval where nitrifying bacteria can compete with phytoplankton for NH_4^+, as the rates of nutrient assimilation are reduced due to light limitation. A sharp peak of nitrification rate is often observed at water column depth where the light intensity is 5–10% of surface light intensity [118, 127-129].

In both deep and shallow sediments, nitrification can be one of the main sinks for oxygen [130, 131]. In continental shelf sediments, nitrification and denitrification are often closely linked. Coupled nitrification/denitrification is invoked to explain the observation that the rate of N_2 flux out of sediments can greatly exceed the diffusive flux of NO_3^- into the sediments [132]. Ammonium is produced during aerobic and anaerobic remineralization of organic matter. It is then oxidized to NO_3^- and subsequently reduced to N_2. Anaerobic oxidation of NH_4^+ can be a factor supporting the imbalance between supply and consumption of NO_3^-. Nitrification can supply up to 100% of the NO_3^-, which is later consumed by denitrification [133, 134].

Although oxygen and NH_4^+ conditions likely differ between planktonic and sediment environments, there is no clear evidence from clone libraries that water column and sediment nitrifying communities are significantly different in composition and regulation.

To measure nitrification rates, isotope dilution can be used. In this method, the isotope dilution of the NO_2^- pool (in the case of NH_4^+ oxidation) or the NO_3^- pool (in the case of NO_2^- oxidation) is monitored over time [135]. Alternatively, in the ^{15}N method $^{15}NH_4^+$ or $^{15}NO_2^-$ is added to a sample and then the appearance of ^{15}N label in the NO_2^- or NO_3^- pools is measured over time [24]. One disadvantage of the method is that the increase in NH_4^+ or NO_2^- concentrations caused by the tracer addition can result in enhanced NH_4^+ or NO_2^- oxidation rates [129]. When small additions of ^{15}N are used, the isotope dilution of the substrate pool must be monitored over time.

4.3. Nitrogen assimilation

Nitrification and nitrogen assimilation are usually found at similar depths in the ocean as they compete for the same reactants (NH_4^+). At the same time, nitrification, producing NO_2^- and NO_3^-, provides assimilation with additional reactants. Nitrogen assimilation is the process of incorporation of reactive nitrogen species (NO_3^-, NO_2^- and NH_4^+) into the bacterial cell. Ammonium, which is produced in the photic zone by heterotrophic processes, is assimilated immediately by phytoplankton and heterotrophic bacteria before it can be nitrified. Ammonium is often the dominant form of dissolved N assimilated in a variety of marine and estuarine systems. This form of N is energetically efficient for cells to use, because it is already reduced and is a common cellular transient in N metabolism, requiring little additional energy for assimilation. Despite the low NH_4^+ concentrations in oceanic systems, uptake and regeneration of NH_4^+ are tightly coupled and result in rapid turnover times [136].

Nitrate concentrations in the surface ocean are usually maintained at low levels as a) phytoplankton assimilate NO_3^- rapidly and b) nitrate can be supplied by mixing or diffusion from the deep NO_3^-reservoir only at the regions with strong upwelling. The assimilation of NO_3^- is more energetically demanding than NH_4^+, because it requires a synthesis of NO_3^- and NO_2^- reductases, associated active transport systems, and the turnover of cellular ATP and nicotinamide adenine dinucleotide phosphate (NADPH) [136]. In order to assimilate NO_3^- for growth, phytoplankton must first possess the genetic capacity to synthesize the necessary enzymes and transport systems- a capacity which not all phytoplankton have. Further, the supply of NO_3^- is limited by nitrification and vertical mixing.

Since nitrite reductase (NiR) is required for NO_3^- assimilation, organisms that assimilate NO_3^- can by default assimilate NO_2^-. Consequently, some organisms that cannot use NO_3^- can still use NO_2^- (e.g., some clones of *Prochlorococcus*; [137]). It has been suggested that this rather unique N physiology has arisen from the evolutionary loss of genes, which are necessary for the assimilation of NO_3^-[138, 139]. The resultant evolution of many ecotypes specifically adapted for unique environments where NO_2^- production can be high.

Nitrogen isotope fractionation can take place at different stages of the processes. Wada and Hattori [140] first argued that N isotope fractionation by phytoplankton occurs during *nitrate reduction*, while Montoya and McCarthy [141] ultimately favored fractionation associated with *nitrate transport* into the cell. The N isotope data presented in Granger at al. [142] suggest that an isotope effect is driven solely by the reductase, and not by uptake at the cell surface and that this effect results in a fractionation factor of 5 – 10‰. Ammonium assimilation is shown to have higher fractionation factor of +14 – +27‰ [143, 144].

In most parts of the ocean, however, assimilation and nitrification occur concurrently, which additionally complicates the isotope analysis. In general, nitrite oxidation, as discussed above, transfers NO_2^- with an elevated $\delta^{15}N$ to the NO_3^- pool, while nitrite assimilation transfers the residual NO_2^- with a lower ^{15}N:^{14}N ratio into the PN pool. If the fractionation factor of nitrite assimilation is 1‰ [144] and fractionation factor of nitrite oxidation is -15‰ [1], then the $\delta^{15}N_{NO2}$ at steady state will be lower than the source of NO_2^-, unless nitrite assimilation is >95% of the NO_2^- sink. This has the opposite effect of the ammonia oxidation/assimilation branching where ammonia oxidation transfers low ^{15}N:^{14}N material into the NO_2^- and NO_3^- pools and higher ^{15}N:^{14}N material into the PN pool [119].

4.3.1. Factors controlling nitrogen assimilation and its distribution

Most nitrogen assimilation takes place in the surface waters of the ocean. The most important factors controlling assimilation are the presence of oxygen for respiration and/or light for photosynthesis and enzymes responsible for metabolic processes. Most of the enzymes involved in the uptake and assimilation of N are bound to energy sources and thus are affected by light, presence of oxygen, and the supply of enzyme co-factors and metabolic substrates. For example, uptake and reduction of NO_3^-, NO_2^-, and urea have been linked to the light supply in phytoplankton due to the necessity for ATP and NADPH from photo-phosphorylation. While uptake and reduction of these compounds is thought to proceed at maximum rates only in the light under nutrient replete conditions, active uptake of these compounds may occur even in the dark [145].

The accumulation of intracellular product pools, on the other hand, can result in the inhibition of uptake and assimilation. The posttranslational modification of enzymes can regulate uptake and assimilation by modulating the number of active sites available for catalysation of specific reactions. Studies examining the transcriptional activation of enzymes involved in the process have begun to demonstrate clear functional relationships between genomes and ecological capabilities. For example, the presence of NH_4^+ represses protein expression (i.e., enzymes) involved in the assimilation of alternative N sources (e.g., NO_3^- and N_2) in some organisms [146]. Overall, most factors controlling assimilation are directly connected with enzymes, which require substrate compounds or energy and oxygen supply for their metabolic activity.

4.4. Nitrogen loss (denitrification and anammox) in the ocean

The term "nitrogen loss" is commonly used for the processes in which fixed nitrogen (nitrate, nitrite or ammonia) is converted into non-bioavailable N_2 in the absence of oxygen. This includes denitrification process, in which combined nitrogen is reduced to gaseous end products, as well as newly discovered process of anammox (anaerobic ammonium oxidation), which converts nitrite and ammonia into N_2 [147]. Another important process of reduction is DNRA (dissimilatory nitrate reduction to ammonia) will be also discussed in this section as it is an important source of ammonia in the ocean. Even though the reduction of nitrate to ammonia is not a N-loss as such, but it is also taking place in suboxic waters, and the ammonia produced can diffuses up to the anoxic waters and may well be oxidized to N_2 by anammox bacteria [148].

4.4.1. Canonical denitrification

Canonical denitrification is a heterotrophic process in which nitrogen oxides serve as the terminal electron acceptor for organic carbon metabolism [149]. The nitrogen oxides are reduced mainly to molecular nitrogen and some nitrous oxide may be formed as a side product. The sequential reduction is enabled by four well-studied enzyme systems: nitrate reductase, nitrite reductase, nitric oxide reductase and nitrous oxide reductase [150] (eq. 15).

$$NO_3^- \xrightarrow{\text{nitrate reductase}} NO_2^- \xrightarrow{\text{nitrite reductase}} NO \xrightarrow{\text{nitric oxide reductase}} N_2O \xrightarrow{\text{nitrous oxide reductase}} N_2 \tag{15}$$

The intermediate NO_2^- is known to escape the cell and is frequently found in denitrifying environments; likewise, N_2O also can accumulate externally [151]. This leads to local maxima of both intermediates, which tend to occur near the boundaries of the suboxic zone. Another intermediate, NO, is actually a free radical and, thus, very reactive and highly toxic to most bacteria including denitrifiers [152]. In order to minimize the accumulation of NO within the cell, nitrite reductase and nitric oxide reductase are controlled interdependently at both the transcriptional and enzyme activity levels [152, 153].

4.4.2. Anammox

Anammox bacteria are chemoautotrophic bacteria that fix CO_2 using NO_2^- as the electron donor [154] and are thought to be strict anaerobes. Recent publication on the tolerance of anammox bacteria to oxygen levels, however, showed that they can perform a significant activity even under ~20 μmol/l O_2 concentrations [155]. All anammox species found have evolved a membrane-bound intra-cytoplasmic compartment called the anammoxasome. The anammoxasome membrane is made up of high-density lipids, called ladderanes because of their ladder-like structure, which are thought to be specific to anammox bacteria. The proposed mechanism of anammox involves a hydrazine hydrolase, which catalyzes the combination of hydroxyl ammine and ammonium to form hydrazine. The hydrazine-oxidizing enzyme subsequently oxidized it to N_2 [156, 157]. The ladderane-lipid membrane is thought to act as a barrier to diffusion thus isolating the toxic intermediates of the anammox reaction within the anammoxasome. Interestingly, for every 1 mol of N_2 produced, 0.3 mol of NO_3^- is also produced alongside from NO_2^-. This reaction is believed to be important for replenishing electrons for the acetyl-CoA carbon fixation pathway [158]. At the same time, these bacteria can perform dissimilatory NO_3^- reduction to NH_4^+ (DNRA), which is subsequently combined with NO_2^- to produce N_2, thus mimicking denitrification [159].

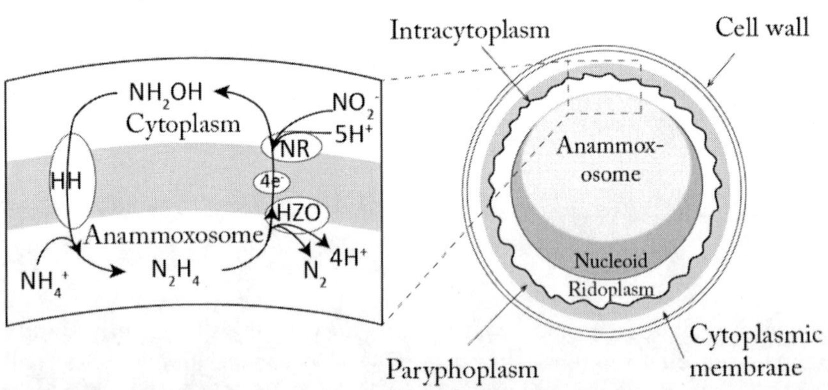

Figure 3. Morphology of the anammox cell and proposed model for the anammox process. HH, hydrazine (N_2H_4) hydrolase; HZO, hydrazine oxidizing enzyme; NR, nitrite reducing enzyme [160].

Anammox bacteria have been identified in many naturally occurring anoxic environments including marine, freshwater, and estuarine sediments [161,

162], oxygen minimum zones [163, 164], soils [165-167], wetlands [168], and anoxic tropical freshwater lakes [169]. These bacteria are not only widespread but may also be very active (e.g. ~67% fixed N-loss [162]). Anammox organisms may live in low abundance but are capable of significant growth and metabolism when a shift in environmental parameters permits anammox communities to grow [170]. The anammox bacteria are slow growing, with a doubling time of ~11 days, even under optimal growth conditions in bioreactors [171]. The state-of-the art method to identify anammox is currently based on isotopic labeling of NH_4^+ or NO_x^-. It relies on the fact that N_2 produced by anammox combines 1:1 atoms from NH_4^+ and NO_2^- [172]. The selective formation of $^{15}N^{14}N$ ($^{29}N_2$) rather than $^{15}N^{15}N$ ($^{30}N_2$) is a clear evidence of anammox activity[162].

4.4.3. Dissimilatory nitrate reduction to ammonium or dnra

The dissimilatory reduction of nitrate to ammonium is a bacteria-mediated heterotrophic process occurring in anaerobic conditions. Nitrate is first reduced to nitrite and then to NH_4^+. Bock et al. [173] also showed that *N. europaea* and *N. eutropha* were able to nitrify and denitrify at the same time when grown under oxygen limitation ([O_2] ~ 0.2 - 0.4 μmol/l). Under these conditions, oxygen and NO_2^- served simultaneously as electron acceptors and both N_2 and N_2O were produced, whereas under anaerobic conditions N_2 was the predominant end product. This type of nitrification–denitrification pathway may help to explain why ammonium oxidizers remain active in nearly suboxic environments and enhanced N_2O production is observed [174, 175].

The reaction has been reported for sediments underlying anoxic waters [176, 177] and sediments with substantial free sulfide, possibly due to sulfide inhibition of nitrification and denitrification [42]. Suboxic NO_3^- reduction to NH_4^+ appears to occur in environments where there is excess available carbon, such as estuarine sediments [178]. Perhaps more important are the NO_3^-fermenting organisms such as *Thioploca* and *Thiomargarita*, that are found in sediments underlying the major suboxic-denitrifying water columns of the Arabian sea, eastern Tropical Pacific and Namibia [179-181]. These organisms are able to couple the reduction of NO_3^- to NH_4^+ with the oxidation of reduced sulfur compounds. Both *Thioploca* and *Thiomargarita* are able to concentrate NO_3^- at up to 500 mM levels in large vacuoles within their cells for subsequent sulfide oxidation [180, 181].

4.4.4. Controlling factors for n-loss processes and its distribution

The controlling factor for all N-loss processes is molecular oxygen. When the concentration of the denitrification intermediate NO_2^- is plotted against dissolved oxygen, we see that NO_2^- does not appear in the water columns of the Eastern Tropical Pacific until oxygen concentrations are reduced below about 2 μM (see figure 4 [182, 183]).

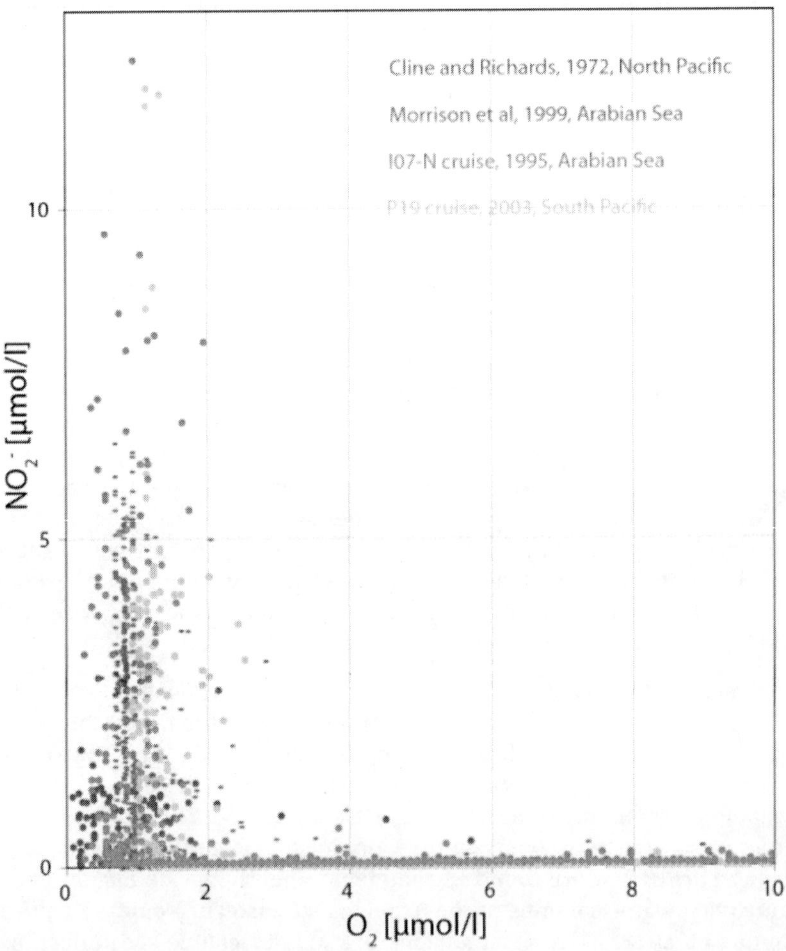

Figure 4. Dissolved oxygen versus nitrite for OMZs. Presented data are taken from publically available cruises I07-N, P19 (http://woce.nodc.noaa.gov/woce_v3/wocedata_2/bathymetry/default.htm) and publications of Cline & Richards [182] and Morisson [183].

Another indication for the denitrification process is consumption of N_2O within the water column, where oxygen concentrations are reduced levels less than c. 5 μM [65]. In oceanic systems the relative importance of heterotrophic denitrification, compared to autotrophic anammox has been debated and believed to depend on the amount of organic matter (OM) available [184, 185]. However, in most of the cases N-loss processes co-exist and it is rather a question which one is playing the first fiddle [186, 187].

About less than half of the current marine denitrification of the global ocean is thought to occur in the three main pelagic Oxygen Minimum Zones (OMZs): Eastern Tropical North Pacific (ETNP), the Eastern Tropical South Pacific (ETSP) and the Northern Arabian Sea (Fig. 4). These zones occur in intermediate waters (~150–1000 m) in locations where the ventilation rate is insufficient to meet the oxygen demand. To calculate nitrogen loss in the water column, the amount of substrate that is consumed in N-loss processes or the amount of N_2 produced should be determined. Similar to estimations of the N-fixation, the Redfield stoichiometry is used to calculate N-deficit or N*. Negative values of N* are interpreted to show the net *N-loss* whereas positive values show the net *nitrogen fixation* (see above).

Devol et al. [188] used N_2:Ar ratios in the Arabian Sea OMZ to determine the amount of N_2produced during denitrification. The amount of N_2 excess in the OMZ was computed from the increase in the N_2:Ar ratio over that present in the source waters. Measurements of N_2 excess predicted a larger nitrogen anomaly than that estimated by nitrate deficit. The discrepancy is said to be due to incorrect assumptions of the Redfield stoichiometry [189]. Inputs of new nitrogen through N-fixation, N_2 contributions from sedimentary denitrification along continental margins, the anammox reaction, or metal catalyzed denitrification reactions all lead to a shift of the N:P ratio from the Redfield stoichiometry.

Due to N-loss processes, i.e. denitrification and anammox, the $\delta^{15}N$ values in the water column increase to 15‰ or more. For the eastern tropical North Pacific OMZ Brandes et al. [190] calculated the fractionation factors between 22‰ (Arabian Sea closed system model) and 30‰ (eastern tropical North Pacific open-system model). Several authors [190-193] have observed high nitrogen isotopic values and nitrate deficits in the Pacific and Arabian Sea OMZ's and estimated fractionation factor, which all cluster around a value of +25‰. The fractionation factor of N-loss processes in the sediments was shown to be significantly lower (~3 ‰) [194]. Thus, in the waters with strong upwelling, the water column signal can be altered via mixing with water originating from the sediments.

4.4.5. Sedimentary denitrification

Another location where oxygen is typically depleted and denitrification takes place is in marine sediments, especially the continental margin and hemipelagic sediments. Respiratory processes within the sediments act as sinks of oxygen, while overlying water deliver nitrate to the sediment. In most continental shelf sediments, oxygen penetrates to less than 1 cm below the sediment–water interface and even in most deep-sea sediments oxygen penetration is restricted to the approximately upper 10 cm. Thus, ample environments exist for N-loss in marine sediments. Due to the fact that marine sediments are usually rich with organic matter, denitrification and DNRA are the prevailing mechanisms in the sediments [176].

5. MODERN OCEAN NITROGEN BUDGET

The discussion of N-budget is usually based on balance or imbalance of nitrogen sources and sinks. Nitrogen sources in this case are N_2-fixation, atmospheric wet and dry deposition, and riverine input. Nitrogen sinks include denitrification and anammox in the water column, sedimentary denitrification, and sedimentation of organic matter. Nitrification and DNRA processes "recycle" nitrogen from ammonia into nitrate and vice versa. Although the primary sources and sinks are relatively well known, the quantification of their fluxes is associated with considerable uncertainty [81, 195-199] (see figure 5). The major problem in calculating global budgets of N sources and sinks is extrapolating measured fluxes from a limited number of oceanic regions to the global ocean.

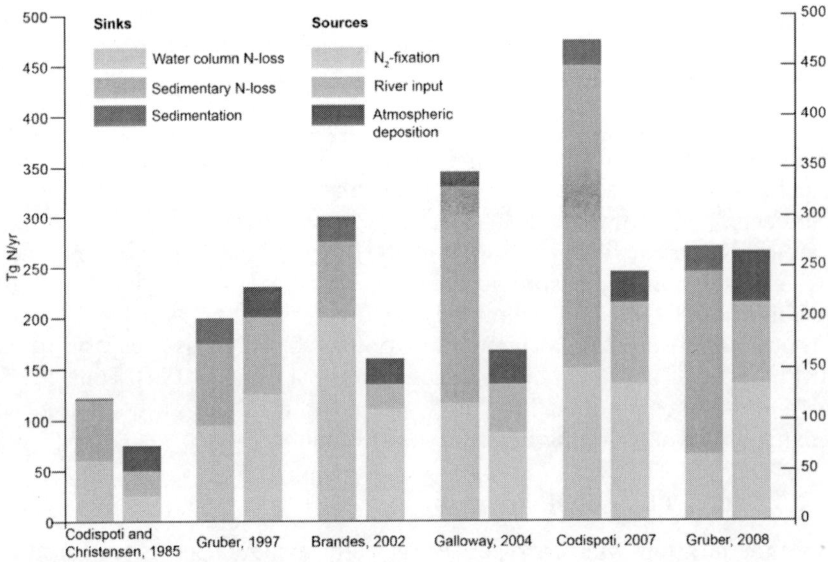

Figure 5. Nitrogen budget estimations in the literature.

Additional complications in calculating N budgets are caused by interplay between and potential coexistence of N cycle processes at the same depth. A schematic water column profile with corresponding processes over O_2, NO_2^-, and N_2O gradients is presented in the figure 6 (based on water column profiles in the South Pacific ocean [65]). In the surface layer, nitrogen assimilation and rapid recycling is taking place, producing nitrite where oxygen and bioavailable N are plentiful enough. Ammonia oxidation (first step of nitrification) lead to efflux of N_2O [200] in the surface waters. Ammonium release and its oxidation in the well-oxygenated waters lead to high production of NO_2^- (primary nitrite maximum). At oxygen concentration ~ 5 μmol/l, denitrification process become more preferential producing to "secondary nitrite maximum" in the OMZ. This NO_2^- can also be a source for anammox process at the top of OMZ, while DNRA process can provide a source of ammonia [148]. N_2O consumption in the water column provides a strong evidence for denitrification producing N_2. At the borders of OMZ nitrification and denitrification processes may coexist [196], leading to production of N_2O at these water depths.

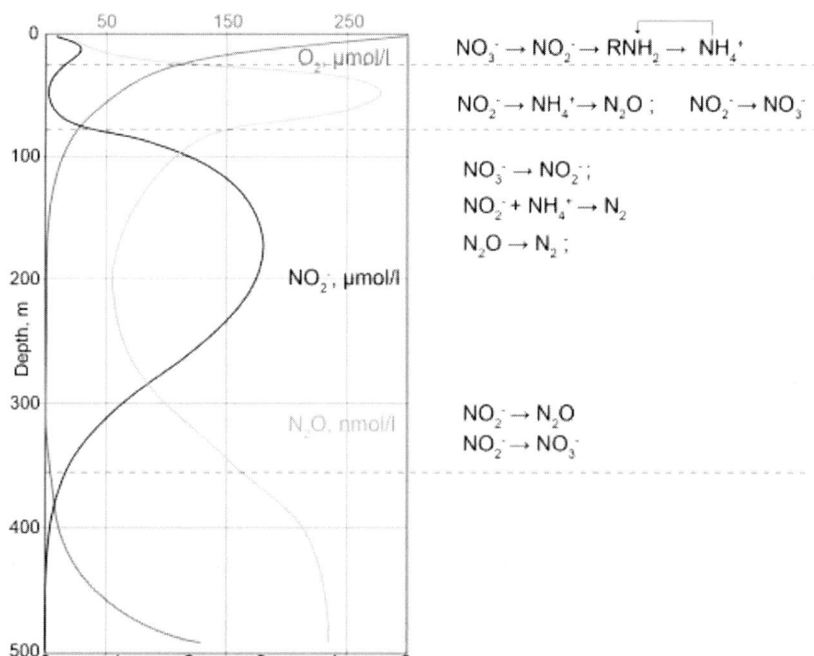

Figure 6. Schematic water column profile with corresponding processes over O_2, NO_2^-, and N_2O gradients.

Nitrogen isotope ratios in the water column and in sedimentary records provide additional constraints for nitrogen budget calculation. Average $\delta^{15}N$ of fixed nitrogen in the deep ocean is close to 5‰. The sedimentary records show that this global average increased during deglaciation periods, when stronger denitrification was taking place. However, the global average $\delta^{15}N$ range for the last 30.000 years was in the range 4-6‰ [201], which suggests that N sources and sinks should be balanced, at least the millennial time scale. In this case, the processes increasing $\delta^{15}N$ (assimilation, N-loss) has to be balanced with those, which are decreasing it (N_2-fixation).

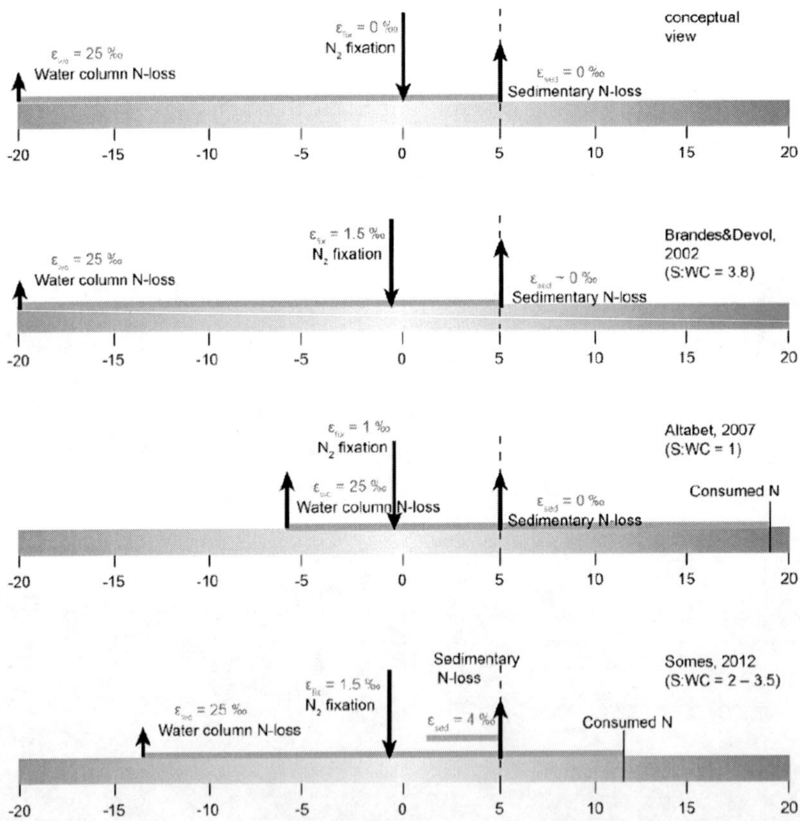

Figure 7. Nitrogen isotope range and processes influencing on the ^{15}N budget. Dashed line at 5‰ represent an oceanic average value [197, 202, 203].

The conceptual model of the nitrogen isotope budget is initially based on two processes: N_2-fixation and denitrification. Sedimentary denitrification was believed not to change the isotopic signature (ε_{sed}) of nitrogen due to N-limitation in the sediments, while water column denitrification (ε_{wc}) has a fractionation factor of ε between 25 and 35 ‰. The N_2- fixation fractionation factor ε_{fix} is close to 0‰ as it incorporates nitrogen from the atmosphere. An early attempt to construct a one-dimensional model was undertaken by Brandes and Devol [197], where they set fractionation factor to ε_{fix} and ε_{sed} to 1.5‰; and ε_{wc} to 25‰. On the figure 8 scale of nitrogen isotope is illustrated. In the Brandes and Devol model, sedimentary N-loss does not change the ^{15}N budget substantially (ε_{sed} in the range of 0 to 1.5‰). Thus, a large amount of sedimentary denitrification is necessary to balance water column denitrification and bring overall isotopic values of the losses into the range of sources, leading to the value of Sediment to Water column denitrification (W:WC) ratio of 3.8. The authors noticed that with fluxes estimations of total N-loss of 400 Tg N/yr and fixation of 100 Tg N/yr, "the combined budget would be out of balance by over -200 Tg N/yr" [197]. This implies that fixation rates should be 2–3 times higher than 100 Tg N/yr, which was an accepted value at that time. Recent observations that N_2- fixation is indeed underestimated by up to two fold support these calculations.

More recent calculation of the nitrogen isotope budget in year 2007 included so-called "consumed-N" into the model, which represents nitrogen utilization in the euphotic zone [202]. During N utilization, nitrogen fractionation takes place, shifting the "initial" value of the ambient NO_3^- from +5‰ to 17 ‰. Water column N-loss leads to production of N_2 gas with $\delta^{15}N$ values 25‰ lighter than the "initial". This process shifts the nitrogen isotope budget significantly and even leads to balance of water column and sedimentary denitrification (S:CW =1). A recent model of the pre-industrial N-budget includes a sedimentary fractionation factor ε_{sed} of 4‰ and ε_{fix} of 1.5‰ [203]. The value of "consumed-N" was set to +12‰, taking into account the fractionation factor of utilization of 7‰. In this model, the S:CW denitrification ratio varies from 2 to 3.5. Current budget estimates can, however, differ from these prior estimates due to an increase of riverine and atmospheric input of nitrogen into the ocean [198].

6. CONCLUSIONS

Of the two approaches whereby nitrogen isotope measurements are used to study marine N cycle processes, stable isotope natural abundance measurements have the great advantage of providing information on integrated processes with little or no manipulation of the system. In contrast tracer-based rate measurements are better used for rate measurements. The natural abundance and tracer approaches are strongly complementary, providing insight into the nitrogen cycle on very different spatial and temporal scales. The magnitude of processes that have distinctive isotopic fractionation patterns can potentially be quantified if the fractionations are known accurately. Thus, there is a significant ongoing need for careful measurement of isotopic fractionation patterns under controlled laboratory conditions as well as under field conditions, which in the best case should be combined. In order to interpret isotopic distributions and trends and to construct meaningful isotopic budgets, the openness of the system (ventilation rates) and the nature of N-exchanges should be known.

The growing set of $\delta^{15}N$ measurements makes it increasingly feasible to incorporate isotopic constraints into ecosystem and biogeochemical models of nitrogen in the ocean. A robust isotope budget for the ocean will require extension of these sampling efforts. Time-series measurements from a variety of locations will also greatly enhance the utility of nitrogen isotope measurements, particularly in quantifying the flux of nitrogen through the biota during blooms and other transient events. Such transient events are effectively natural tracer experiments created by isotopic fractionation during phytoplankton consumption of inorganic nitrogen and could provide an ecosystem-level analog to the short-term tracer experiments typically carried out in small volumes aboard ship. Complementing nitrogen stable isotope measurements with focused rate measurements may ultimately provide the best strategy for studying the dynamics of the nitrogen cycle across a variety of temporal and spatial scales.

REFERENCES

1. K. L Casciotti, *Inverse kinetic isotope fractionation during bacterial nitrite oxidation.* Geochimica et Cosmochimica Acta, 200920612076
2. O Nier, *A redetermination of the relative abundances of the isotopes of carbon, nitrogen, oxygen, argon, and potassium.* Physical Review, 1950789793
3. T. B Coplen, H. R Krouse, and J. K Böhlke, *Reporting of nitrogen isotope abundances-(Technical report).* Pure and Applied Chemistry, 1992907908
4. J Hoefs, *Stable Isotope Geochemistry* 6ed2009Berlin: Springer Verlag.
5. N. J. P Owens, and A. P Rees, *Determination of nitrogen-15 at sub-microgram levels of nitrogen using automated continuous flow isotope ratio mass spectrometry.* Analyst, 198916551657
6. E Ryabenko, M. A Altabet, and D. W. R Wallace, *Effect of chloride on the chemical conversion of nitrate to nitrous oxide for δ 15N analysis.* Limnology and Oceanography-Methods, 2009545552
7. M. R Mcilvin, and M. A Altabet, *Chemical conversion of nitrate and nitrite to nitrous oxide for nitrogen and oxygen isotopic analysis in freshwater and seawater.* Analytical Chemistry, 200555895595

8. D. M Sigman, et al*A bacterial method for the nitrogen isotopic analysis of nitrate in seawater and freshwater.* Analytical Chemistry 200141454153
9. K. L Casciotti, and M. R Mcilvin, *Isotopic analyses of nitrate and nitrite from reference mixtures and application to Eastern Tropical North Pacific waters.* Marine Chemistry, 2007184201
10. K. L Casciotti, et al*Measurement of the oxygen isotopic composition of nitrate in seawater and freshwater using the denitrifier method.* Analytical Chemistry, 200249054912
11. J Granger, et al*A method for nitrite removal in nitrate N and O isotope analyses.* Limnology and Oceanogrophy-Methods, 2006July): 205212
12. P. M Glibert, et al*Isotope-dilution models of uptake and remineralization of ammonium by marine plankton.* Limnology and Oceanography, 1982639650
13. W. G Harrison, and L. R Harris, *Isotope-dilution and its effects on measurements of nitrogen and phosphorus uptake by oceanic microplankton.* Marine Ecology-Progress Series, 1986253261
14. W Mohr, et al*Methodological underestimation of oceanic nitrogen fixation rates.* Plos One, 2010e12583
15. N Frew, *The role of organic films in air-sea gas exchange*, in *The sea surface and globale change*, R.A. Duce and P.S. Liss, Editors. 1997Cambridge University Press: Cambridge, UK. 121172
16. W. E Asher, and J. F Pankow, *Prediction of gas/water mass transport coefficients by a surface renewal model* Environmental Science & Technology, 199112941300
17. A Mariotti, *Atmospheric nitrogen is a reliable standard for natural 15N abundance measurements.* Nature, 1983685687
18. Montoya, J. P.: Nitrogen stable isotopes in marine environments, in: Nitrogen in Marine Environment, edited by: Capone, D. G., Bronk, D. A., Mulholland, M. R., and Carpenter, E. J., Elsevier Inc, Amsterdam, 1277-1302, 2008.
19. F. P Wilkerson, and G Grunseich, *Formation of blooms by the symbiotic ciliate Mesodinium rubrum- The significance of nitrogen uptake* Journal of Plankton Research, 1990973989
20. P Raimbault, G Slawyk, and N Garcia, *Comparison between chemical and isotopic measurements of biological nitrate utilization: further evidence of low new-production levels in the equatorial Pacific.* Marine Biology, 200011471155
21. S. P Libby, and P. A Wheeler, *Particulate and dissolved organic nitrogen in the central and eastern equatorial Pacific.* Deep Sea Research Part I: Oceanographic Research Papers, 1997345361
22. S Taguchi, and E. A Laws, *On the microparticles which pass through glass-fiber filter type GF/F in coastal and open waters.* Journal of Plankton Research, 19889991008
23. H Maske, and E Garcia-mendoza, *Adsorption of dissolved organic matter to the inorganic filter substrate and its implications for 14C uptake measurements.* Applied and Environmental Microbiology, 199438873889
24. P Raimbault, and G Slawyk, *A Semiautomatic wet-oxidation method for the determination of particlulate organic nitrogen collected on filters.* Limnology and Oceanography, 1991405408
25. R. J Olson, *Differential photoinhibition of marine nitrifying bacteria: A possible mechanism for the formation of the primary nitrite maximum.* Journal of Marine Research, 1981227238
26. H Yakushiji, and J Kanda, *Determination of experimentally enriched 15N in nitrate nitrogen based on an improved method of azo dye formation.* Journal of Oceanography, 1998337342
27. D. A Bronk, and B. B Ward, *Gross and net nitrogen uptake and DON release in the euphotic zone of Monterey Bay, California.* Limnology and Oceanography, 1999573585
28. D. M Sigman, et al*Natural abundance-level measurement of the nitrogen isotopic composition of oceanic nitrate: an adaptation of the ammonia diffusion method.* Marine Chemistry, 1997227242
29. T Tanaka, and T Saino, *Modified method for the analysis of nitrogen isotopic composition of oceanic nitrate at low concentration.* Journal of Oceanography, 2002539546
30. N Risgaard-petersen, et al*Application of the isotope pairing technique in sediments where anammox and denitrification coexist.* Limnology and Oceanography-Methods, 20036373

31. K. M Jensen, M. H Jensen, and R. P Cox, *Membrane inlet mass spectrometric analysis of N-isotope labelling for aquatic denitrification studies.* Fems Microbiology Ecology, 1996101109
32. N. A Feast, and P. F Dennis, *A comparison of methods for nitrogen isotope analysis of groundwater.* Chemical Geology, 1996167171
33. T Preston, et al*Isotope dilution analysis of combined nitrogen in natural waters.1. Ammonium.* Rapid Communications in Mass Spectrometry, 1996959964
34. M. F Lehmann, S. M Bernasconi, and J. A Mckenzie, *A method for the extraction of ammonium from freshwaters for nitrogen isotope analysis.* Analytical Chemistry, 200147174721
35. Deen, W. A. a. n. d L. K Porter, *Devarda's alloy reduction of nitrate and tube diffusion of the reduced nitrogen for indophenol ammonium and nitrogen-15 determinations.* Analytical Chemistry, 198011641166
36. R. M Holmes, et al*Measuring 15N-NH4+ in marine, estuarine and fresh waters: An adaptation of the ammonia diffusion method for samples with low ammonium concentrations.* Marine Chemistry, 1998235243
37. N Risgaard-petersen, N. P Revsbech, and S Rysgaard, *Combined microdiffusion-hypobromite oxidation method for determining nitrogen-15 isotope in ammonium.* Soil Science Society of America Journal 199510771080
38. M Sebilo, et al*The use of the 'Ammonium diffusion' method for δ15N-NH4+ and δ15N-3-measurementsComparison with other techniques.* Environmental Chemistry, 200499103
39. N Dudek, M. A Brzezinski, and P. A Wheeler, *Recovery of ammonium nitrogen by solvent extraction for the determination of relative 15N abundance in regeneration experiments.* Marine Chemistry, 19865969
40. L Zhang, et al*Sensitive measurement of NH4+ 15N/14N (δ15NH4+) at natural abundance levels in fresh and saltwaters.* Analytical Chemistry, 2007
41. S. M Steingruber, et al*Measurement of denitrification in sediments with the N-15 isotope pairing technique.* Applied and Environmental Microbiology, 200137713778
42. D. G Capone, and J. P Montoya, *Nitrogen fixation and denitrification*, in *Methods in Marine Microbiology*, J. Paul, Editor 2001Academic Press: New York. 501515
43. S. M An, W. S Gardner, and T Kana, *Simultaneous measurement of denitrification and nitrogen fixation using isotope pairing with membrane inlet mass spectrometry analysis.* Applied and Environmental Microbiology, 200111711178
44. T. M Kana, et al*Denitrification in estuarine sediments determined by membrane inlet mass spectrometry.* Limnology and Oceanography, 1998334339
45. H. E Hartnett, and S. P Seitzinger, *High-resolution nitrogen gas profiles in sediment porewaters using a new membrane probe for membrane-inlet mass spectrometry.* Marine Chemistry, 20032330
46. T. M Kana, et al*Membrane Inlet Mass Spectrometer for rapid high-precision determination of N2, O2, and Ar in environmental water samples.* Analytical Chemistry, 199441664170
47. S Punshon, and R. M Moore, *Nitrous oxide production and consumption in a eutrophic coastal embayment.* Marine Chemistry, 20043751
48. K. L Casciotti, M Mcilvin, and C Buchwald, *Oxygen isotopic exchange and fractionation during bacterial ammonia oxidation* Limnology and Oceanography, 2010753762
49. C. H Frame, and K. L Casciotti, *Biogeochemical controls and isotopic signatures of nitrous oxide production by a marine ammonia-oxidizing bacterium.* Biogeosciences 201030193059
50. S Toyoda, and N Yoshida, *Determination of nitrogen isotopomers of nitrous oxide on a modified isotope ratio mass spectrometer.* Analytical Chemistry, 199947114718
51. R. L Sutka, et al*Distinguishing nitrous oxide production from nitrification and denitrification on the basis of isotopomer abundances.* Applied and Environmental Microbiology, 2006638644
52. N Ostrom, and P Ostrom, *The isotopomers of nitrous oxide: Analytical considerations and application to resolution of microbial production pathways*, in *Handbook of Environmental Isotope Geochemistry*, M. Baskaran, Editor 2011Springer Berlin Heidelberg. 453476

53. B. B Ward, and D. A Bronk, *Net nitrogen uptake and DON release in surface waters: importance of trophic interactions implied from size fractionation experiments.* Marine Ecology-Progress Series, 20011124
54. J. A Brandes, A. H Devol, and C Deutsch, *New developments in the marine nitrogen cycle.* Chemical Reviews, 2007577589
55. J. J Mccarthy, and E. J Carpenter, *Oscillatoria (Trichodesmium) thiebautii (Cyanophyta) in the central north Atlantic Ocean.* Journal of Phycology, 19797582
56. S. P Seitzinger, R. W Sanders, and R Styles, *Bioavailability of DON from natural and anthropogenic sources to estuarine plankton.* Limnology and Oceanography, 2002353366
57. J. P Montoya, E. J Carpenter, and D. G Capone, *Nitrogen fixation and nitrogen isotope abundances in zooplankton of the oligotrophic North Atlantic.* Limnology and Oceanography, 200216171628
58. J Zehr, and H Paerl, *Biological nitrogen fixation in the marine environment.*, in *Microbial Ecology of the Oceans.*, D.L. Kirchman, Editor 2008Wiley-Liss, Inc.: New York. 481525
59. E. J Carpenter, *Nitrogen fixation by Oscillatoria (Trichodesmium) thiebautii in the southwestern Sargasso Sea.* Deep-Sea Research, 1973285288
60. T Saino, and A Hattori, *Diel variation in nitrogen fixation by a marine blue-green alga, Trichodesmium thiebautii.* Deep-Sea Research, 197812591263
61. E. J Carpenter, *Nitrogen fixation by a blue-green epiphyte on pelagic Sargassum.* Science, 197212071209
62. T. H Mague, F. C Mague, and O Holmhansen, *Physiology and chemical composition of nitrogen-fixing phytoplankton in the central north Pacific ocean.* Marine Biology, 1977213227
63. T. A Villareal, *Widespread occurrence of the Hemiaulus-cyanobacterial symbiosis in the southwest north Atlantic ocean.* Bulletin of Marine Science, 199417
64. I Berman-frank, et al*Segregation of nitrogen fixation and oxygenic photosynthesis in the marine cyanobacterium Trichodesmium.* Science, 200115341537
65. E Wada, and A Hattori, *Natural abundance of 15N in particulate organic matter in the North Pacific Ocean.* Geochimica et Cosmochimica Acta, 1976249251
66. E Ryabenko, et al*Contrasting biogeochemistry of nitrogen in the Atlantic and Pacific oxygen minimum zones.* Biogeosciences 2012203215
67. M. F Lehmann, et al*Origin of the deep Bering Sea nitrate deficit: Constraints from the nitrogen and oxygen isotopic composition of water column nitrate and benthic nitrate fluxes.* Global Biogeochemical Cycles, 2005GB4005
68. C. C Delwiche, et al*Nitrogen isotope distribution as a presumptive indicator of nitrogen-fixation.* Botanical Gazette, 1979S65S69
69. J. P Zehr, E. J Carpenter, and T. A Villareal, *New perspectives on nitrogen-fixing microorganisms in tropical and subtropical oceans.* Trends in Microbiology, 20006873
70. D. G Capone, et al*Basis for diel variation in nitrogenase activity in the marine planktonic cyanobacterium Trichodesmium thiebautii.* Applied and Environmental Microbiology, 199035323536
71. R. W Howarth, R Marino, and J. J Cole, *Nitrogen-fixation in fresh-water, estuarine, and marine ecosystems.2. Biogeochemical controls.* Limnology and Oceanography, 1988688701
72. R. W Howarth, et al*Nitrogen fixation in freshwater, estuarine, and marine ecoystems. 1. rates and importance.* Limnology and Oceanography, 1988469687
73. D Karl, et al*Dinitrogen fixation in the world's oceans.* Biogeochemistry, 20024798
74. E. J Carpenter, *Physiology and ecology of marine Oscillatoria (Trichodesmium).* Marine Biology Letters, 19836985
75. J Lehtimaki, et al*Growth, nitrogen fixation, and nodularin production by two Baltic sea cyanobacteria.* Applied and Environmental Microbiology, 199716471656
76. J. A Sohm, E. A Webb, and D. G Capone, *Emerging patterns of marine nitrogen fixation.* Nature Reviews Microbiology, 2011499508
77. J. R Gallon, *N2 fixation in phototrophs: adaptation to a specialized way of life.* Plant and Soil, 20013948

78. I Berman-frank, et al*Iron availability, cellular iron quotas, and nitrogen fixation in Trichodesmium.* Limnology and Oceanography, 200112491260
79. LaRocheJ. and E. Breitbarth, *Importance of the diazotrophs as a source of new nitrogen in the ocean.* Journal of Sea Research, 20056791
80. L. A Codispoti, and F. A Richards, *An analysis of the horizontal regime of denitrification in the eastern tropical North Pacific.* Limnology and Oceanography, 1976379388
81. C Deutsch, et al*Denitrification and N2 fixation in the Pacific Ocean.* Global Biogeochemical Cycles, 2001483506
82. N Gruber, and J. L Sarmiento, *Global patterns of marine nitrogen fixation and denitrification.* Global Biogeochemical Cycles, 1997235266
83. F Michaels, et al*Inputs, losses and transformations of nitrogen and phosphorus in the pelagic North Atlantic Ocean.* Biogeochemistry, 1996181226
84. D. A Hansell, and R. A Feely, *Atmospheric intertropical convergence impacts surface ocean carbon and nitrogen biogeochemistry in the western tropical Pacific.* Geophysical Research Letters, 200010131016
85. D. A Hansell, and T. Y Waterhouse, *Controls on the distributions of organic carbon and nitrogen in the eastern Pacific Ocean.* Deep-Sea Research Part I-Oceanographic Research Papers, 1997843857
86. M Vidal, C. M Duarte, and S Agusti, *Dissolved organic nitrogen and phosphorus pools and fluxes in the central Atlantic Ocean.* Limnology and Oceanography, 1999106115
87. J Abell, S Emerson, and P Renaud, *Distributions of TOP, TON and TOC in the North Pacific subtropical gyre: Implications for nutrient supply in the surface ocean and remineralization in the upper thermocline.* Journal of Marine Research, 2000203222
88. D Karl, et al*The role of nitrogen fixation in biogeochemical cycling in the subtropical North Pacific Ocean.* Nature, 1997533538
89. D. M Karl, et al*Ecosystem changes in the North Pacific subtropical gyre attributed to the 1991-92 El-Nino.* Nature, 1995230234
90. B. A Bergquist, and E. A Boyle, *Dissolved iron in the tropical and subtropical Atlantic Ocean.* Global Biogeochemical Cycles, 2006
91. S Blain, S Bonnet, and C Guieu, *Dissolved iron distribution in the tropical and sub tropical South Eastern Pacific.* Biogeosciences, 2008269280
92. E. A Boyle, et al*Iron, manganese, and lead at Hawaii Ocean Time-series station ALOHA: Temporal variability and an intermediate water hydrothermal plume.* Geochimica et Cosmochimica Acta, 2005933952
93. J Wu, et al*Soluble and colloidal iron in the oligotrophic North Atlantic and North Pacific.* Science, 2001847849
94. C Mahaffey, et al*Biogeochemical signatures of nitrogen fixation in the eastern North Atlantic.* Geophysical Research Letters, 20031300
95. K. S Johnson, et al*Surface ocean-lower atmosphere interactions in the Northeast Pacific Ocean Gyre: Aerosols, iron, and the ecosystem response.*Global Biogeochemical Cycles, 20031063
96. J. M Lenes, et al*Iron fertilization and the Trichodesmium response on the West Florida shelf.* Limnology and Oceanography, 200112611277
97. M. M Mills, et al*Iron and phosphorus co-limit nitrogen fixation in the eastern tropical North Atlantic.* Nature, 2004292294
98. R Baker, et al*Atmospheric deposition of nutrients to the Atlantic Ocean.* Geophysical Research Letters, 200322962300
99. C Ridame, and C Guieu, *Saharan input of phosphate to the oligotrophic water of the open western Mediterranean Sea.* Limnology and Oceanography, 2002856869
100. J. L Ramos, F Madueno, and M. G Guerrero, *Regulation of nitrogenase levels in Anabaena sp. ATCC 33047 and other filamentous cyanobacteria* Archives of Microbiology, 1985105111
101. J Martín-nieto, A Herrero, and E Flores, *Control of nitrogenase mRNA levels by products of nitrate assimilation in the cyanobacterium Anabaena sp. Strain PCC 7120.* Plant Physiology, 1991825828

102. S Sanzalferez, and F. F Delcampo, *Relationship between nitrogen fixation and nitrate metabolism in the Nodularia strains M1 and M2.* Planta, 1994339345
103. M. R Mulholland, K Ohki, and D. G Capone, *Nutrient controls on nitrogen uptake and metabolism by natural populations and cultures of Trichodesmium (Cyanobacteria).* Journal of Phycology, 200110011009
104. C. M Holl, and J. P Montoya, *Interactions between nitrate uptake and nitrogen fixation in continuous cultures of the marine diazotroph Trichodesmium (Cyanobacteria).* Journal of Phycology, 200511781183
105. P Raimbault, and N Garcia, *Evidence for efficient regenerated production and dinitrogen fixation in nitrogen-deficient waters of the South Pacific Ocean: impact on new and export production estimates.* Biogeosciences, 2008323338
106. C Fernandez, L Farias, and O Ulloa, *Nitrogen fixation in denitrified marine waters.* Plos One, 2011e20539
107. E. J Carpenter, and D. G Capone, *Nitrogen fixation in the marine environment,* in *Nitrogen in the marine environment,* D.G. Capone, et al., Editors. 2008Elsevier. 141198
108. V. J Bertics, and W Ziebis, *Bioturbation and the role of microniches for sulfate reduction in coastal marine sediments.* Environmental Microbiology, 201030223034
109. F Gilbert, et al*Hydrocarbon influence on denitrification in bioturbated Mediterranean coastal sediments.* Hydrobiologia, 19976777
110. R. A Herbert, *Nitrogen cycling in coastal marine ecosystems.* FEMS Microbiology Reviews, 1999563590
111. I Schmidt, R. J. M Van Spanning, and M. S. M Jetten, *Denitrification and ammonia oxidation by Nitrosomonas europaea wild-type, and NirK- and NorB-deficient mutants.* Microbiology-Sgm, 200441074114
112. D Zart, and E Bock, *High rate of aerobic nitrification and denitrification by Nitrosomonas eutropha grown in a fermentor with complete biomass retention in the presence of gaseous 2or NO.* Archives of Microbiology, 1998282286
113. B Hooper, and K. R Terry, *Hydroxylamine oxidoreductase of Nitrosomonas: Production of nitric oxide from hydroxylamine.* Biochimica et Biophysica Acta (BBA)- Enzymology, 19791220
114. B Hooper, et al*Enzymology of the oxidation of ammonia to nitrite by bacteria.* Antonie Van Leeuwenhoek International Journal of General and Molecular Microbiology, 19975967
115. M Poth, and D. D Focht, *N kinetic analysis of N2O production by Nitrosomonas europaea: an examination of nitrifier denitrification.* Applied and Environmental Microbiology, 198511341141
116. A Remde, and R Conrad, *Production of nitric oxide in Nitrosomonas europaea by reduction of nitrite.* Archives of Microbiology, 1990187191
117. A Mariotti, et al*Experimental determination of nitrogen kinetic isotope fractionation- Some principles- Illustration for the denitrification and nitrification processes.* Plant and Soil, 1981413430
118. K. L Casciotti, D. M Sigman, and B. B Ward, *Linking diversity and stable isotope fractionation in ammonia-oxidizing bacteria.* Geomicrobiology Journal, 2003335353
119. R. L Sutka, et al*Stable nitrogen isotope dynamics of dissolved nitrate in a transect from the North Pacific Subtropical Gyre to the Eastern Tropical North Pacific.* Geochimica et Cosmochimica Acta, 2004517527
120. K. L Casciotti, and C Buchwald, *Insights on the marine microbial nitrogen cycle from isotopic approaches to nitrification.* Frontiers in Microbiology, 2012
121. H. L Schmidt, et al*Is the isotopic composition of nitrous oxide an indicator for its origin from nitrification or denitrification? A theoretical approach from referred data and microbiological and enzyme kinetic aspects.* Rapid Communications in Mass Spectrometry, 200420362040
122. S Toyoda, et al*Fractionation of N2O isotopomers during production by denitrifier.* Soil Biology & Biochemistry, 200515351545
123. R. L Sutka, et al*Nitrogen isotopomer site preference of N2O produced by Nitrosomonas europaea and Methylococcus capsulatus Bath.* Rapid Communications in Mass Spectrometry, 2003738745

124. D Nevison, et al*Interannual and seasonal variability in atmospheric N2O.* Global Biogeochemical Cycles, 2007GB3017
125. Y Cohen, and L. I Gordon, *Nitrous oxide in the oxygen minimum of the eastern tropical North Pacific: evidence for its consumption during denitrification and possible mechanisms for its production.* Deep Sea Research, 1978509524
126. T Yoshinari, *Nitrous oxide in the sea.* Marine Chemistry, 1976189202
127. C Nevison, J. H Butler, and J. W Elkins, *Global distribution of N2O and the delta N2O-AOU yield in the subsurface ocean.* Global Biogeochemical Cycles, 20031119
128. F Lipschultz, et al*Bacterial transformations of inorganic nitrogen in the oxygen-deficient waters of the Eastern Tropical South Pacific Ocean.* Deep Sea Research Part A. Oceanographic Research Papers, 199015131541
129. B Ward, *Nitrogen transformations in the Southern California Bight.* Deep-Sea Research Part A-Oceanographic Research Papers, 1987785805
130. B. B Ward, M. C Talbot, and M. J Perry, *Contributions of phytoplankton and nitrifying bacteria to ammonium and nitrite dynamics in coastal waters.* Continental Shelf Research, 1984383398
131. T. H Blackburn, and N. D Blackburn, *Coupling of cycles and global significance of sediment diagenesis* Marine Geology, 1993101110
132. V Grundmanis, and J. W Murray, *Nitrification and denitrification in marine sediments from Puget Sound.* Limnology and Oceanography, 1977804813
133. H Devol, and J. P Christensen, *Benthic fluxes and nitrogen cycling in sediments of the continental margin of the eastern North Pacific* Journal of Marine Research, 1993345372
134. E Laursen, and S. P Seitzinger, *The role of denitrification in nitrogen removal and carbon mineralization in Mid-Atlantic Bight sediments.* Continental Shelf Research, 200213971416
135. M. F Lehmann, D. M Sigman, and W. M Berelson, *Coupling the 15N/14N and 18O/16O of nitrate as a constraint on benthic nitrogen cycling.* Marine Chemistry, 2004120
136. R Clark, A. P Rees, and I Joint, *A method for the determination of nitrification rates in oligotrophic marine seawater by gas chromatography/mass spectrometry.* Marine Chemistry, 20078496
137. M. R Mulholland, and M. W Lomas, *Nitogen uptake and assimilation*, in *Nitrogen in the Marine Environment*, D.G. Capone, et al., Editors. 2008Elsevier Inc.: Amsterdam et al. 303384
138. J. K Moore, et al*An intermediate complexity marine ecosystem model for the global domain.* Deep-Sea Research Part II-Topical Studies in Oceanography, 2002403462
139. J. M Garcia-fernandez, N. T De Marsac, and J Diez, *Streamlined regulation and gene loss as adaptive mechanisms in Prochlorococcus for optimized nitrogen utilization in oligotrophic environments.* Microbiology and Molecular Biology Reviews, 2004630638
140. J. M Garcia-fernandez, and J Diez, *Adaptive mechanisms of nitrogen and carbon assimilatory pathways in the marine cyanobacteria Prochlorococcus.* Research in Microbiology, 2004795802
141. E Wada, and A Hattori, *Nitrogen isotope effect in the assimilation of inorganic nitrogenous compounds by marine diatoms.* Geomicrobiology Journal, 197885101
142. J. P Montoya, and J. J Mccarthy, *Isotopic fractionation during nitrate uptake by phytoplankton grown in continuous culture.* Journal of Plankton Research, 1995439464
143. J Granger, et al*Coupled nitrogen and oxygen isotope fractionation of nitrate during assimilation by cultures of marine phytoplankton.* Limnology and Oceanography, 200417631773
144. M. P Hoch, M. L Fogel, and D. L Kirchman, *Isotope fractionation associated with ammonium uptake by a marine bacterium.* Limnology and Oceanography, 199214471459
145. N. A Waser, et al*Nitrogen isotope fractionation during nitrate, ammonium and urea uptake by marine diatoms and coccolithophores under various conditions of N availability.* Marine Ecology-Progress Series, 19982941
146. N. J Antia, P. J Harrison, and L Oliveira, *The role of dissolved organic nitrogen in phytoplankton nutrition, cell biology and ecology.* Phycologia, 1991189

147. R. W Eppley, and W. H Thomas, *Comparison of half-saturation constants for growth and nitrate uptake of marine phytoplankron* Journal of Phycology, 1969375379
148. M Strous, J. G Kuenen, and M. S. M Jetten, *Key physiology of anaerobic ammonium oxidation.* Applied and Environmental Microbiology, 199932483250
149. P Lam, et al*Revising the nitrogen cycle in the Peruvian oxygen minimum zone.* Proceedings of the National Academy of Sciences of the United States of America, 200947524757
150. L. A Codispoti, et al*The oceanic fixed nitrogen and nitrous oxide budgets: Moving targets as we enter the anthropocene?* Scientia Marina, 200185105
151. W. G Zumft, and H Korner, *Enzyme diversity and mosaic gene organization in denitrification.* Antonie Van Leeuwenhoek International Journal of General and Molecular Microbiology, 19974358
152. L. A Codispoti, T Yoshinari, and A. H Devol, *Suboxic respiration in the oceanic water column.*, in *Respiration in Aquatic Ecosystems*, P.A. del Giorgio and P.J.L.B. Williams, Editors. 2005Blackwell Scientific: Oxford. 225247
153. W. G Zumft, *Cell biology and molecular basis of denitrification.* Microbiology and Molecular Biology Reviews, 1997533616
154. S. J Ferguson, *Denitrification and its control.* Antonie Van Leeuwenhoek International Journal of General and Molecular Microbiology, 199489110
155. D Güven, et al*Propionate oxidation by and methanol inhibition of anaerobic ammonium oxidizing bacteria* Applied and Environmental Microbiology, 200510661071
156. T Kalvelage, et al*Oxygen sensitivity of anammox and coupled N-cycle processes in oxygen minimum zones.* Plos One, 2011e29299
157. M. S. M Jetten, et al*Anaerobic ammonium oxidation by marine and freshwater planctomycete-like bacteria.* Applied Microbiology and Biotechnology, 2003107114
158. L Van Niftrik, et al*The anammoxasome: an intracytoplasmic compartment in anammox bacteria.* FEMS Microbiology Letters, 2004713
159. M Strous, et al*Deciphering the evolution and metabolism of an anammox bacterium from a community genome.* Nature, 2006790794
160. B Kartal, et al*Anammox bacteria disguised as denitrifiers: nitrate reduction to dinitrogen gas via nitrite and ammonium.* Environmental Microbiology, 2007635642
161. M. M. M Kuypers, et al*Anaerobic ammonium oxidation by anammox bacteria in the Black Sea.* Nature, 2003608611
162. O. R Dale, C. R Tobias, and B Song, *Biogeographical distribution of diverse anaerobic ammonium oxidizing (anammox) bacteria in Cape Fear River estuary.* Environmental Microbiology, 200911941207
163. B Thamdrup, and T Dalsgaard, *Production of N2 through anaerobic ammonium oxidation coupled to nitrate reduction in marine sediments.* Applied and Environmental Microbiology, 200213121318
164. T Dalsgaard, et al*N2 production by the anammox reaction in the anoxic water column of Golfo Dulce, Costa Rica.* Nature, 2003606608
165. A Podlaska, et al*Microbial community structure and productivity in the oxygen minimum zone of the eastern tropical North Pacific.* Deep Sea Research Part I: Oceanographic Research Papers, 20127789
166. B Hu, l., et al*New anaerobic, ammonium-oxidizing community enriched from peat soil.* Applied and Environmental Microbiology, 2011966971
167. S Humbert, et al*Molecular detection of anammox bacteria in terrestrial ecosystems: distribution and diversity.* ISME J, 2009450454
168. G Zhu, et al*Anaerobic ammonia oxidation in a fertilized paddy soil.* ISME J, 201119051912
169. V Erler, B. D Eyre, and L Davison, *The contribution of anammox and denitrification to sediment N2 production in a surface flow constructed Wetland.* Environmental Science & Technology, 200891449150
170. J Schubert, et al*Anaerobic ammonium oxidation in a tropical freshwater system (Lake Tanganyika).* Environmental Microbiology, 200618571863
171. M Hannig, et al*Shift from denitrification to anammox after inflow events in the central Baltic Sea.* Limnology and Oceanography, 200713361345

172. M Strous, et al*The sequencing batch reactor as a powerful tool for the study of slowly growing anaerobic ammonium-oxidizing microorganisms.* Applied Microbiology and Biotechnology, 1998589596
173. van de GraafA.A., et al., *Autotrophic growth of anaerobic ammonium-oxidizing micro-organisms in a fluidized bed reactor.* Microbiology, 199621872196
174. E Bock, et al*Nitrogen loss caused by denitrifyling Nitrosomonas cells using ammonium or hydrogen as electron-donors and nitrite as electron-acceptor.* Archives of Microbiology, 19951620
175. M Blackmer, J. M Bremner, and E. L Schmidt, *Production of nitrous oxide by ammonia oxidizing chemoautotrophic microorganisms in soil.* Applied and Environtal Microbiology, 198010601066
176. T. J Goreau, et al*Production of 2-andN2O by nitrifying bacteria at reduced concentrations of oxygen.* Applied and Environmental Microbiology, 1980526532
177. L Bohlen, et al*Benthic nitrogen cycling traversing the Peruvian oxygen minimum zone.* Geochimica et Cosmochimica Acta, 201160946111
178. M. M Jensen, et al*Intensive nitrogen loss over the Omani Shelf due to anammox coupled with dissimilatory nitrite reduction to ammonium.* Isme Journal, 201116601670
179. S Rysgaard, N Risgaard, Petersen, and N.P. Sloth, *Nitrification, denitrification, and nitrate ammonification in sediments of two coastal lagoons in Southern France.* Hydrobiologia, 1996p. 133141
180. L Farías, *Potential role of bacterial mats in the nitrogen budget of marine sediments: the case of Thioploca spp.* Marine ecology progress series, 1998291292
181. H Fossing, et al*Concentration and transport of nitrate by the mat-forming sulphur bacterium Thioploca.* Nature, 1995713715
182. H. N Schulz, et al*Dense populations of a giant sulfur bacterium in Namibian shelf sediments.* Science, 1999493495
183. J. D Cline, and F. A Richards, *Oxygen deficient conditions and nitrate reduction in Eastern Tropical North-Pacific Ocean.* Limnology and Oceanography, 1972885900
184. J. M Morrison, et al*The oxygen minimum zone in the Arabian Sea during 1995.* Deep-Sea Research Part II-Topical Studies in Oceanography, 199919031931
185. W Koeve, and P Kähler, *Heterotrophic denitrification vs. autotrophic anammox-quantifying collateral effects on the oceanic carbon cycle.* Biogeosciences, 201023272337
186. M Voss, and J. P Montoya, *Nitrogen cycle. Oceans apart.* Nature, 20094950
187. S. E Bulow, et al*Denitrification exceeds anammox as a nitrogen loss pathway in the Arabian Sea oxygen minimum zone.* Deep-Sea Research Part I-Oceanographic Research Papers, 2010384393
188. A Galan, et al*Anammox bacteria and the anaerobic oxidation of ammonium in the oxygen minimum zone off northern Chile.* Deep-Sea Research Part II-Topical Studies in Oceanography, 200911251135
189. H Devol, et al*Denitrification rates and excess nitrogen gas concentrations in the Arabian Sea oxygen deficient zone.* Deep-Sea Research Part I-Oceanographic Research Papers, 200615331547
190. A. S Van Mooy, R. G Keil, and A. H Devol, *Impact of suboxia on sinking particulate organic carbon: Enhanced carbon flux and preferential degradation of amino acids via denitrification.* Geochimica et Cosmochimica Acta, 2002457465
191. J. A Brandes, et al*Isotopic composition of nitrate in the central Arabian Sea and eastern tropical North Pacific: A tracer for mixing and nitrogen cycles.* Limnology and Oceanography, 199816801689
192. M Sigman, et al*The δ15N of nitrate in the Southern Ocean: Consumption of nitrate in surface waters.* Global Biogeochemical Cycles, 199911491166
193. M Voss, J. W Dippner, and J. P Montoya, *Nitrogen isotope patterns in the oxygen-deficient waters of the Eastern Tropical North Pacific Ocean.* Deep-Sea Research Part I-Oceanographic Research Papers, 200119051921
194. J. D Cline, and I. R Kaplan, *Isotopic fractionation of dissolved nitrate during denitrification in the eastern tropical north Pacific Ocean.* Marine Chemistry, 1975271299

195. M. F Lehmann, et al*The distribution of nitrate 15N/14N in marine sediments and the impact of benthic nitrogen loss on the isotopic composition of oceanic nitrate.* Geochimica et Cosmochimica Acta, 200753845404
196. L. A Codispoti, *An oceanic fixed nitrogen sink exceeding 400 Tg Na-1 vs the concept of homeostasis in the fixed-nitrogen inventory.* Biogeosciences, 2007233253
197. L. A Codispoti, and J. P Christensen, *Nitrification, denitrification and nitrous-oxide cycling in the eastern tropical South-Pacific Ocean.* Marine Chemistry, 1985277300
198. J. A Brandes, and A. H Devol, *A global marine-fixed nitrogen isotopic budget: Implications for Holocene nitrogen cycling.* Global Biogeochemical Cycles, 200211201134
199. J. N Galloway, et al*Nitrogen Cycles: Past, Present, and Future.* Biogeochemistry, 2004153226
200. N Gruber, *The marine nitrogen cycle: Overview and challanges*, in *Nitrogen in Marine Environment*, D.G. Capone, et al., Editors. 2008Elsevier Inc. 151
201. L. A Codispoti, *Interesting times for marine N2O.* Science, 201013391340
202. D Galbraith, et al*Global nitrogen isotopic constraints on the acceleration of oceanic denitrification during deglacial warming.* Nature Geosciences, 2012submitted.
203. M. A Altabet, *Constraints on oceanic N balance/imbalance from sedimentary 15N records.* Biogeosciences, 20077586
204. C Somes, *Nitrogen Isotopes in the Global Ocean: Constraints on the pre-industrial fixed-N budget*, in *Biogeochemical modelling* 2012Christian-Albrechts-Universität zu Kiel: Kiel.

CHAPTER 15

Acoustic identification of marine species using a feature library

Rolf J. Korneliussen[a,b,*], Yngve Heggelund[b], Gavin J. Macaulay[a], Daniel Patel[b], Espen Johnsen[a], Inge K. Eliassen[b]

[a] *Institute of Marine Research, P.O. Box 1870, 5817 Bergen, Norway*

[b] *Christian Michelsen Research AS, P.O. Box 6031, NO-5892 Bergen, Norway*

ABSTRACT

Sonars and echosounders are widely used for remote sensing of life in the marine environment. There is an ongoing need to make the acoustic identification of marine species more correct and objective and thereby reduce the uncertainty of acoustic abundance estimates. In our work, data from multi-frequency echosounders working simultaneously with nearly identical and overlapping acoustic beams are processed stepwise in a modular sequence to improve data, detect schools and categorize acoustic targets by means of the Large Scale Survey System software (LSSS). Categorization is based on the use of an acoustic feature library whose main components are the relative frequency responses. The results of the categorization are translated into acoustic abundance of species. The method is tested on acoustic data from the Barents Sea, the Norwegian Sea and the North Sea, where the target species were capelin (Mallotus villosus L.), Atlantic mackerel (Scomber scombrus L.) and sandeel (Ammodytes marinus L.), respectively. Manual categorization showed a high conformity with automatic categorization for all surveys, especially for schools.

Keywords: Relative frequency response; School-detection; Acoustic-based categorization; Acoustic feature library; Large Scale Survey System; LSSS

1. INTRODUCTION

The sustainable management of commercially exploited marine species requires reliable knowledge of their abundance as well as of that of their prey and predator species. This is often informed by spatially extensive surveys using a variety of sampling tools that catch the targeted species, or remotely detect their presence (King, 2007). A commonly used remote sensing method is acoustics, where echosounders are used to generate directed pulses of sound and the resulting backscatter from marine organisms is measured and mapped (Simmonds and MacLennan, 2005). Given knowledge of the backscatter and other characteristics for species of interest, it is possible to assign species or species groups to the backscatter. Then, when combined with a relationship between backscatter amplitude and species size one can derive an estimate of the species biomass in the surveyed area (Simmonds and MacLennan, 2005). Estimating and understanding the sources of uncertainty in biomass estimates is an important means to reduce uncertainty and improve the inputs to the fisheries management process.

A key procedure in acoustic biomass estimation is the correct assigning of backscatter to species or species-group level (MacLennan and Holliday, 1996 ; Horne, 2000), a process typically called scrutinizing or categorization. Categorization, supported by biological sampling and human experience is time-consuming and subjective and there is a need to make the process more objective so as to reduce human-generated bias and uncertainty in acoustic abundance estimates. It has been estimated that the systematic error associated with the categorization of acoustic backscatter can be as high as ±80% (Simmonds and MacLennan, 2005). Automated species identification can be a valuable input to the categorization process and offers the potential for a more objective result.

The potential for objective classification of targets from backscatter was recognized in the 1970s (Holliday, 1977 ; Deuser et al., 1979). The acoustic scattering characteristics of some species may not be reliably distinguished from others by acoustic means only due to similarities in morphology and behaviour. Tilt distribution, i.e. swimming upwards or downwards, usually results in decreasing backscatter with increasing frequency. However, groups of similar species may be distinguished from other species or scattering groups (e.g. fish with swimbladder vs. zooplankton or fish without swimbladder).

The categorization methods can be divided into those that use multiple frequencies and those that primarily use a single frequency to identify school species. In some specific cases, school characteristics from a single frequency may provide sufficient information for species identification, but these techniques have not been universally applied because temporal changes in school shape and location (on scales from hours to years) can affect the results. For example, Scalabrin et al. (1996) concluded that their use of school structure and species-related spatial distribution did not give a high probability of species identification.

The attempts to utilize multiple echosounder frequencies to derive automated categorizing routines have been more successful. In the 1990s artificial neural feedback networks (ANN) were used to categorize acoustic data and are improved and expanded by the use of ground-truthed data for training the network (Haralabous and Georgakaros, 1996). A major problem with ANNs is that weights involved in the neural network are "hidden" (i.e. cannot be interpreted by humans), which makes systematic improvement to species identification difficult. A problem for many methods (Kloser et al., 2002; Korneliussen and Ona, 2002; Korneliussen and Ona, 2003; Fernandes et al., 2002 ; Korneliussen, 2010) is the use of properties that are pre-defined and difficult to modify. The average relative frequency responses, $r(f)$, (Korneliussen and Ona, 2002), were measured for some species (Korneliussen and Ona, 2002; Korneliussen and Ona, 2003; Fernandes et al., 2002 ; Korneliussen, 2010), and acoustic categories (representing the species) were then hardcoded into the BEI (Bergen Echo Integrator) postprocessing system (Korneliussen, 2004) to identify the acoustic categories. Pedersen and Korneliussen (2009) corrected for the beam-patterns of the transducer by using split-beam data of single-fish tracks so that species could be reliably distinguished from fewer samples than were needed from school measurements. De Robertis et al. (2010) used the Z-score as a summary to classify backscatter measurements based on $r(f)$, and used it in real-time with reasonable success. Although not explicitly stated, the Z-score implementation is likely to contain predefined features (given in their Table 2) in a similar manner to Fernandes et al. (2002) and Korneliussen (2010).

A more flexible solution is to use a range of features, such as $r(f)$, but let the statistical properties be derived from an expandable knowledge database. Many species occur in monospecific registrations that are the preferred type of acoustic registrations used to populate an acoustic library. A library could also contain non-acoustic features, such as geographical areas, depth and temperature ranges. Korneliussen et al. (2009a) organized $r(f)$ features in a library, but their focus was to show how aggregating and averaging acoustic measurements inside schools could improve the ability to distinguish species with similar acoustic properties. That work did not describe the details of the categorization method, so their results would be hard to reproduce and verify. Furthermore, they applied the method to a limited dataset where the school shapes were used as supporting information to distinguish between capelin and herring only, making the reliability of the method difficult to assess.

For some species it is difficult to find monospecific registrations and the categorization can be based on scattering models. The use of scattering models for zooplankton (Stanton et al., 1994 ; Demer and Conti, 2003) in a framework for estimating zooplankton specimen size (Holliday, 1977; Holliday et al., 1989 ; Fernandes et al., 2002) overcomes the problem of finding monospecific aggregations of zooplankton in-situ. Some large schools of euphausiids **can be identified both from scattering models as well as from an acoustic** library (for example, Euphausia superba L. in the southern ocean (Korneliussen et al., 2009b), or Meganyctiphanes norwegica L. in the North Atlantic Ocean), but such schools of zooplankton are the exception rather than the rule (Kaartvedt et al., 2005).

In this paper we describe categorization methods, methods to improve data quality, techniques for efficient processing of multi-frequency acoustic data, and automatic identification of biologically generated backscatter. These are all steps that make categorization of acoustic data practicable on operational surveys and provide a more objective scrutinizing process. The methods allow for processing large volumes of high resolution data from echosounders, which in turn makes it possible for a limited number of experts to analyse such data within restricted time periods. Special emphasis is given to methods for automatic categorization of acoustic features by means of statistical properties with reference to an acoustic feature library. We describe the realization of these methods for the efficient scrutinization of acoustic surveys. The reliability of the methods is illustrated with data from three surveys targeting Atlantic mackerel (Scomber scombrus L.), sandeel (Ammodytes marinus L.).

2. CONCEPTS AND BACKGROUND

2.1. Overview of analysis system

Data were scrutinized on a daily basis by a team of experts within 24 h of data collection (Fig. 1(A)). The result of the scrutiny was nautical area scattering coefficient values (s_A, m^2 nmi^{-2}, MacLennan et al., 2002) allocated to categories and stored in a relational database. s_A is a measure of the backscattering intensity per unit surface area over some depth interval. The results of echogram categorization (Fig. 2) are used to estimate fish stock abundance for each of the surveys and are stored in scrutiny files together with the acoustic data for additional inspection.

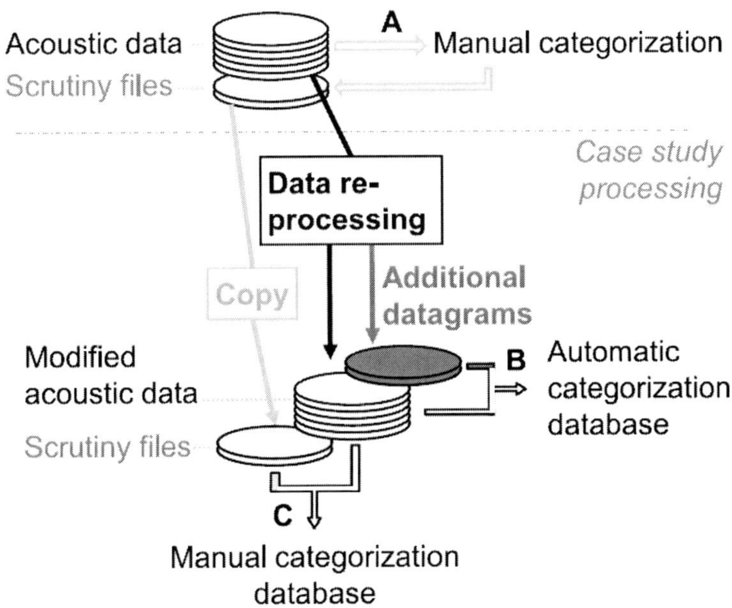

Figure. 1. Workflow used in the analysis. (A) Data were manually categorized and the results stored in scrutiny files. (B) Data were reprocessed, e.g. to remove noise. The scrutiny files were used with reprocessed acoustic data to generate the manual categorization database. (C) The acoustic data were further processed to add datagrams (grey in figure), e.g. automatically identified acoustic library categories and school-detections. The school detections were used only to improve acoustic library categorization in schools. The acoustic library categories were mapped to manual scrutiny categories so they could be stored in a database, in this case the automatic categorization database. capelin (Mallotus villosus L.) in three different ecosystems (Norwegian Sea, North Sea, and Barents Sea, respectively).

2.2. The LSSS software

The Large Scale Survey System (LSSS) software was developed for the analysis of acoustic backscatter from marine organisms such as zooplankton, pelagic and demersal fishes. LSSS is a software package programmed in Java, developed specifically for the needs of the Institute of Marine Research (IMR), Norway. It is designed for efficient post-processing of large amounts of echosounder and sonar data, and contains tools to isolate echo-traces by creating regions and assigning categories to these regions. A design aim of LSSS was to achieve the categorization of acoustic data collected over 24 h within 2 h of operator time. LSSS has a pre-processing facility (called KORONA) that carries out time-intensive automated processing of acoustic data, such as removing noise, aligning data due to transducers offsets, and other operations that make manual categorization actions faster (such as school detection, single fish track detection and bottom detection; Anon, 1990). Furthermore, through automatic categorization KORONA provides information that allows for more objective scrutiny of the echograms. Some of this information is not accessible without pre-processing, e.g. the identification of specified acoustic categories from completely mixed acoustic aggregations. The information from KORONA is available during categorization, but it is the scrutinizing team that decides if any, parts or all the information from KORONA is used. The primary GUI of LSSS focusses on the acoustic data, with overlays providing additional information to the operator (Fig. 2).

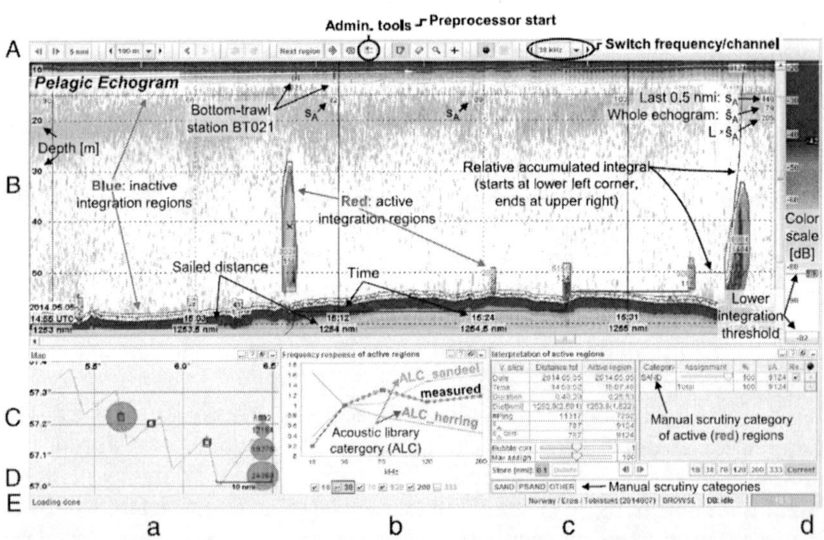

Figure. 2. Graphical user interface (GUI) of LSSS for interpreting backscatter data. The numbers (red for the active and black for inactive scatter-regions) on the echograms (B) show values of s_A. The s_A-values against each vertical black line refer to the whole water column within the block; the mean of these for the whole echogram is shown in the top right corner. Various scatter-regions outlined in red appear on the echogram, along with the corresponding s_A. The blue numbers, when present, show the mean s_A multiplied by the corresponding length of the scatter region. (A)

Toolbar; (B) Pelagic echogram, i.e. showing backscatter with surface at the top. The areas indicated with red outlines are the regions being scrutinized. These are linked to the windows at position Cb and Cc. (Bd) Echogram colour scale showing volume backscattering strength, S_v. (Ca) Cruise track on a latitude/longitude grid; (Cb) The relative frequency response, $r(f)$. (Cc) Summary of interpretation results and general info; (Dc) Manual scrutiny categories (MSC). (E) Status bar that includes ship and survey name.

2.3. Echogram pre-processing

Repeated echoes from isolated single organisms, such as fish, are detected automatically (Handegard, 2007). Schools are detected as described below, where a school is defined by the boundary that encloses an aggregation of individually unresolved organisms with some minimum backscatter intensity.

In general, backscatter from two or more frequencies is combined to detect schools. Here we use the combined frequency channel $[s_v(38)+3s_v(200)]/4$ for all case studies, where $s_v(38)$ and $s_v(200)$ are the volume backscattering coefficients at 38 kHz and 200 kHz. Although any combination of frequencies could be used, this combination is particularly well suited to the case studies used here: both 38 and 200 kHz are good for detecting swimbladdered fish such as herring and capelin, while 200 kHz is particularly useful for the detection of the two swimbladderless fish species used in the case studies (mackerel and sandeel). The use of combined frequency data, which is only used to detect the extent of the school, results in a smoother school boundary that is less dependent on frequency-dependent scattering. The combined frequency channel is smoothed with a 2D-Gaussian window (the normalized weights at ?±12.5 m horizontally and ±0.5 m vertically are half the value of the centre weight). Median filters followed by erosion filters and dilation filters (Dougherty, 1992) are then applied to remove small scale variability while preserving the main features of the dataset.

Next, measured samples are grouped together in school regions and averaged. The school detection module of the pre-processor KORONA traces the borders of each region as defined by a mean volume backscattering strength (S_v, MacLennan et al., 2002) threshold of View the MathML source re View the MathML source for the examples presented here, and joins or splits borders at the correct location as soon as a new ping arrives, thereby providing near real-time detection of schools (i.e. within a few seconds). The traced borders are merged to form closed curves which constitute outlines of potential schools. The potential schools must match user-defined criteria to be accepted as a school. These are:

- Scattering strength: upper and lower limits on the average S_v inside the region;
- Overlap: the values inside the school border are compared with those just outside the region by fitting each to a Gaussian distribution. The percentage overlap of these distributions is calculated and used as a criterion. The thickness of this band is user-defined.
- Lower limits on the length, thickness and area of the region.

The density interval and overlap criteria make it possible to detect regions with subtle data variations, such as seen in plankton concentrations.

3. CATEGORIZATION METHODS

The LSSS software provides three types of acoustic backscatter categories. These are:

1) Acoustic library categories (**ALC**), which are obtained from reference to a library of ground-truthed acoustic backscatter characteristics.
2) Model-based categories (**MBC**), which are obtained by comparison to backscatter frequency spectra obtained from acoustic backscattering models.
3) Manual scrutiny categories (**MSC**), where the operator manually attaches categories to acoustic backscatter.

Only the manual scrutiny categories can be used for echo integration and subsequent biomass estimation. The two other categories are either used to inform the manual categorization, or are mapped to manual scrutiny categories via an operator defined mapping matrix. Manual scrutiny categories correspond closely to species or species groups. Acoustic library categories and model-based categories do not necessarily correspond to species or species groups, but are more focussed on the distinguishable characteristics of the backscatter. The conversion of ALC and MBC to the manual scrutiny categories of relevance to particular surveys is an important function of the LSSS operator. There are currently 118 MSC, 25 ALC and five MBC available in LSSS. Acoustic categories should not be confused with species, which have a unique scientific name.

3.1. Manual interpretation of echograms (in general)

The backscatter data are visualized as echograms, and manually categorized (e.g. Jech and Michaels, 2006) by means of a computer program. IMR used the BEI software (Korneliussen, 2004) for surveys prior to 2007, and then the Large Scale Survey System (LSSS) software. A range of features are considered during manual categorization . First, unwanted areas are excluded, then unwanted backscatter is removed (e.g. bottom echoes and noise spikes). The operators then identify and select regions with similar acoustic properties. Some fish species typically appear as single individuals, while other species are usually seen in schools, layers or loose aggregations. The spatial distribution of the targets and the distribution of target strength aid identification of single-fish. The size, density and shape of a school also help to indicate the species. The average $r(f)$ of an encircled region may be used in differentiating between species. Active use of the S_v threshold (volume backscattering strength, Fig. 2(Bd)) is useful to differentiate some species of zooplankton and fish, but also to partition s_A between those species. Some species of fish and zooplankton are associated with sharp changes in temperature profile. The geographical area and proximity to sea surface and bottom is also used to identify species. However, the results of directed sampling of the biology are the most important supportive information used when manually categorizing echograms.

3.2. Acoustic feature library

The acoustic feature library contains ground-truthed data from selected acoustic backscatter that has been processed in a standardized manner. The processing involves noise removal and smoothing.

The statistical properties of each of the features of the acoustic library are calculated on several sample collection levels as described below. The library is organized by ALC, which for convenience are named after species since they are based on ground-truthed species. They are referred to as ALC due to the fact that the features of one species may also be used to identify species of similar acoustic properties (for example, a category derived from herring echoes may also identify the physically similar sardine). The average values of the acoustic measures $r(f)$ and s_u (38khz) are influenced by morphology and tilting behaviour.

3.2.1. Populating the acoustic feature library

The ALC are identified by their multi-frequency properties measured over some scattering volume. Currently, these are the volume backscattering coefficients, s_v, at 38 kHz and the ratio of s_v at the other frequencies to that at 38 kHz, that is $r(f)=s_v(f)/s_v(38)$. Thus, the variation of backscatter with frequency is covered by $r(f)$, and the scattering strength by $s_v(38)$. The logarithmic version $R(f)=10\log_{10}(r(f))$ is used in most figures for convenience.

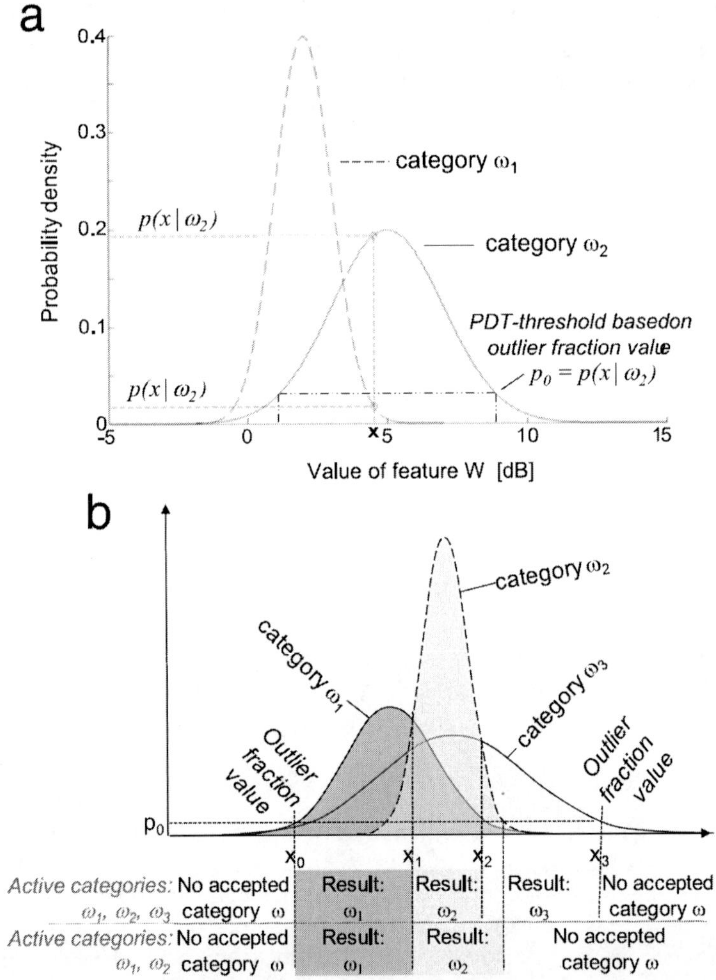

Figure. 3. (a) Probability density function of one acoustic feature *W* (i.e. one dimension) for two hypothetical ALC, ω_1 and ω_2, that is used to decide if measurement *x* (of acoustic feature *W*) is most likely to originate from category ω_1 or ω_2. Here $p(x|\omega_2) > p(x|\omega_1)$ indicates category ω_2 as the most likely. In practice many such comparisons between different features and categories contribute to the chosen category. (**b**) Example of results for value *x* of feature *W* when three acoustic-library categories ω_1, ω_2 and ω_3 are used. At values less than x_1, ω_1 is the most probable category (for probabilities above the threshold), but if $x < x_0$ the probability $p(x|\omega_1)$ is too small, being less than the threshold $p_0 = p(x_0|\omega_1)$, as determined by the specified outlier fraction (cf. 3.2.2). If ω_1 was not used, ω_3 would be accepted for $x < x_1$ (but with a different minimum value for *x*).

The acoustic feature library for each category contains a multi-dimensional normal probability density function (PDF) for the set of features, $W=W(R(18),S_v(38),R(70),\ldots,R(f),\ldots)$. The parameters of the PDFs are estimated from the training data by the maximum likelihood method (Theodoridis and Koutroumbas, 1999). Fig. 3(a) shows the fitted PDF for two hypothetical ALC, $\omega 1$ and $\omega 2$, in one dimension. The covariances among the individual acoustic features are considered as well as the variances. This means that the contours of the distributions will be hyper-ellipses, whose axes are not necessarily parallel to those of the property-space axes (Fig. 4, lower panel). The use of covariance in addition to variance provides a more comprehensive set of distribution functions and better separation between the different ALC.

The acoustic feature library used to identify categories is based on high quality acoustic measurements combined with trustworthy species identification, such as that obtained from targeted trawls that caught only one species, or were dominated by one species. Although trawl catches may not be representative of the local species mix, if one species dominates the catch then its abundance is clearly significant in that area. Thus, given some basic knowledge of biota scattering properties, the use of such ground-truthed measurements reduces the probability of assigning calculated acoustic features to the wrong ALC.

To develop an acoustic feature such as the $R(f)$, multi-frequency S_v training data are manually selected from rectangular sections of the echogram, and only values simultaneously above specified S_v thresholds at all available and chosen frequencies are used further in the training process. Next, unwanted data are identified and removed from the training dataset, such as observations near the edges of a school as these values may include sampling volume that are not representative of the acoustic target (Fig. 4(a)). Then, a statistical outlier fraction of the remaining data that deviates most from the multi-dimensional normal distribution (Fig. 4(b)) is removed from the training dataset. The remaining data (Fig. 4(c)) are stored in the feature library at different resolutions, referred to as pixel, cell, school, or track. The narrowest curves with the highest peak in Fig. 5 show the currently selected samples (in black) and the best-fit normal distribution for one feature, R(200), for the same data (as in Fig. 4(c)). The acoustic library may be edited or expanded at any time, should additional high quality ground-truthed data become available.

The pixels represent the original and highest spatial resolution, but they show the poorest separation between ALC due to having the largest variance and covariance between the acoustic features R(f) and Sv(38) (compare Fig. 6(a)–(c)). Here, a cell is defined as 16 pings horizontally by 50 pixels vertically, giving 800 samples covering approximately View the MathML source of the echogram, given a ping rate of View the MathML source, a cruising speed of 10 knots and a vertical resolution of View the MathML source. The cells are required to be at least 50% full, that is with 400–800 samples in each cell with View the MathML source re View the MathML source being available simultaneously at all frequencies. Data at school resolution are averaged over a school, and data at track resolution are averaged over a single fish track.

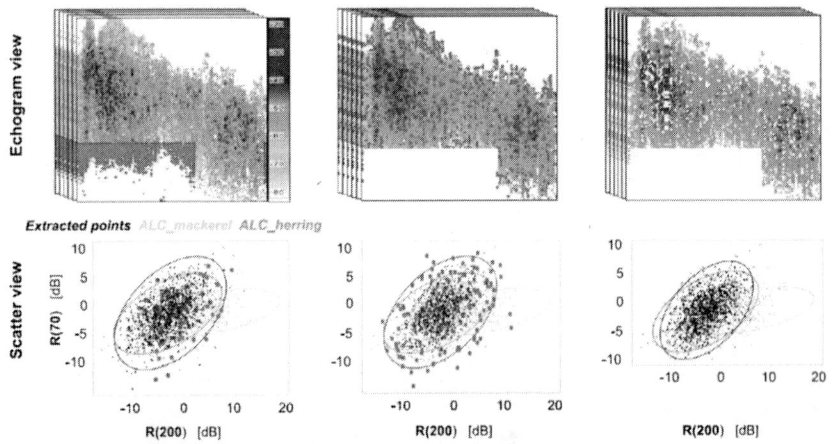

(a) Spatial marking: lower left corner in echogram.

(b) Statistical marking: 0.1 outlier fraction.

(c) Final result (black) after removal of unwanted samples.

Figure. 4. Stepwise removal of unwanted samples from a school when populating the acoustic feature library. Echogram data from upper panel is shown as small black points in the lower panels. Fitted probability density functions (PDF) are shown as ellipses in the lower panels. *ALC_mackerel* (orange) and *ALC_herring* (purple) from the acoustic feature library are shown for comparison (Fig. 5). (a) Spatial selection of data: data to be removed is selected in the echogram. The selected data are visualized in red. (b) Statistical selection of data: statistical outliers are marked in red, being the data outside the hyper-ellipsoid encapsulating the 0.9 fraction of the data, i.e. 90% of the data should be inside. (c) Final result to be stored in the feature library.

The acoustic features of several ALC may not be well separated (e.g. ALC_herring and ALC_capelin, Fig. 6(a)). Backscatter from a single fish track clearly belongs to a particular species, and those from a region identified as a school are likely to belong to just one species (Pitcher, 1993). Averaging the acoustic data over school cells or entire schools reduces the variance of the sv(38) and r(f) data, thus the acoustic features of the ALC become better separated as the data aggregation level goes from pixel to cell to school (Fig.

6(a)–(c)). The splitting of a school into cells where each cell contains a fixed number of samples is partly used to check if a school seems to contain more than one species (Korneliussen et al., 2009a), and partly to standardize the number of samples used to calculate the acoustic features.

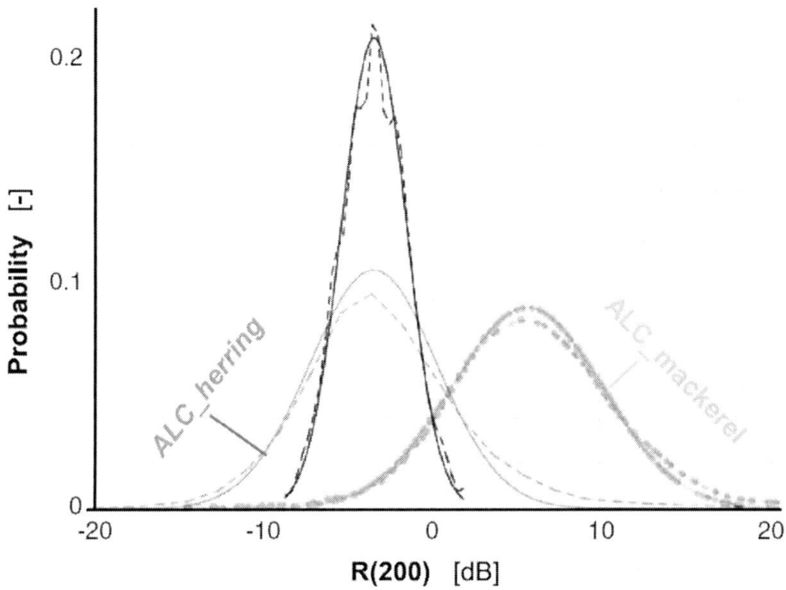

Figure. 5. Final result of data in Fig. 4(c) (black) after removal of unwanted samples. Normal distributions (continuous lines) fitted to real data (stippled). Selected data in narrow curves left, ALC_herring in grey wide curves to left, ALC_mackerel in curves to right.

Fig. 6 shows two of the hyper-ellipse dimensions, R(70) and R(200), in the training dataset for ALC_mackerel, ALC_herring and ALC_capelin, and compares the results at three resolutions—pixels, cells, and entire schools. The variances (the 10% percentile) that appear as sizes of the ellipses are clearly larger at pixel resolutions than at cell and school. Several of the ALC can be completely separated when visualized at cell resolution, in which case the automatic separation of these categories is considered reliable (Fig. 7). The reliable identification of some ALC depends on particular features. For example the category ALC_mackerel needs R(200) and ALC_resonant_18 needs R(18), so there is no attempt to identify ALC_mackerel beyond the maximum range accessible at 200 kHz. Similarly, there is no attempt to identify ALC_resonant_18 if 18 kHz data is not available. ALC such as ALC_herring may be identified beyond 300 m by using the features based on frequencies still valid at larger ranges. To compensate for this range limitation, KORONA can merge data from hull-mounted systems and deep towed systems, in which case data at higher frequencies may be used at any depth within the range limitations of the frequencies from the towed vehicle.

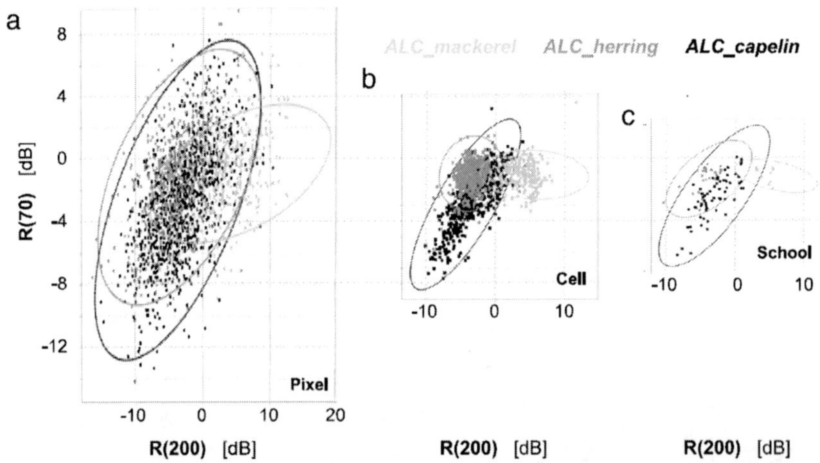

Figure. 6. Acoustic features of three ALC and the R(70) and R(200) different spatial resolutions: (a) Pixel resolution; (b) Cell resolution (800 pixels averaged spatially); (c) School resolution (one value per identified school). The separation of the categories improves as the resolution decreases.

The content of some of the ALC can be visualized as scatter plots (Fig. 7(a)) or as r(f) (Fig. 7(b)

3.2.2. Identification of ALC in small volumes

The estimation of the most probable ALC to be associated with a pixel is to some extent subjective, but we still use the term probability even if it is not a rigorous statistical probability. The subjectivity comes from, for example, operator decisions about how strong the requirement of belonging to the same category as the adjacent pixels should be (cf. step 7 below), and how much (if any) adjustment of the á priori probability (cf. step 5) there should be (e.g. that capelin is less likely to be found in the Norwegian Sea than in the Barents Sea, and thereby subjectively setting the maximum possible probability $P_{apriori}<1$), although $P_{apriori}=1$ for all case studies here.

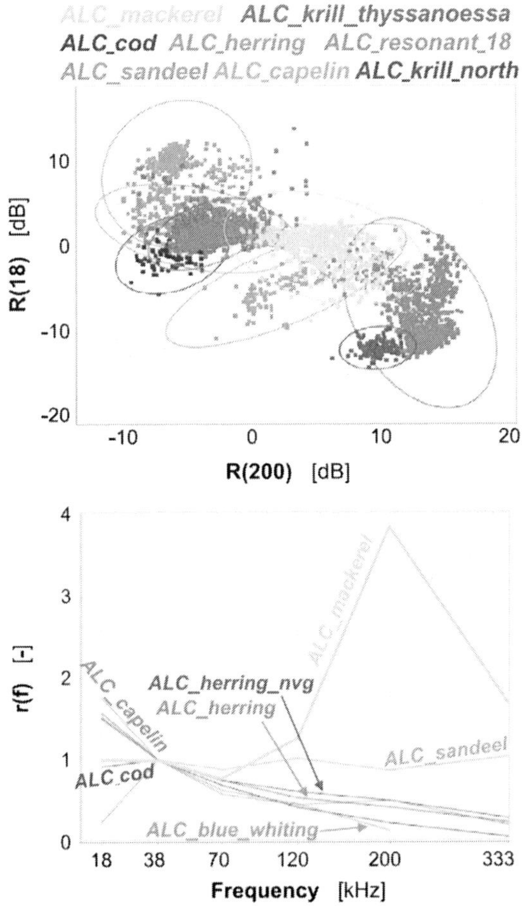

Figure. 7. (a) Relative frequency responses R(18) vs. R(200) at cell resolution (800 samples averaged spatially) for the ALC of importance for the case studies. The ellipses are statistical boundaries on each category that contain 90% of the observations. The plot shows that for the features R(18) vs. R(200)ALC_mackerel is almost completely separated from ALC_resonant_18, while ALC_herring has partial overlap with both ALC_mackerel and ALC_resonant_18 . The use of more features in addition to R(18) and R(200) improves the separation of categories. (b) The categories are generated from ground-truthed backscatter of Scomber scombrus L. (ALC_mackerel), Ammodytes marinus L. (ALC_sandeel), Mallotus villosus L. (ALC_capelin), Gadus morhua L. (ALC_cod), Micromesistius poutassou L. (ALC_blue_whiting) and Clupea harengus L. (ALC_herring and ALC_herring_nvg). The ALC_herring_nvg category is trained on Norwegian spring spawning herring, while the ALC_herring category also includes North Sea herring.

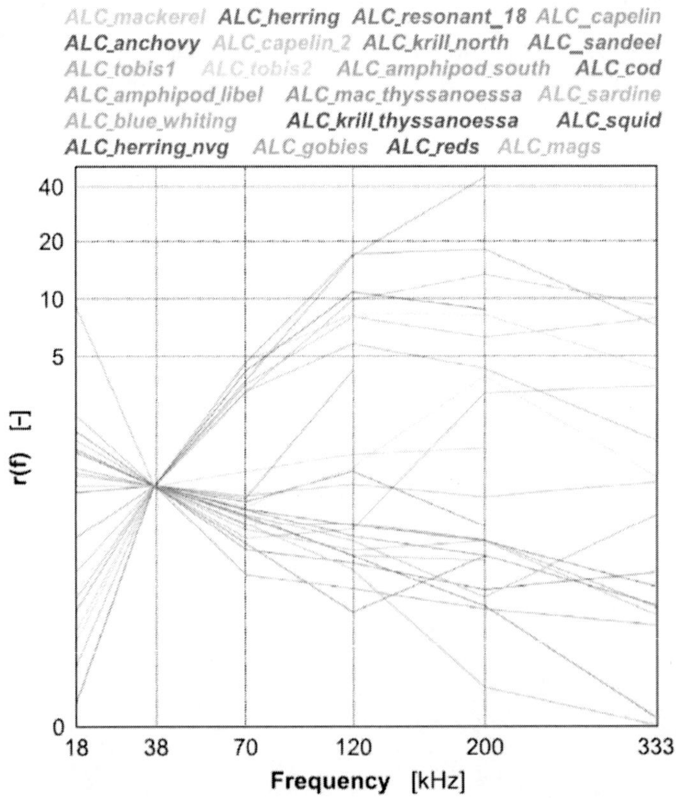

Figure. 8. r(f) of acoustic library categories currently available in LSSS.

The categorization method is based on Bayes decision theory (Theodoridis and Koutroumbas, 1999). New backscatter measurements are compared to the content of the feature library to decide which, if any, of the existing ALC the new measurements are most similar to. The categorization of $R(f)$ and $S_v(38)$ measurements from a pixel proceeds as described below. The ALC with the highest probability is allocated to a pixel (but the other candidates are also kept), provided the probability is above a specified minimum value (cf. step 2).

The outlier fraction sets the cut-off for the measured pixel. When there are several ALC active, the most likely category (as shown in Fig. 3(b)) is associated with the pixel. This is done for all pixels. Further, a pixel is more likely to be occupied by the same species as the majority of nearby pixels, which is why the probability is adjusted according to the categories of the nearest spatial neighbours (cf. step 7).

Algorithm

Multi-frequency data from each pixel are tested against the properties of the ALC. The most probable ALC is associated with that pixel. This is determined by the probability that the tested data belongs to a particular ALC. The ALC categorization proceeds as follows for all pixels by means of features selected and valid at the current range:

(0) Find the subset of useable ALC, $\Omega=\{\omega_1,\omega_2,\ldots,\omega_n\}$, based on user specified criteria in terms of available features. For example, categorization of *ALC*_mackerel needs $R(200)$, which is not valid for hull-mounted transducers beyond a depth of about 300 m.

(1) Assign special ALC to pixels that should not be categorized using the acoustic features. These special categories include transducer near-field, return from the seafloor, noise, etc.

(2) Find the probability density function (PDF) threshold for each ALC based on an outlier fraction, normally 0.1. Fig. 4, Fig. 6 ; Fig. 7 show the borders corresponding to the probability minima visualized as ellipses.

(3) Compute the feature vector x, for example $[R(18),S_v(38),R(70),R(200)]$ in case of a 4-dimensional feature space.

(4) For each ω_i (i.e. ALC), calculate the ALC-dependent probability density $p(x|\omega_i)$. Fig. 3(a) shows probability density functions (PDF)s for the feature W for categories ω_1 and ω_2 for a 1-dimension feature space.

(5) Calculate a discriminant function $g(\omega_i|x)=P(\omega_i)\cdot p(x|\omega_i)$, where $P(\omega_i)$ is a user-defined á priori probability for the category ω_i. For the case studies $P(\omega_i)=1$.

(6) Assign to the pixel the ALC ω_i with the largest discriminant g. Leave the volume unassigned if the measurement is not above the PDF-threshold from step (2) of any category.

(7) Update the á priori probability of each category by considering neighbouring pixels (Besag, 1986). The updated probability, $P'(\omega_i)$, is calculated using the formula $P'(\omega_i)=P_0(\omega_i)e^{\beta(ni-3)}$, where $P_0(\omega_i)$ is the original á priori probability, ni is the number of neighbouring pixels assigned category ω_i (maximum of 8), and β, with default value 0.1, is a parameter that determines how much the probability should change. Thus, the probability is modified by the factor $e^{\beta(ni-3)}$, which is greater than unity if more than three neighbouring pixels have the same category.

(8) Steps (5)–(7) may be repeated several times (usually three). New categorizations hence take into account previous changes to the categorization of neighbouring pixels.

3.2.3. Identification of ALC in sample collections

The algorithm for identification of ALC in sample collections is the same as for small volumes except that step (7) is omitted. Categorization of data inside detected schools and single fish tracks are more highly averaged than at pixel resolution. The categories ALC_capelin and ALC_herring are hardly distinguishable at pixel level (Fig. 6(a)), but are better distinguished at the cell level (Fig. 6(b)) are even better distinguished at the school level (Fig. 6(c)). To ensure that a school is monospecific, it is split into cells and each cell is categorized as one unit (as described by Korneliussen et al., 2009a). The school categorization uses all pixels inside an aggregation, (e.g. School 1 in Fig. 9(b)), including those that would be uncategorized at pixel level. If the cell categorization indicates a multiple species aggregation, then pixel categorization is used as for School 2 in Fig. 9(b). Trawl samples targeting the schools in Fig. 9 contained both mackerel and herring, and therefore supported the results of a two species aggregation (or two schools in close proximity).

A disadvantage of testing all schools for multiple species as described above is that the result may be inconclusive, in which case the pixel-categorization will be used for the school, which in turn results in many uncategorized pixels inside the schools. Examination of echograms indicates that schools in Norwegian waters almost always contain only one species. Therefore, the schools are considered to be monospecific in the analysis.

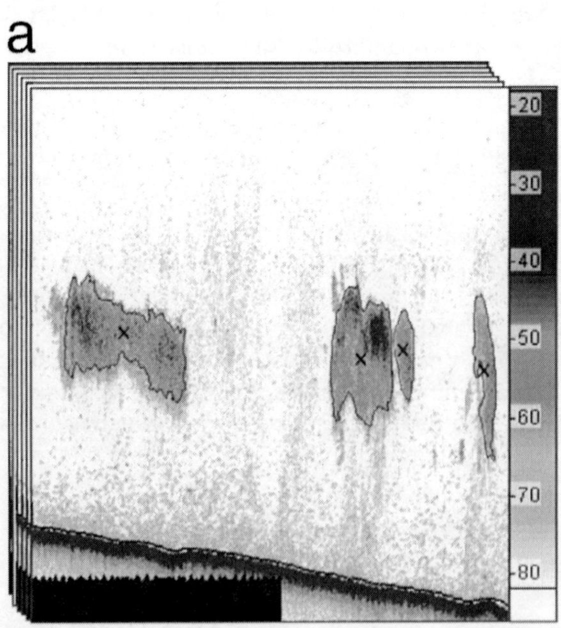

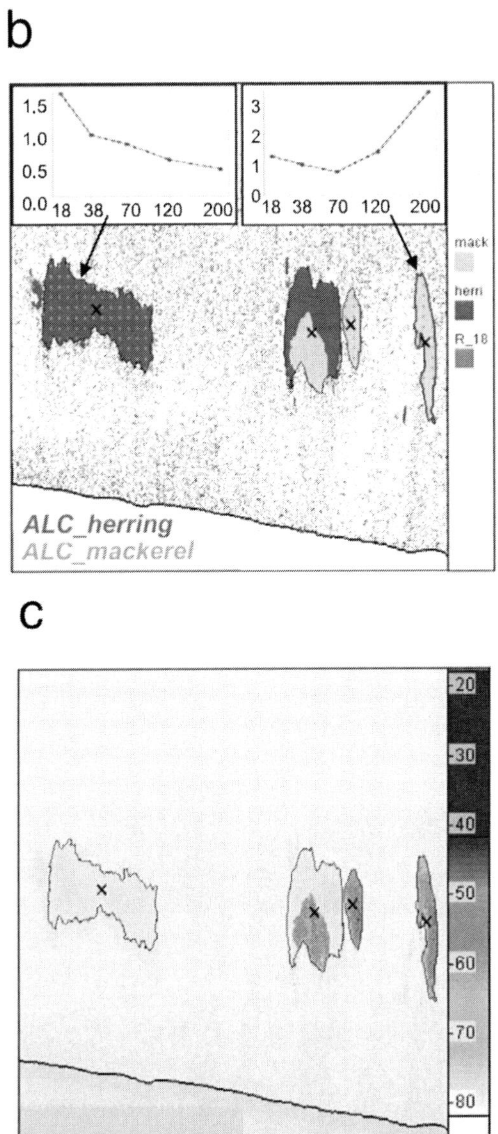

Figure. 9. Results of acoustic library categorization. (a) Echograms at six frequencies 18–364 kHz on overlaid sheets, 38 kHz on top. (b) Identified ALC shown with solid colours (grey for ALC_herring, light grey for ALC_mackerel) with relative frequency responses, *r*(*f*), for two schools shown. (c) Synthetic 38 kHz echogram with ALC_mackerel kept and all others removed.

The analysis of data from a single fish track is similar to that of schools, but the analysis is based on the backscattering cross section, σ_{bs}, instead of s_v. The knowledge that a track clearly contains a single specimen, combined with the use of within-beam location data to correct for the transducer beam pattern, reduces the variance of the relative frequency responses. Schools require more samples than single fish tracks for good separation between acoustic categories, but a school contains far more samples than a single fish track.

3.2.4. Model-based categorization of aggregated samples

The MBC algorithm uses several scattering models to estimate backscatter characteristics from zooplankton, such as euphausiids and amphipods. These currently include the randomly oriented fluid bent cylinder model (Stanton et al., 1993) and the stochastic distorted-wave Born approximation (Demer and Conti, 2003). Those models have been implemented in an optimized framework (Fernandes et al., 2002, sections 6.2.3 to 6.2.4). Zooplankton classification may be applied to pixels, but is preferably done on cells or schools to avoid large variations in estimated sizes. We use the term pixel to refer to this minimal volume being the highest resolution of the echosounder/transducer combination. To speed up the processing, data allocated to ALC that are clearly not zooplankton are ignored. Note that the results of model-based categorization are not used in the case studies in this paper.

4. CASE STUDIES

4.1. Material and methods

Multi-frequency acoustic data were collected according to existing protocols (Korneliussen et al., 2008), with the intention of combining and analysing co-incident observations from several echosounders operating at different frequencies. This requires the use of calibrated systems (Foote et al., 1987 ; Demer et al., 2015) operating at the same pulse-duration (in these data 1 ms) and ensonifying nearly the same volume of water at all frequencies. Data were collected during three surveys using Simrad EK60 echosounders mounted in a tightly-packed configuration on a protruding keel (Korneliussen, 2010). The 18, 38, 70, 120, and 200 kHz echosounder frequencies were used on all surveys. In addition, a 364 kHz echosounder was used in 2004 and a 333 kHz on the 2014 sandeel survey. The 18 kHz system provided a transducer beam width of 11° between the half-power points, while the other frequencies provided 7° beamwidths. All surveys were used for stock abundance estimation based on the manual categorization. The surveys were:

1. RV G.O. Sars in October–November 2004 in the northern part of the North Sea and southern part of the Norwegian Sea targeting Atlantic mackerel. The survey was selected to compare the acoustic library categorization method with an earlier method (Korneliussen, 2010).

2. FV Eros in 2014 in the North Sea targeting sandeel. The survey was selected because it observed small schools on and just above the seabed, which provides a challenge to the categorization process.

3. RV G.O. Sars in the Barents Sea in 2014 for investigation of the major components of the ecosystem. The survey was selected for identification of capelin, which is acoustically similar to herring and cod in the same waters.

The acoustic data were manually categorized during the surveys (Fig. 1(A)) and were reprocessed with a standardized setup (Fig. 1(B) and (C)). LSSS used the same processing setup in all cases, to remove noise spikes and ambient noise (Korneliussen, 2000). Spatial offsets between the transducers and pulse-transmission delay between frequencies were corrected (Korneliussen et al., 2008) for the purpose of comparing data at different frequencies in small volumes. These reprocessed data were used during the manual categorization (Fig. 1(C)).

The manual categorization was carried out daily within 24 h after it was collected and then stored in scrutiny files. Those scrutiny files were loaded into LSSS and applied to the reprocessed acoustic data (Fig. 1(C)) and stored into a manual categorization database. The results of the automatic categorization were ignored.

The reprocessed data were also used during the automatic acoustic categorization (Fig. 1(B)), but the processing continued by detecting ALC at pixel-level, and by detecting the extent of echo-traces. These are investigated as a unit when detecting ALC. Finally, the MBC are detected. An echo-trace based ALC is used before a pixel-based ALC. Each pixel becomes associated with an ALC thus:

- All pixels: ALC with probability based on pixel-data (even within echo-traces);
- Pixels within echo-traces: ALC with probability based on echo-trace data;
- All pixels: A single MBC with probability (not used in case studies).

The settings for pixel categorization and school categorization were the same for all surveys, except that the selection of ALC was different due to geographical considerations (see Table 1).

The process of automatic acoustic categorization is automated (i.e. no operator intervention), and the resulting categories (MSC) stored in an automatic categorization database (Fig. 2(B)). The process starts with automatic acoustic categorization, which results in ALC and MBC for each pixel. Furthermore, those categories are mapped into manual scrutiny categories (MSC) chosen by the operator prior to the start of the process. For example, in the 2014 Barents Sea survey, ALC_nvg_herring was mapped to the herring MSC.

The s_A was stored to the databases at a horizontal resolution of 0.1 nautical miles both for the original manual scrutiny and for the automatic categorization. Table 1 shows the ALC used by the categorization modules in the different case studies.

4.2. Results

The s_A from the automatic categorization and the manual categorization for the three case studies are similar (Fig. 10), especially for large s_A. A notable result especially for capelin, is that the values from the automatic categorization are equal to or slightly smaller than the manual categorization except for small values of sA. Furthermore, there are many small to moderate values from the automatic categorization that were given zero sA for the manual categorization, especially for sandeel. Subsequent inspection of the manually categorized sandeel survey echograms showed there were some sandeel schools that were clearly categorized incorrectly. This objective check of the manual categorization is one advantage of the automatic process. There was high spatial conformity between the manual categorization and the automatic categorization (Fig. 10(b)) except for low values of sv where the conformity varied considerably.

The overall Pearson correlation coefficients between manual scrutiny and automatic categorization are very high for all surveys (Table 2). For mackerel and sandeel the total acoustic abundance of the automatic categorization and the manual categorization were within 2% of each other. This slight difference is explained by the observation that manually drawn school integration regions are always slightly larger than the automatically drawn school regions. The capelin data correlation is close to unity (Table 2), but the automatic categorization resulted in almost 15% lower sA than the manual categorization. This may be due to school categorization failing so that some schools are not being categorized as a unit but as individual pixels. This in turn leaves many pixels uncategorized and some of the pixels being wrongly identified as cod or herring instead of capelin, which is quite acoustically similar. In that case, masking the measurements with pixels identified as capelin will give sA values that are too low. This problem may be reduced if additional features are used in the categorization process although some uncertainty will always remain.

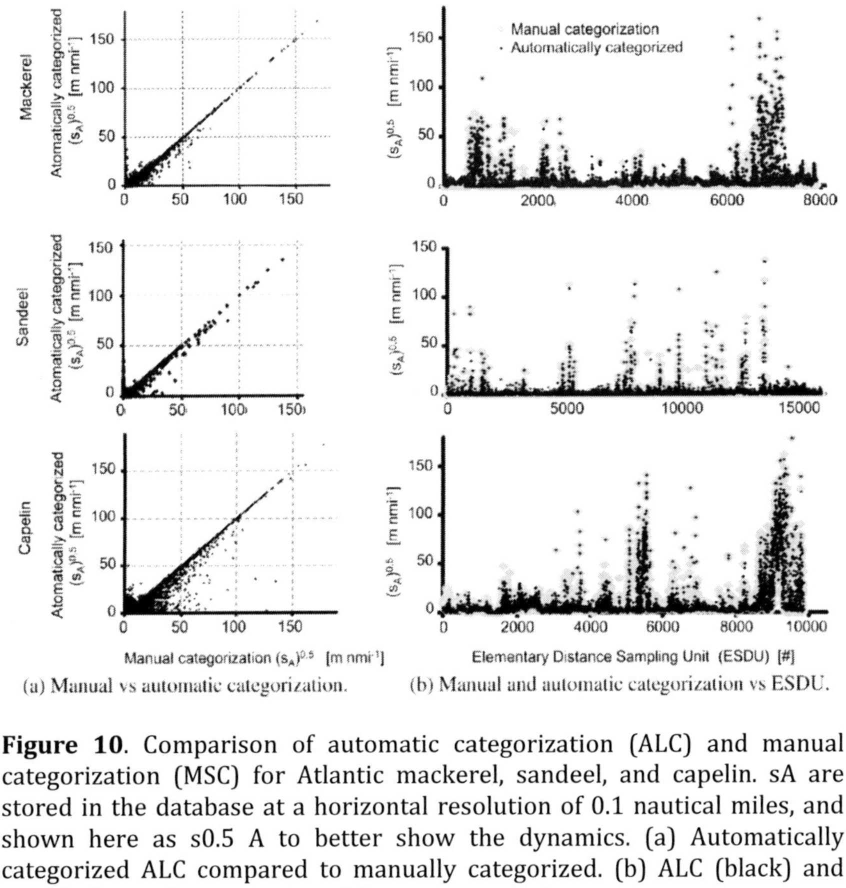

(a) Manual vs automatic categorization. (b) Manual and automatic categorization vs ESDU.

Figure 10. Comparison of automatic categorization (ALC) and manual categorization (MSC) for Atlantic mackerel, sandeel, and capelin. sA are stored in the database at a horizontal resolution of 0.1 nautical miles, and shown here as s0.5 A to better show the dynamics. (a) Automatically categorized ALC compared to manually categorized. (b) ALC (black) and MSC (light grey) as a unction of distance travelled.

The high correlation coefficients (Table 2) show that, at least for these surveys, the s_A (and thereby also the biological abundance) from the target species can be calculated with high conformity using automatic categorization method. For sandeel and Atlantic mackerel, there is insignificant difference in s_A, while automatic categorization of capelin was 15% less than manual categorization.

Table 1. Categories and species compared in the case studies. The second column is the target species, and the third and fourth columns are the similar scrutinize category and ALC believed to originate from the target species. The final column contains ALC that were identified simultaneously with the target library category. If the backscatter of a pixel is not most likely to originate from the target ALC, or if the probability is below the probability threshold, sv=0 for that pixel.

Case study survey number	Target species	MSC[g]	ALC[g]	Number of training schools	Training schools collection year	ALC used to remove backscatter
2004113	*Scomber scombrus* L.	Mackerel	*ALC_mackerel*	35	1999 2005	*ALC_resonant_18*[a] *ALC_krill_north*[b] *ALC_krill_thyssanoessa*[c] *ALC_herring*[d]
2014807	*Ammodytes marinus* L.	Sandeel	*ALC_sandeel*	44	2008 2009	*ALC_resonant_18*[a] *ALC_krill_north*[b] *ALC_krill_thyssanoessa*[c] *ALC_herring*[d]
2014116	*Mallotus villosus* L.	Capelin	*ALC_capelin*	67	2008 2014	*ALC_resonant_18*[a] *ALC_krill_north*[b] *ALC_krill_thyssanoessa*[c] *ALC_herring_nvg*[e] *ALC_cod*[f]

[a] *ALC_resonant_18* were trained from 4 shoals of zooplankton that were resonant at 18 kHz.
[b] *ALC_krill_north* were trained from 24 schools of *Meganictiphanes norwegica* L.
[c] *ALC_krill_thyssanoessa* were trained from 8 shoals of *Thyssanoessa inermis* L.
[d] *ALC_herring* were trained from 40 schools of *Clupea harengus* L.
[e] *ALC_herring_nvg* were trained from 9 schools of Norwegian Spring Spawning herring.
[f] *ALC_cod* were trained from 10 schools of *Gadus morhua* L. These were actually schools of cod.
[g] ALC are always in italic and preceded by *ALC_*, MSC are not italic.

Table 2. Comparison of manual scrutiny and automatic categorization. The sA(200) samples are averaged over 0.1 nautical miles. Column 6 gives number of samples, column 7 the total s_A of automatic detection relative to s_A of manual scrutiny.

Survey	Target species	MSC	ALC	Pearson correlation coefficient	$s_A(200)$ samples	$\Sigma s_A(200)_{auto} / \Sigma s_A(200)_{man}$
2004113	*Scomber scombrus* L.	Mackerel	*ALC_mackerel*	0.996	7913	0.985
2014807	*Ammodytes marinus* L.	Sandeel	*ALC_sandeel*	0.998	17925	0.994
2014116	*Mallotus villosus* L.	Capelin	*ALC_capelin*	0.976	9869	0.855

4.3. Discussion

The target ALC (ALC_mackerel, ALC_sandeel and ALC_capelin) were used in the three case studies, along with ALC_herring, ALC_cod, ALC_blue_whiting that have features with properties not very different from ALC_capelin . The mean $r(f)$ values of ALC_mackerel (Korneliussen, 2010) and ALC_sandeel (Johnsen et al., 2009) are quite different from the other ALC (Fig. 7), which made reliable automatic categorization for schools a not unexpected outcome. However, the sample variation of $R(f)$ illustrated by the ellipses of the scatterplots (Fig. 4, Fig. 6 ; Fig. 7) is much larger than the variation of mean $R(f)$ at school, cell and pixel resolution. This makes occasional misclassifications likely even for ALC_mackerel and ALC_sandeel , especially at pixel level. In other words, misclassification is more likely if the scattering regions cannot be grouped into single fish tracks or schools. The sample correlation coefficient is high for large s_A, thus the total correlation also becomes large. The less common alternative, the rank correlation coefficient (Walpole et al., 2002) gives equal weight to all values, and results in a somewhat lower correlation. However, it is the large values that have most influence on the abundance estimates, and it is therefore felt that the use of sample correlation is more appropriate.

The high correlation between manual categorization and automatic categorization of capelin was unexpected, especially since the $R(f)$ of ALC_capelin in the feature library is similar to that of other fish with swimbladder, such as ALC_cod and ALC_herring_nvg (Fig. 7(b) and Fernandes et al., 2002). The manual categorization of capelin is considered to be reliable in this case due to the low water temperature in which capelin is often found, the location in the water column, the schooling behaviour and the results of the biological sampling. It was expected that additional, non-acoustic characteristics would be needed for a high correlation between manual categorization and automatic categorization. Differences in school shapes have previously been used to improve automatic categorization (Scalabrin et al., 1996 ; Korneliussen et al., 2009a), but that was not used in this case as the scattering regions mostly formed large layers for capelin in the 2014 Barents Sea cruise, and not schools of limited size, so morphology was not considered to be reliable. A slight reduction in the á priori probability of herring and cod in the 2014 Barents Sea case study, for example from 1 to 0.9, while keeping the á priori probability of capelin as 1, may be defended due to the low water temperature. Such a reduction in the á priori probability did in fact make the s_A resulting from automatic categorization of capelin closer to manual categorization. However, the quantification of the á priori probability is not easy, and the á priori probability is therefore set to P=1 for all ALC in all case studies.

The statistical properties of the features should be calculated from a sufficiently large quantity of data for each ALC, and preferably from more than one survey, to make the calculated values representative. Poor training of one ALC in the acoustic feature library can lead to biased results, as would the use of data from uncalibrated echosounders. Uncalibrated instruments result in biased s_v, and therefore an incorrect $r(f)$ (unless all $s_v(f)$ are equally wrong at all frequencies f). The uncritical use of fully automated categorization during operational surveys is inadvisable. Although the case studies showed good agreement between automatic categorization and manual categorization, the $r(f)$ may potentially vary with water temperature, fat content, specimen size and species behaviour, so that similarly good results may not be found in other cases. Ona (1990) and Ona et al. (2001) showed that TS (and thereby s_v) vary with physiology and with season, and unless s_v(f) varies similarly at all frequencies, $r(f)$ will also vary somewhat. A solution to this problem is to build an acoustic library that allows for seasonal variations of scattering properties. Alternatively, one species may be used to build up several ALC, such as ALC_mackerel_spring, ALC_mackerel_summer, etc.

A comprehensive study of acoustic features, such as $r(f)$, was made by Fernandes et al. (2002) for several species, and those $r(f)$ may be used as a reference during scrutiny of multi-frequency echograms. De Robertis et al. (2010) also have a list of scattering classes that could be used during categorization (their Fig. 4), but they were able to integrate all $r(f)$ into a single Z-score, and used that to make an operational system for multi-frequency backscatter classification. Like De Robertis et al. (2010), we also integrate many features into a discriminant function used to identify ALC in real-time, and like them we may expand the features we use with additional $r(f)$ for additional frequencies, f. Our method has additional flexibility in that we can also use $s_v(38)$, and may expand the features used (for example depth). Unlike De Robertis et al. (2010), we extract the statistical properties of the features directly from measured data, and can easily update the statistical properties during the survey in an operational sense.

A future enhancement of our method will incorporate more of the information that the operator has at hand, such as the local geography, and use it to automatically adjust the á priori probability (e.g. reducing the probability of finding Atlantic mackerel in the Barents Sea during the winter). Based on geography (and an ALC mapped to a species), the á priori probability could be multiplied by 1 within the main residential area of a species, and gradually reduced outside that area. The methods for acoustic categorization presented here can be applied to wideband acoustic data, although this has not been shown here. Wideband data can improve range resolution and hence improve single fish detection and tracking. Wideband backscatter can also increase the number of points in the r(f) curve, potentially leading to improved species identification. Here wideband is defined to consist of one or several continuous sub-bands, where each sub-band (called broadband) obeys approximately to the ratio bandwidth/center_frequency ≥ 0.5.

The two methods both give high conformity between manual categorization and automatic categorization, although the presented results used different Elementary Sampling Distance Unit (ESDU). A natural application of the high conformity between manual categorization and automatic categorization shown here and by De Robertis et al. (2010) is that automatic categorization may be used to reanalyse data from old surveys to obtain a new and more objective view of those data. Furthermore, our method may be used to analyse data that has no temporally relevant ground-truthed information available, such as backscatter data collected by autonomous underwater vehicles or ocean observatories.

Although no comprehensive comparison for a survey has been done, the 2014 sandeel survey typically required 10 h per 24 h data for manual scrutiny of echograms for the few days with the most complex registrations, while it was carried out in less than 1 h when manual categorization was assisted with automatic categorization, and 5 min (excluding 20 min processing) when fully automated. The 5 min is time needed for loading data and storing the results of the categorization to the database.

5. SUMMARY AND FUTURE EXPECTATIONS

Automatic categorization is reliable for the large s_A of the ALC (or species) investigated in the case studies. Although the manual categorizations, are kept separate from automatic categorization for these case studies, that would not be the case for most future survey analyses. The automatic acoustic categorization provides a more objective view that assists the manual categorization and creates a common baseline for categorization. The speed and objectivity of the categorization naturally also improves with the use of reliable echo trace detection, as the acoustic library categorization becomes more reliable, and because less time is needed to draw school regions. Thus, more time is available for the most difficult part of the categorization. During routine categorization it is a design aim of the software that the operators (not the software) decide if the automatic acoustic categorization should be used or whether the categorization should be wholly manual, or a combination of the two. One reason for this is that automatic acoustic categorization may identify species in small volumes that are difficult to see or too labour intensive for manual categorization.

Automatic categorization can provide a more objective view to assist manual categorization. Efficient and effective categorization of acoustic data in a resource limited environment requires that attention be paid to the time required to produce the estimates of acoustic abundance, s_A, used to estimate fish stock size. The use of automatic categorization contributes to this aim for surveys which require estimates of fish stock size at completion.

The improved scrutiny objectivity obtained from automatic or semi-automatic data analysis and the ability to process, in a timely manner, the increased volume of acoustic data during surveys is advantageous. Automatic analysis produces consistent and reproducible results, and enables improved assessment and understanding of the spatial distribution of ecosystem components. Such automatic processing makes practical the timely analysis of the very large quantities of data that will be collected by autonomous vehicles (Fernandes et al., 2003; Ghani et al., 2014 ; Suberg et al., 2014); and stationary systems (Pawlowicz and McClure, 2010). Much of these data can be processed automatically as described here, with the manual work being simply verification of the processing.

REFERENCES

1. Anon 1990. SIMRAD EK500 Scientific echo sounder. Instruction manual. P2172E, 1990–1997. Contents: Operators manual, P2170E Installation Manual P2180E, Service Manual 2161E.
2. Besag, J., 1986. On the statistical analysis of dirty pictures. J. R. Stat. Soc. Ser. B Stat. Methodol. 48 (3), 259–302.
3. De Robertis, A., McKelvey, D.R., Ressler, P.H., 2010. Development and application of an empirical multifrequency method for backscatter classification. Can. J. Fish. Aquat. Sci. 67, 1459–1474.
4. Demer, D.A., Berger, L., Bernasconi, M., Bethke, E., Boswell, K., Chu, D., Domokos, R., et al., 2015. Calibration of acoustic instruments. ICES Cooperative Research Report No. 326. p. 130.
5. Demer, D.A., Conti, S.G., 2003. Reconciling theoretical versus empirical target strengths of krill: effects of phase variability on the distorted-wave Born approximation. ICES J. Mar. Sci. 60, 429–434.
6. Deuser, L., Middleton, D., Plemons, T., Vaughan, J., 1979. On the classification of underwater acoustic signals. II. Experimental applications involving fish. J. Acoust. Soc. Am. 65, 444–445.
7. Dougherty, E.R., 1992. An Introduction to Morphological Image Processing. ISBN: 0-8194-0845-X.
8. Fernandes, P.G., Korneliussen, R.J., Lebourges-Dhaussy, A., Masse, J., Iglesias, M., Diner, N., Ona, E., et al., 2002. The SIMFAMI project: species identification methods from acoustic multi-frequency information. Final Report to the EC Number Q5RS-2001-02054 (Report accessible from any of the authors).
9. Fernandes, P.G., Stevenson, P., Brierley, A.S., Armstrong, F., Simmonds, E.J., 2003. Autonomous underwater vehicles: future platforms for fisheries acoustics. ICES J. Mar. Sci. 60, 684–691.

10. Foote, K.G., Knudsen, H.P., Vestnes, G., MacLennan, D.N., Simmonds, E.J., 1987. Calibration of acoustic instruments for fish density estimation: a practical guide. ICES Cooperative Research Report, 144. p. 69.
11. Ghani, M.H., Hole, L.R., Fer, I., Kourafalou, V.H., Wienders, N., Kang, H., Drushka, K., Peddie, D., 2014. The SailBuoy remotelycontrolled unmanned vessel: Measurements of near surface temperature, salinity and oxygen concentration in the Northern Gulf of Mexico. Methods Oceanogr. 10, 104–121.
12. Handegard, N.O., 2007. Observing individual fish behavior in fish aggregations: Tracking in dense fish aggregations using a split-beam echosounder. J. Acoust. Soc. Am. 122 (1), 177–187.
13. Haralabous, J., Georgakaros, S., 1996. Artificial neural network as a tool for species identification fish schools. ICES J. Mar. Sci.53, 173–180.
14. Holliday, D.V., 1977. Extracting bio-physical information from the acoustic signature of marine organisms. In: Andersen, N.R., Zahuranec, B.J. (Eds.), Oceanic Sound Scattering Prediction. Plenum, NY, pp. 619–624.
15. Holliday, D.V., Pieper, R.E., Kleppel, , 1989. Determination of zooplankton size and distribution with multifrequency acoustic technology. J. Int. Counc. Explor. Sea 46, 52–61.
16. Horne, J.K., 2000. Acoustic approaches to remote species identification: a review. Fisheries Oceanography 9 (4), 356–371.
17. Jech, J.M., Michaels, W.L., 2006. A multifrequency method to classify and evaluate fisheries acoustics data. Can. J. Fish. Aquat. Sci. 63 (10), 2225–2235.
18. Johnsen, E., Pedersen, R., Ona, E., 2009. Size-dependent frequency response of sandeel schools. ICES J. Mar. Sci. 66, 1100–1105.
19. Kaartvedt, S., Røstad, A., Fiksen, Ø., Melle, W., Torgersen, T., Breien, M.T., Klevjer, T.A., 2005. Piscivorous fish patrol krill swarms. Mar. Ecol. Prog. Ser. 299, 1–5.
20. King, M., 2007. Fisheries Biology, Assessment and Management. Wiley-Blackwell, Oxford, p. 400.
21. Kloser, R.J., Ryan, T., Sakov, P., Williams, A., Koslow, J.A., 2002. Species identification in deep water using multiple acoustic frequencies. Can. J. Fish. Aquat. Sci. 59, 1065–1077.
22. Korneliussen, R.J., 2000. Measurement and removal of echo integration noise. ICES J. Mar. Sci. 57, 1204–1217.
23. Korneliussen, R.J., 2004. The Bergen echo integrator post-processing system, with focus on recent improvements. Fish. Res. 68, 159–169.
24. Korneliussen, R.J., 2010. The acoustic identification of Atlantic Mackerel. ICES J. Mar. Sci. 67, 1749–1758.
25. Korneliussen, R.J., Diner, N., Ona, E., Berger, L., Fernandes, P.G., 2008. Proposals for the collection of multifrequency acoustic data. ICES J. Mar. Sci. 65, 982–994.
26. Korneliussen, R.J., Heggelund, Y., Eliassen, I.K., Johansen, G.O., 2009a. Acoustic species identification of schooling fish. ICES J. Mar. Sci. 66, 1111–1118.
27. Korneliussen, R.J., Heggelund, Y., Eliassen, I.K., Øye, O.K., Knutsen, T., Dalen, J., 2009b. Combining multibeam-sonar and multifrequency echosounder data: examples of the analysis and imaging of large euphausiid schools. ICES J. Mar. Sci. 66, 991–997.
28. Korneliussen, R.J., Ona, E., 2002. An operational system for processing and visualising multi-frequency acoustic data. ICES J. Mar. Sci. 59, 293–313.
29. Korneliussen, R.J., Ona, E., 2003. Synthetic echograms generated from the relative frequency response. ICES J. Mar. Sci. 60, 636–640.
30. MacLennan, D.N., Fernandes, P.G., Dalen, J., 2002. A consistent approach to definitions and symbols in fisheries acoustics. ICES J. Mar. Sci. 59, 365–369.
31. MacLennan, D.N., Holliday, D.V., 1996. Fisheries and plankton acoustics: past, present, and future. ICES J. Mar. Sci. 53, 513–516.
32. Ona, E., 1990. Physiological factors causing natural variations in acoustic target strength of fish. J. Mar. Biol. Assoc. UK 70, 107–127.
33. Ona, E., Zhao, X., Svellingen, I., Fosseidengen, J.E., 2001. Seasonal Variation in Herring Target Strength. Herring: Expectations for a New Millennium 461 Alaska Sea Grant College Program AK-SG-01-04, 461-487.

34. Pawlowicz, R., McClure, B., 2010. Inverted echosounder for continuous high-resolution water column profiling from the NEPTUNE (Canada) ocean observatory, OCEANS 2010 MTS/IEEE SEATTLE. Presented at the OCEANS 2010 MTS/IEEE SEATTLE, pp. 1–8.
35. Pedersen, G., Korneliussen, R.J., 2009. The relative frequency response derived from individually separated targets of northeast Arctic cod (Gadus morhua), saithe (Pollachius virens), and Norway pout (Trisopterus esmarkii). ICES J. Mar. Sci. 66, 1149–1154.
36. Pitcher, T.J., 1993. Behaviour of Teleost Fishes. In: Fish and Fisheries Series, vol. 7. Chapman and Hall, London, p. 715.
37. Scalabrin, C., Diner, N., Weill, A., Hillion, A., Mouchot, M.-C., 1996. Narrowband acoustic identification of monospecific fish shoals. ICES J. Mar. Sci. 53, 181–188.
38. Simmonds, E.J., MacLennan, D.N., 2005. Fisheries Acoustics: Theory and Practice, second ed. Blackwell, Oxford, p. 437.
39. Stanton, T.K., Chu, D., Wiebe, P.H., Benfield, M.C., Clay, C.S., 1993. Average echoes from randomly oriented random-length finite cylinders: Zooplankton models. J. Acoust. Soc. Am. 94 (6), 3463–3472.
40. Stanton, T.K., Wiebe, P.H., Chu, D., Benfield, M.C., Scanlon, L., Martin, l., Eastwood, R.L., 1994. On acoustic estimates of zooplankton biomass. ICES J. Mar. Sci. 51, 505–512.
41. Suberg, L., Wynn, R.B., Kooij, J. van der, Fernand, L., Fielding, S., Guihen, D., Gillespie, D., Johnson, M., Gkikopoulou, K.C., Allan, I.J., Vrana, B., Miller, P.I., Smeed, D., Jones, A.R., 2014. Assessing the potential of autonomous submarine gliders for ecosystem monitoring across multiple trophic levels (plankton to cetaceans) and pollutants in shallow shelf seas. Methods in Oceanogr. 10, 70–89.
42. Theodoridis, S., Koutroumbas, K., 1999. Classifiers based on Bayes theory. In: Pattern Recognition, third ed. Academic press, Oval Road, London NW1 7DX, UK, pp. 24–28 (Chapter 2). Classifiers based on Bayes decision theory. http://www.amazon.com/Pattern-Recognition-Sergios-Theodoridis/dp/0126861404.
43. Walpole, R.E., Myers, R.H., Myers, S., Ye, K., 2002. Probability & Statistics for Engineers & Scientists. Prentice-Hall, New Jersey, p. 730.

INDEX